# Developments in Hydrobiology 111

*Series editor*
H. J. Dumont

# Advances in Littorinid Biology

Proceedings of the Fourth International Symposium on Littorinid Biology,
held in Roscoff, France, 19–25 September 1993

*Edited by*

## P.J. Mill and C.D. McQuaid

*Reprinted from Hydrobiologia, vol. 309 (1995)*

SPRINGER-SCIENCE+BUSINESS MEDIA, B.V.

**Library of Congress Cataloging-in-Publication Data**

A C.I.P. Catalogue record for this book is available from the Library of Congress.

ISBN 978-94-010-4194-2    ISBN 978-94-011-0435-7 (eBook)
DOI 10.1007/978-94-011-0435-7

*Printed on acid-free paper*

# Contents

vi

*Hydrobiologia* **309**, 1995.
*P.J. Mill & C.D. McQuaid (eds), Advances in Littorinid Biology.*

# Preface

Members of the gastropod family Littorinidae are common throughout the world. They form a very abundant component of many intertidal and shallow subtidal ecosystems and, through their grazing effects, often play a central role in shaping these communities. They also display a wide range of life history strategies and many are polymorphic, making them attractive model organisms for ecologists, evolutionary biologists and physiologists alike. Over the last 20 years several major fields of littorinid research have developed. These include ecological interactions with other animals and with algae; the effects of pollutants and the use of littorinids as sentinel species for monitoring pollution; the effects of parasites on growth and ecology; taxonomy and the study of genotypic/phenotypic responses to environmental factors. There is still much to be done in all of these fields and the littorinids are proving to be an ideal group on which to work.

The Fourth International Symposium on Littorinid Biology was held at the Marine Station at Roscoff, France from 19 to 25 September 1993 and was attended by over forty participants. This Proceedings contains 20 papers, including one which was presented as a poster and a companion paper to one of those presented. All of the papers have been subjected to normal refereeing procedures and a considerable number of referees were involved who were not associated with the meeting; we wish to thank all of the referees for the time and effort which they put into this work. Abstracts of the papers presented at the meeting have been published in volume 35 (1994) of Cahiers de Biologie Marine; not all of these were submitted for this volume.

Previous Symposia were held at the National History Museum in London, England (1986), the Tjärnö Marine Biological Laboratory, Sweden (1988) and the Dale Fort Field Centre, Wales (1990). The Proceedings of the second symposium were published as volume 193 of *Hydrobiologia* (and as Developments in Hydrobiology volume 56) (1990); those of the third by the Malacological Society of London (1992).

The meeting was felt to be a great success by all those who attended. We are indebted to Prof. Pio Fioroni, his wife Esther and Dr David Reid, who organised the symposium and did so much to make us all feel welcome; also, of course, to the Director of the Roscoff Marine Station, Professor Pierre Lasserre, for providing us with such excellent facilities. We also thank Kluwer Academic Publishers for agreeing to publish this volume. Last, but by no means least, we thank all of the participants for their presentations, for the many discussions and for their company, all of which made the meeting so enjoyable. The fifth symposium is planned for Cork, Ireland in September 1996.

PETER MILL
CHRISTOPHER MCQUAID

*Caption for photograph:*

*Back row*: Jörg Oehlmann, Martial Huet, Rupert Lewis, Frank Gentil, Gray Williams, Hans De Wolf, Yiu Ming Mak, Yoshitake Takada, Kevin Caley, Roger Hughes, John Grahame, Peter Mill, Bo Johannesson, Kerstin Johannesson.

*Middle row*: Elaine Rumbak, Sue Hull, Barbara Bauer, Delmont Smith, Johan Erlandson, Bob McMahon, Mark Davies, Sue Crossland, Thierry Backeljau, Nadezhda Zaslavskaya, Henrike Schmidtberg, Daniela Uthe, Stefanie Liebe, Gurutze Calvo Ugarteburu, Imke Ide, Ruth O'Riordan, Charles Fletcher, Emilio Rolan-Alvarez, Clara Johannesson.

*Front row*: Joe Britton, Jon Sigurdsson, Christopher McQuaid, Esther Fioroni, Pio Fioroni, David Reid, Roger Stamm, Helen Hughes, Andrey Tatarenkov.

*Not present on photograph*: Becky Britton, Pat Fletcher, Colette McMahon, Cesare Sacchi, Mireille Sacchi, Min Ho Son, Elly Stamm.

*Hydrobiologia* **309**: 1–14, 1995.
*P. J. Mill & C. D. McQuaid (eds), Advances in Littorinid Biology.*
©1995 *Kluwer Academic Publishers.*

1

# Resource allocation, demography and the radiation of life histories in rough periwinkles (Gastopoda)

R. N. Hughes
*School of Biological Sciences, University of Wales, Bangor, Gwynedd LL57 2UW, UK*

*Key words:* allocation trade offs, oviparity/brooding, body size, niche diversification

## Abstract

Applicability of life-history theory to higher levels of comparison (from populations, through ecotypes to sibling species) was investigated in rough periwinkles, whose life histories have diversified since colonization of the North Atlantic by an oviparous ancestor in the upper Pliocene. Comparisons were made among populations of the ovoviviparous *Littorina saxatilis*, between *L. saxatilis* and its ecotype, *L. neglecta* (with an annual life history) and between the sibling species *L. saxatilis* and *L. arcana*, the latter of which retains the ancestral oviparity.

Resource-allocation priority, reproductive effort and related trade offs were compared between the ecotypes and the sibling species by measuring changes in flesh mass and reproductive output in snails subjected to different degrees of food deprivation, and by measuring mortality rate of snails stressed by desiccation, high temperature and low salinity. Body size had a marked effect on all parameters, but after statistically removing this effect there remained no significant differences in allocation among ecotypes or species.

Published demographical data were reviewed for correlations between habitat, mortality regime and life-history characteristics. Populations of *L. saxatilis* varied principally in size at birth and in adult size. Theoretical premises based on density-dependent versus density-independent mortality regimes could not explain these trends. Instead, size at birth may have reflected the mechanical, physiological or biological nature of mortality risk rather than its density dependence or independence. Adult size reflected the available sizes of crevices used for shelter and perhaps also the quality of feeding conditions.

Radiation of life histories within the rough periwinkles is interpreted as a series of adaptations to a progressively wider range of habitats. The transition from oviparity to ovoviviparity allows colonization of estuaries, saltmarshes and pebble beaches too hazardous for naked egg masses. The transition from a perennial to an annual life history in barnacle ecotypes follows from allometric re-scaling of morphological and physiological parameters, enabling reproduction and brooding to occur at the small body size necessary for life within empty barnacle tests. This suite of adaptations allows exploitation of a relatively benign microhabitat that occurs almost ubiquitously on exposed rocky shores of the temperate North Atlantic. The persistence of oviparous forms, presumably in the face of competition from sympatric ovoviviparous forms, remains unexplained.

## Introduction

Physiologically based life-history theory rests upon the principle of allocation, that given a limited budget increased expenditure of resources on one function entails a corresponding decrease in expenditure on others (Sibly & Calow, 1986). This principle generates a set of possible trade offs between pairs of life-history characters (Sibly & Calow, 1986; Stearns, 1992). Exposure of such trade offs is experimentally

demanding but has met with success in some cases where the energy budget could be manipulated (e.g. Ernsting & Isaaks, 1991). There is little doubt, therefore, that such life-history theory captures the truth of nature at certain levels, particularly regarding the fine-tuning of resource allocation. But how far can the theory be extended to other levels of comparison? In moving from individuals, through populations to species and higher taxa, gross adaptations to habitat may overshadow matters of allocation and ultimately there must come a point where phylogeny exerts an overriding influence on life history. I address this question by experimentally examining the applicability of the allocation principle to the radiation of life histories in rough periwinkles, representing grades of differentiation from ecotypes to sibling species. Other aspects of life-history theory are centred upon mortality regimes, particularly density dependence versus density independence (Pianka, 1970), and I address this aspect by reviewing previously published data on the intra- and inter-specific demography of rough periwinkles.

Rough periwinkles populate rocky shores, harbours, estuaries and saltmarshes throughout the temperate North Atlantic. Collectively referred to as the *Littorina saxatilis* species complex, the component taxa *Littorina saxatilis* (Olivi), *L. arcana* Hannaford Ellis, *L. neglecta* Bean and *L. nigrolineata* Gray comprise a genetically recognizable (Ward, 1990) yet ecologically diverse group (Raffaelli, 1982). Probably, the group has ancestry in the North-East Pacific, having arisen from the trans-Arctic invasion of the Atlantic by an *arcana*-like ancestor, following the opening of the Bering Strait in the Upper Pliocene (Reid, 1990). Whereas some members of the group clearly are specifically distinct (*L. saxatilis*, *L. arcana*, *L. nigrolineata*), others are less easily categorized (*L. saxatilis*, *L. neglecta*) (Johannesson & Johannesson, 1990a,b). Whether the latter position reflects a young group in the process of radiation or a polymorphism held back from speciation by population-genetical constraints is debatable (Johannesson & Sundberg, 1992; Reid, 1993).

Taxonomic problems notwithstanding, the *L. saxatilis* group exhibits a remarkable diversity of life histories among closely related clades. Moreover it is reasonably clear which major life-history characters are ancestral and which are derived, for example oviparity is ancestral to ovoviviparity and a perennial ancestral to an annual life history (Reid, 1990). I sought evidence of allocation priorities and trade offs, expected to be relevant to the contrasted life histories, by comparing resistance to physiological stressors.

Two pairs of sympatric taxa were studied: *L. arcana* / *L. saxatilis*, and *L. neglecta* / *L. saxatilis*. *L. arcana* and *L. saxatilis* are often morphologically indistinguishable except for details of the reproductive tract (Hannaford Ellis, 1979). *L. arcana*, however, is oviparous, depositing egg masses in crevices and beneath stones on the upper shore. Each egg mass consists of a batch of eggs embedded in protective jelly. Accordingly, the reproductive tract is equipped with a large jelly gland. *L. saxatilis*, a sibling species to

*L. arcana*, is ovoviviparous, the ancestral jelly gland having become modified to serve as a brood chamber. The young leave the mother at the same developmental stage as young *L. arcana* crawl free of the egg mass.

*L. neglecta* is an ecotype (Reid, 1993), almost obligately associated with barnacles. The maximum size of adult *L. neglecta* is small enough for individuals to fit within the protective cavities afforded by the tests of dead barnacles (Raffaelli, 1978). Growth characteristics, size at maturation and brood capacity have become modified, enabling reproduction to proceed at sizes ancestrally typical of juveniles. Indeed, *L. neglecta* has been regarded as a neotenous derivative of *L. saxatilis* (Raffaelli, 1976).

The above differences in life history suggest that trade offs among competing functions may have operated during the radiation of taxa and that shifts in resource-allocation priority may have occurred. Trade offs can influence both the proportion and priority of allocation to competing functions. The proportion of allocation, for example of assimilated energy to reproduction (reproductive effort), is broadly determined by phylogeny and natural selection. Priority, while also susceptible to evolutionary forces, defines preferential allocation among functions when resources become limiting on a physiological time scale and does not necessarily reflect the proportion of allocation broadly set by inheritance.

This investigation is in three parts. The first part concerns the proportion and priority of allocation to somatic and reproductive functions in rough periwinkles. The change from oviparity to ovoviviparity morphologically involves only slight adjustments to the reproductive tract (Hannaford Ellis, 1979) and this, in itself, should not require alteration of the resource-allocation policy as reflected in reproductive effort. Survival of the brood, however, depends on survival of the mother. Moreover, packing constraints limit the size of the brood and so the loss of any young should be a relatively heavy penalty. These considerations predict that ovoviviparity should demand greater somatic priority over resources in times of stress than oviparity.

The change from a perennial to an approximately annual life history in *L. neglecta* has involved drastic reduction in adult size, with corresponding reduction in brood size. Allometric adjustments have enabled brooding to begin at a smaller size and earlier age, but the reproductive life span is still much shorter than in the ancestral condition. As a result of these factors, lifetime reproductive output in *L. neglecta* is an order

3

of magnitude less than that of sympatric *L. saxatilis*. Short adult life and limited energy reserves associated with small size lead to the prediction that allocation priority when under stress should be shifted more in favour of reproduction. Moreover, the shorter length of the adult phase relative to that of the juvenile in *L. neglecta* leads one to expect greater reproductive effort than in *L . saxatilis*.

The second part of the investigation is concerned with trade offs between somatic and reproductive functions and between subcategories within each of these. Trade offs amenable to investigation with the present material, and appropriate comparisons for revealing them, are listed in Table 1.

The third part reviews the ability of life-history theory to explain patterns observed in previous demographic studies on the *L. saxatilis* group. A model based on morphological and physiological adaptation to habitat is proposed to account for intraspecific variations in life-history characters. Major differences in reproductive mode among species are ascribed to phylogeny.

## Methods

### General maintenance

*Littorina arcana* (8.10-14.7 mm shell height, base to apex) and *L. saxatilis* (7.8-14.75 mm) were collected from the upper intertidal zone on exposed rocks at Abraham's Bosom, Anglesey, Wales (British National Grid reference SH 215814), where they were intermingled. *L. neglecta* (3.65-4.90 mm) and *L. saxatilis* (7.15-9.45 mm) were collected from an exposed cliff at Cable Bay (Porth Trecastell) Anglesey (SH 330707). At this site *L. neglecta* was collected from the upper barnacle zone and *L. saxatilis* from crevices and rock pools immediately above the barnacle zone, no more than 3 m from the source of *L. neglecta*.

Snails were maintained routinely in the laboratory at 12 °C. Following Warwick (1983), they were housed in plastic boxes furnished with native, algal covered stones as a source of food. Enough seawater was added to barely submerge the stones and any snails that crawled out of the water were reimmersed daily. Stones were replaced by fresh material every three days.

### Part 1. Allocation priority

Allocation priority was investigated by comparing the relative responses of growth and reproductive output to food deprivation. Mature females, whose growth and reproductive output was to be monitored, were placed individually in separate boxes, each accompanied by four males to ensure fertilization. Food ration was manipulated by controlling exposure to native, algal covered stones as used in routine maintenance. There were three treatments: full ration, where snails were transferred at 3 d intervals to clean boxes furnished with fresh algal-covered stones; half ration, where snails were transferred alternately at 3 d intervals to clean boxes with or without algal covered stones; and zero ration, where snails were transferred at 3 d intervals to clean boxes without algal covered stones. After a conditioning period of 36 days exposure to the experimental food ration, each female was weighed both in air and when immersed in seawater, enabling flesh mass and shell mass to be estimated separately (Palmer, 1982). This was repeated after a further 90 days, and the difference in flesh mass, corrected for changes in mass of the reproductive tract by subtracting the final mass from the estimated initial mass of the tract, was used to estimate the relative growth rate as $\log_e[2w_2/(w_1+w_2)+1]$. Also, reproductive output was measured at each 3 d transfer by counting and weighing any crawling young in the case of *L. saxatilis* and *L. neglecta* or eggs in the case of *L. arcana*. Reproductive output was finally expressed as the total over 90 d.

Each experiment was repeated, with a different batch of 10 snails per treatment per species, at two times of the year (January-March, June-August) to account for any seasonal effects on performance.

### Statistical analyses

Data were analysed using the GLM command of Minitab (Ryan *et al.*, 1985). $\log_e$ shell height, taken as an independent measure of size, was used as covariate to detrend for body size. F-ratios were calculated from sequential mean squares in the order length, ration, and reproductive activity or species, as appropriate. The interaction term for species and ration was used to test for differences in resource-allocation priority. Snails that failed to reproduce during both the conditioning period and the experiment (126 days) were excluded from the analysis of reproductive output.

3

Table 1. Possible life-history trade offs in the Littorina saxatilis group and comparisons used to test for their existence.

| Trade-off | Comparison |
|---|---|
| Somatic/reproductive allocation | Growth of fertilized and unfertilized L. saxatilis |
| Parental investment/juvenile survivorship | Survivorship under stress of juvenile L. arcana and L. saxatilis |
| Parental investment/adult survivorship | Survivorship under stress of adult L. arcana and L. saxatilis |

Table 2. Biochemical analysis of the flesh of snails maintained for 36 days on three levels of food supply. Analyses (see Methods) are based on an homogenized sample of 10 snails per treatment. Data are means of 3 determinations, with standard errors. FR = full ration, HR = half ration, ZR = zero ration, as described in Methods. All units are in J mg$^{-1}$

| Taxon | Protein | | | Carbohydrate | | | Lipid | | |
|---|---|---|---|---|---|---|---|---|---|
| | FR | HR | ZR | FR | HR | ZR | FR | HR | ZR |
| L. arcana | 29.0 (1.0) | 29.5 (0.5) | 29.8 (1.0) | 7.1 (0.3) | 11.4 (0.3) | 6.2 (1.5) | 16.3 (0.4) | 13.6 (0.6) | 14.1 (1.0) |
| L. saxatilis | 26.2 (1.0) | 27.6 (0.5) | 30.4 (0.5) | 9.8 (0.8) | 11.2 (2.4) | 2.9 (0.2) | 17.9 (0.6) | 15.9 (0.5) | 13.0 (1.4) |
| L. saxatilis | 26.1 (0.6) | 30.1 (0.6) | 36.0 (0.8) | 30.9 (1.0) | 15.9 (0.6) | 13.5 (0.2) | 17.1 (0.8) | 16.1 (1.5) | 15.8 (0.7) |
| L. neglecta | 31.7 (0.6) | 32.8 (0.9) | 34.6 (0.8) | 30.8 (0.2) | 28.5 (3.0) | 20.9 (0.8) | 17.7 (1.4) | 18.8 (0.4) | 17.9 (1.5) |
| egg masses | 31.3 (0.0) | | | 5.5 (0.2) | | | 3.8 (0.2) | | |

## Reproductive effort

Data on growth and reproduction obtained from the allocation-priority experiments (above) were converted to energy units first by analysing for gross biochemical composition, then applying energy equivalents of 23.645 J mg$^{-1}$ for protein, 17.159 J mg$^{-1}$ for carbohydrate and 39.548 J mg$^{-1}$ for lipid (Brody, 1945). Protein was assayed using Sigma Diagnostics Protein Assay Kit no. P5656 (Sigma Chemical Company Ltd., Poole, Dorset, England), carbohydrate using Boehringer Mannheim Diagnostic Kit no. 124010 (Boehringer Mannheim UK Ltd., Lewes, East Sussex, England) and lipid using Boehringer-Manheim Diagnostics Kit no. 124303. Results of the analyses are summarized in Table 2. Newly released young of L. saxatilis and L. neglecta were assigned energy equivalents of 26 J mg$^{-1}$ (Holland et al., 1975; Hughes & Roberts, 1980) and the egg masses of L. arcana were analysed biochemically, as for flesh.

Measurements of respiration rate were made on additional sets of 10 individuals per source population. These sets were subjected to the experimental food rations for 36 days. Individual snails were then placed in 5 ml vials, immersed in a constant-temperature bath held at 12 °C. The decline in oxygen tension of the seawater, stirred by magnetic flea, was measured for 1 h using oxygen electrodes. Regressions of respiration rate on dry flesh mass, after logarithmic transformation of the data (Table 3), were used together with an oxycalorific equivalent of 20.3601 J ml$^{-1}$ (Elliott & Davidson, 1975), to estimate metabolic energy loss at each food ration.

Growth, reproduction and metabolism were standardized to units of J mg$^{-1}$ d$^{-1}$ and reproductive effort was then calculated as reproduction/growth+reproduction+metabolism.

## Part 2. Trade-offs

### (i) Somatic/reproductive allocation
L. saxatilis were collected from Cable Bay. Those bearing a penis were discarded and the remainder placed individually in plastic boxes according to general maintenance procedures (above). Individuals producing young within a week were then assigned to two groups. One group was maintained in a communal aquarium without males, while the other was maintained in an aquarium along with four males per female, the latter being marked with paint. After six months on

Table 3. Regressions of oxygen-consumption rate ($\mu$l $O_2$ $h^{-1}$) on dry body mass (mg), after logarithmic transformation of data. F = full ration, SF = half ration, S = starved. n = 10 for all regressions.

| Species | Diet | Range of body size (mm) | Slope | SD | Intercept | SD | P | $R^2$ |
|---------|------|-------------------------|-------|-----|-----------|-----|---|-------|
| L. arcana | P | 10.0–21.4 | −0.0930 | 0.0956 | 0.6528 | 0.1121 | 0.359 | 0.106 |
| | SF | 3.5–22.4 | 0.0645 | 0.1032 | 0.4076 | 0.1199 | 0.549 | 0.047 |
| | S | 10.2–20.0 | −0.1122 | 0.0914 | 0.6317 | 0.1051 | 0.254 | 0.159 |
| L. saxatilis | P | 10.2–18.6 | −0.3287 | 0.1311 | 0.9780 | 0.1519 | 0.037 | 0.440 |
| | SF | 9.3–20.0 | −0.0817 | 0.0970 | 0.5741 | 0.1124 | 0.424 | 0.081 |
| | S | 10.2–19.5 | −0.3330 | 0.1407 | 0.8909 | 0.1629 | 0.046 | 0.412 |
| L. saxatilis | F | 1.2–4.3 | 0.5280 | 0.0541 | 0.1978 | 0.0221 | 0.001 | 0.923 |
| | SF | 1.7–4.2 | 0.6246 | 0.1158 | 0.0464 | 0.0479 | <0.001 | 0.784 |
| | S | 1.5–3.9 | 0.1908 | 0.1197 | 0.1970 | 0.0489 | 0.150 | 0.241 |
| L. neglecta | F | 1.4–3.9 | 0.2907 | 0.0726 | 0.3121 | 0.0296 | 0.004 | 0.667 |
| | SF | 1.5–4.1 | 0.1010 | 0.0819 | 0.2966 | 0.0356 | 0.252 | 0.160 |
| | S | 1.4–3.5 | 0.6170 | 0.1168 | 0.0395 | 0.0459 | <0.001 | 0.777 |

Table 4. Energy-allocation budgets, listing specific rates of somatic growth, reproduction and respiration (J $mg^{-1}$ $d^{-1}$) among species and diets. RE = reproduction/(reproduction+growth+respiration), RE' = reproduction-growth.

| Species | Diet | Growth | Reproduction | Respiration | RE | RE' |
|---------|------|--------|--------------|-------------|-----|-----|
| L. arcana | F | 0.0043 | 0.0343 | 0.1651 | 0.17 | 0.03 |
| | SF | 0.0022 | 0.0100 | 0.1462 | 0.06 | 0.01 |
| | S | −0.0558 | 0.0016 | 0.1753 | 0.01 | 0.06 |
| L. saxatilis | F | 0 | 0.0123 | 0.1250 | 0.09 | 0.01 |
| | SF | −0.0008 | 0.0095 | 0.1193 | 0.07 | 0.01 |
| | S | −0.0686 | 0.0062 | 0.1511 | 0.07 | 0.07 |
| L. saxatilis | F | 0.0040 | 0.0463 | 0.3601 | 0.11 | 0.04 |
| | SF | 0.0030 | 0.0332 | 0.2927 | 0.10 | 0.03 |
| | S | −0.0909 | 0.0209 | 0.3287 | 0.08 | 0.11 |
| L. neglecta | F | 0.0038 | 0.0345 | 0.5733 | 0.06 | 0.03 |
| | SF | 0.0039 | 0.0120 | 0.4479 | 0.03 | 0.01 |
| | S | −0.0738 | 0.0252 | 0.4386 | 0.07 | 0.10 |

full food ration, the reproductive output of the group lacking males had dwindled to zero and these females were deemed unfertilized, even though it is known that sperm can be stored for at least this length of time (D.G. Reid personal communication). The reproductive output of the group containing males continued at a high level and these females were deemed fertilized.

Ten individuals from each of the fertilized and unfertilized groups were then maintained on each of the three food rations and their growth monitored, as described for allocation priority (above). Using GLM,

the significance level of the main effect, reproductive activity, was used as one criterion for evaluating any trade off between somatic and reproductive allocation. Reproductive output was also plotted as a function of time during the experiment, since this revealed trends obscured by the gross comparisons made with GLM.

*Parental investment/juvenile survivorship*
The change from oviparity to ovoviviparity may be regarded as an increase in parental investment per

offspring because brooding increases embryonic residence time in the reproductive tract. It remains to be shown whether or not brooding entails a greater energetic investment per offspring. Early embryos can be raised in vitro, although the success rate may be low due to bacterial infection (Roberts, 1979). Moreover, gestation times are comparable with those of oviparous species (Roberts, 1979; Hughes, 1980; Warwick *et al.*, 1990). Despite these considerations, it remains possible that embryos receive nourishment from within the brood chamber.

Egg masses of *L. arcana* (low parental investment) and embryos within the brood chamber of *L. saxatilis* (high parental investment) were subjected to the potential stressors desiccation, high temperature and low salinity, using the general protocol described in Cannon & Hughes (1991) and summarized as follows.

Desiccation for 3 h at 2% RH at 15 °C (approximately 14 mm Hg aqueous vapour pressure deficiency) was applied to five *L. arcana* egg masses and five brooding *L. saxatilis*. The egg masses of *L. arcana* were replaced in seawater and the percentage of embryos eventually hatching compared with undesiccated controls. The embryos of *L. saxatilis* were dissected from the brood chamber and examined for signs of mortality (partial disintegration of early stages, lack of movement in later stages), compared with undesiccated controls.

One *L. arcana* egg mass and one brooding *L. saxatilis* were allocated to each of ten temperatures from 26-35 °C and exposed to the experimental temperature for 24 h. Mortality was estimated as for desiccation.

One *L. arcana* egg mass and one brooding *L. saxatilis* were allocated to each of five salinities (0, 5, 10, 15 and 35 ‰) and immersed in the medium for 5 d. Mortality was estimated as before.

Young, newly emerged from the egg mass of *L. arcana* and from the brood chamber of *L. saxatilis*, were subjected to the stressors, as above. Batches of juveniles of 2–3 mm shell height (1–2 months after emergence from the egg mass or brood chamber) per species were treated similarly. Equivalent experiments for *L. saxatilis* and *L. neglecta* are described in Cannon & Hughes (1992).

*Parental investment/adult survivorship*
Batches of 20 adult *L. arcana* and of *L. saxatilis* were subjected to the stressors, as above, using three replicate batches per treatment (see Cannon & Hughes (1992) for *L. saxatilis* and *L. neglecta*).

## Results

### Part 1. Allocation priority

Season had no significant effect on growth or reproductive output and so data from replicate experiments were pooled for further analysis.

### (i) L. arcana and L. saxatilis

Flesh growth was close to zero at full and half rations, becoming strongly negative at zero ration (Fig. 1a). Thus, whereas the main effect, ration, was highly significant ($F_{2,108} = 94.4$, $P<0.0001$), there was no significant difference between species ($F_{1,108} = 0.04$, $P = 0.85$). The interaction between ration and species was significant ($F_{2,108} = 4.1$, $P = 0.02$), but this was due to the opposite signs of growth rate at half ration (Fig. 1a) and not to any meaningful trend.

Only 19 of the 60 *L. arcana* produced egg masses, severely limiting the power of the experiment. Thus, although reproductive output was diminished at half and zero rations (Fig. 1b), the effect of ration was barely statistically significant ($F_{2,13} = 3.7$, $P = 0.054$). All the *L. saxatilis* liberated young, but at a progressively declining rate (Fig. 1b). The effect of ration on total reproductive output was not statistically significant ($F_{2,54} = 0.8$, $P = 0.46$).

Since food deprivation strongly depressed flesh growth but not reproductive output, it appears that priority of resource allocation was given to reproduction. Lack of a consistent inter-specific difference in the response of growth to ration suggests that there was no difference in the degree of priority given to reproduction by *L. arcana* and *L. saxatilis*.

### (ii) L. saxatilis and L. neglecta

For both ecotypes lower rations increasingly depressed flesh growth, but whereas the flesh growth of *L. saxatilis* became negative only at zero ration, that of *L. neglecta* was negative at all rations (Fig. 2a). Even at full ration, therefore, conditions must have been stressful to *L. neglecta*, perhaps through suppression of feeding activity. The inverse relationship between ration and flesh growth, however, followed a similar trend in both ecotypes. Thus, ration had a highly significant effect on growth ($F_{2,108} = 122.0$, $P<0.0001$) and there was a highly significant difference in growth between ecotypes ($F_{1,108} = 51.2$, $P<0.0001$), but the

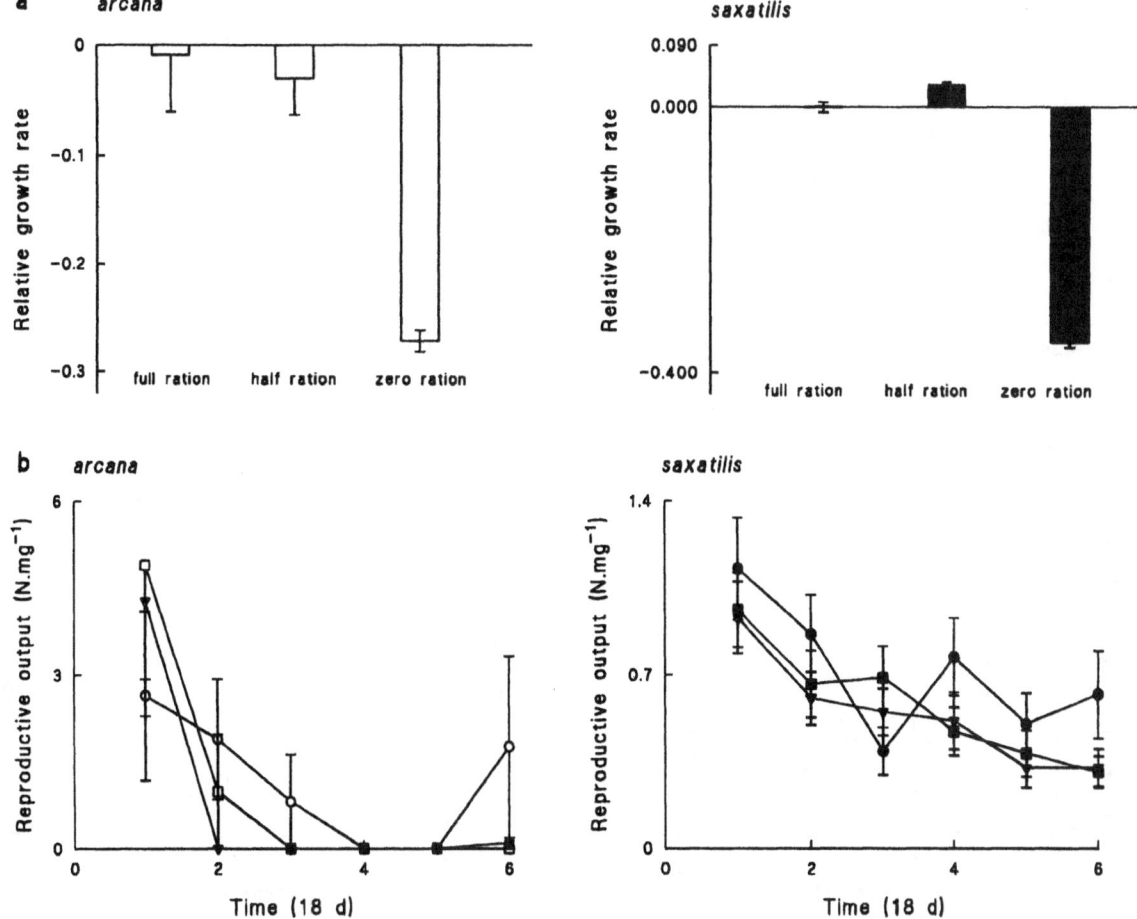

*Fig. 1.* Resource allocation, at three levels of food supply, in the sibling species *Littorina arcana* and *L. saxatilis* (a) mean flesh growth, with standard errors. Adjustment of the means to remove the effect of body size was not done, since the regression slopes were significantly different ($F_{1,108} = 4.1$, $p = 0.045$). (b) mean number of young (*L. saxatilis*) or eggs within egg masses (*L. arcana*) produced per 18 d interval, with standard errors. ○ full ration, □ half ration, ▽ zero ration

interaction between diet and ecotype was not significant ($F_{2,108} = 0.3$, $P = 0.74$).

Reproductive output of both species declined during the experiment, the level of output tending to be lower at increasing food deprivation, particularly in the case of *L. neglecta* (Fig. 2b). The effect of ration on total reproductive output over the experimental period was not statistically significant for *L. saxatilis* ($F_{2,51} = 1.4$, $P = 0.26$), but was highly significant for *L. neglecta* ($F_{2,42} = 5.6$, $P = 0.007$). The greater depression of reproductive output by starvation in *L. neglecta* probably reflected the limitation of energy reserves caused by smaller body size, and perhaps also by unfavourable feeding conditions (discussed above). These possibilities, together with the non-significant interaction between the effects of ration and ecotype

on growth (above), suggested that there was no difference in allocation priority between ecotypes.

### Reproductive effort

Reproductive output at full ration accounted for some 10% of assimilated energy in *L. saxatilis*, 17% in *L. arcana* and 6% in *L. neglecta* (Table 4). The relatively high estimate for *L. arcana* and low estimate for *L. neglecta* were caused by the allometry between respiration rate and body size. When expressed as the difference between reproductive output and flesh growth, reproductive effort was similar in these taxa (Table 4). Only in the case of *L. arcana* was there a marked reduction in reproductive effort at decreased ration. This result should be regarded with caution since it was

8

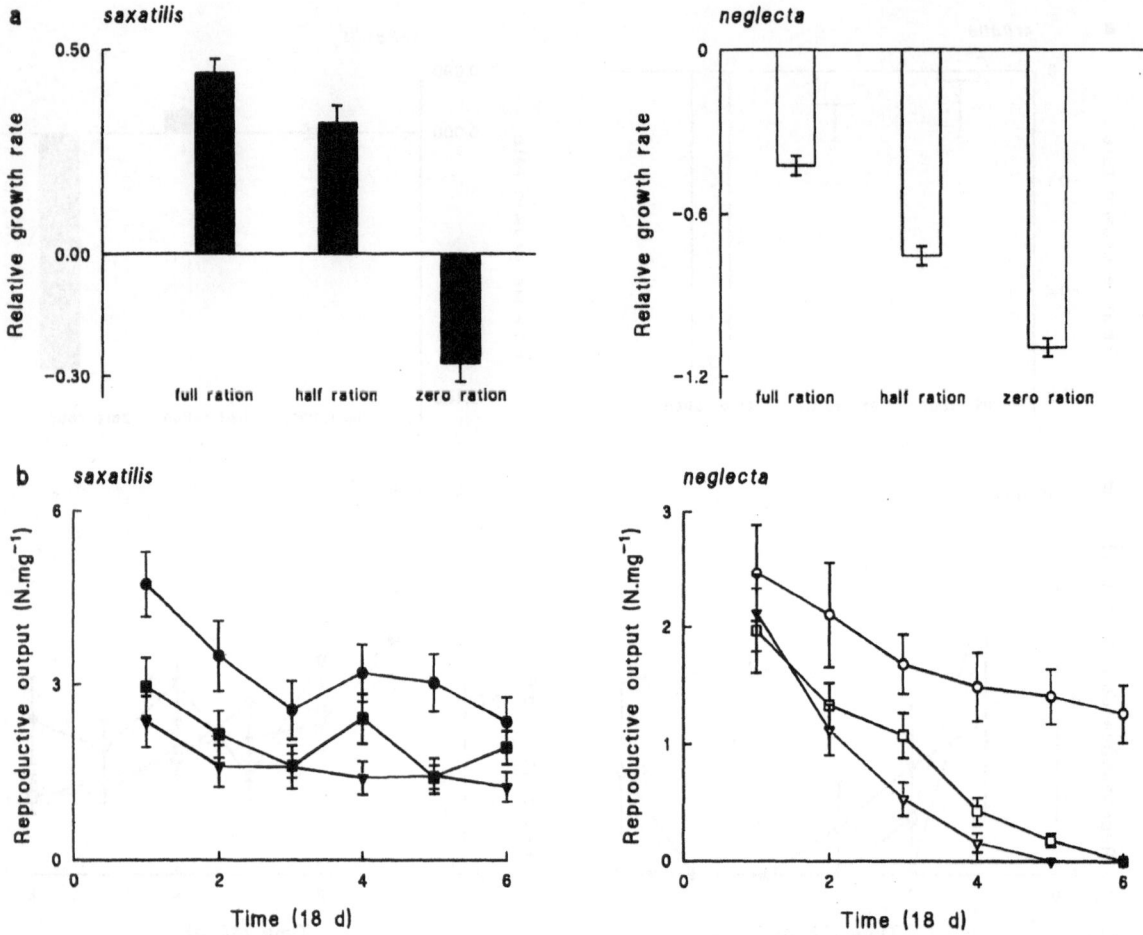

*Fig. 2.* Resource allocation, at three levels of food supply, in the ecotypes *Littorina saxatilis* and *L. neglecta* (a) mean flesh growth, adjusted by GLM to remove the effect of body size, with standard errors. The regression slopes for the effect of body size were not significantly different ($F_{1,108} = 0.1$, $P = 0.75$). - (b) mean number of young released per 18 d interval, with standard errors. ○ full ration, □ half ration, ▽ ration

based on very small samples of reproductively active snails. As the estimates of reproductive effort were based on mean values for the constituent variables, it was not possible to derive standard errors and so no statistical comparison among taxa could be made.

### Part 2. Trade-offs

#### (i) Somatic/reproductive allocation

At all food rations, flesh growth of unfertilized, nonreproductive *L. saxatilis* was significantly greater than that of fertilized, reproductive individuals (Fig. 3) ($F_{1,48} = 61.3$, $P < 0.0001$). This result suggested that somatic and reproductive functions were competing for resources, as envisaged by the concept of a trade-off. The difference in growth between reproductive

and non-reproductive snails, however, was uncorrelated with ration, contrary to expectation if allocation is prioritized when resources become limited.

#### (ii) parental investment/juvenile survivorship

Embryos of *L. arcana* suffered a mortality of 65%±6% (SE) when the egg masses were desiccated, whereas no mortality occurred among the brooded embryos of *L. saxatilis* when the mothers were desiccated. Embryos within the egg masses of *L. arcana* sustained 7% and 12% mortality at 26 °C and 27 °C, whereas there was no mortality at these temperature among the brooded embryos of *L. saxatilis*. Conversely, at 34 °C and 35 °C all *L. saxatilis* embryos died whereas those of *L. arcana* suffered only 66% and 84% mortality. Between the range 29-33 °C, the mortality of

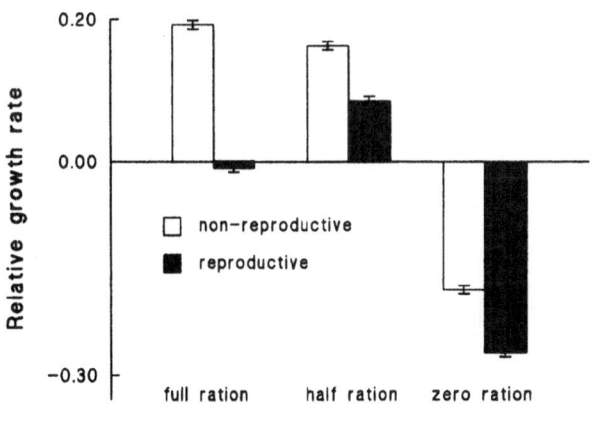

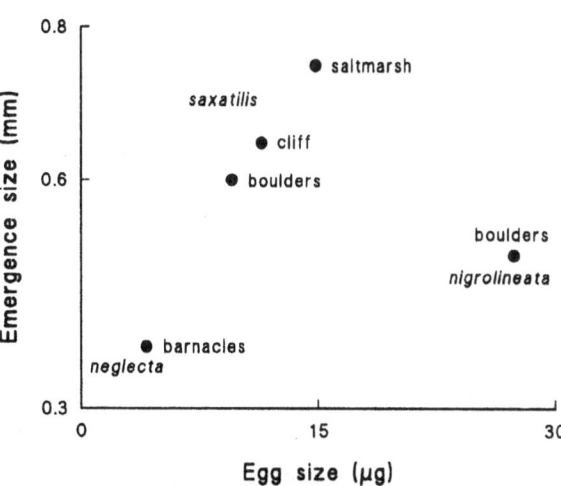

**Diet**

*Fig. 3.* Allocation trade off. Flesh growth of fertilized (reproductive) and unfertilized (non-reproductive) *Littorina saxatilis* at three levels of food supply. Data are means, adjusted by GLM to remove the effect of body size, with standard errors.

*Fig. 5.* Size of the newborn emerging from egg masses (*Littorina nigrolineata*) or from the brood chamber (*L. saxatilis*). Emergence size is the distance from base to apex of the protoconch and egg size is the ash-free dry weight of the young (including egg-mass jelly in the case of *L. nigrolineata*). Species are referred to by their trivial names. (original data from Hughes & Roberts, 1980; Fish & Sharp, 1985).

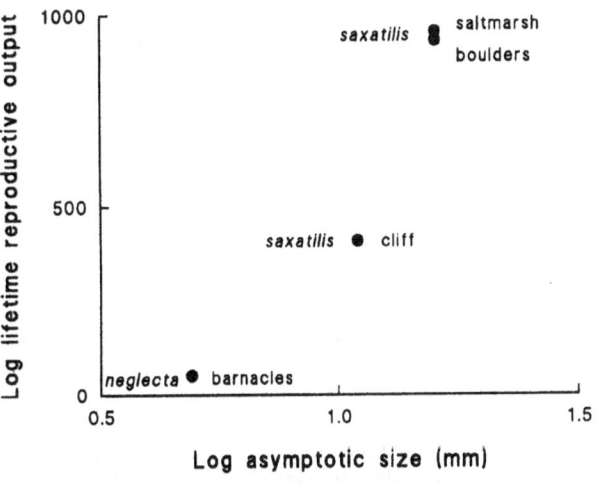

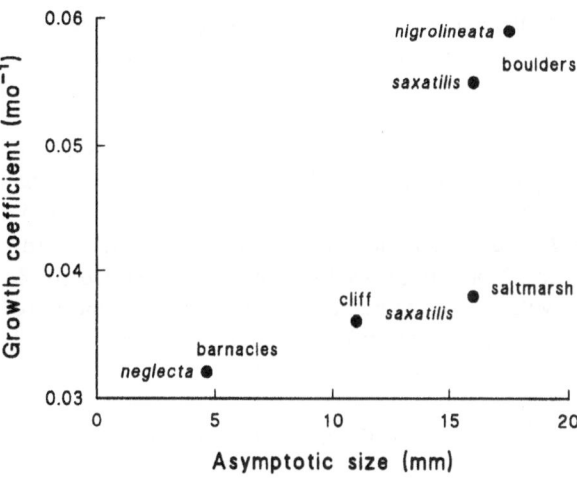

*Fig. 4.* Lifetime reproductive output plotted as a function of adult size, after logarithmic transformation of the data. Reproductive output was calculated from life tables by summing the total numbers of young produced at successive age classes. Size is expressed as the maximum size recorded in the population. Species are referred to by their trivial names. (original data from Hughes, 1980; Hughes & Roberts, 1981; Fish & Sharp, 1985).

*Fig. 6.* Growth rate of the shell plotted as a function of final, adult size. Growth rate is expressed as the coefficient, k, of the Bertalanffy equation and asymptotic size is estimated by the maximum size found in the population. Species are referred to by their trivial names. (original data from Roberts & Hughes 1980; Fish & Sharp, 1985).

*L. saxatilis* embryos always exceeded that of *L. arcana* embryos. The poorer survival of brooded embryos under high-temperature stress probably reflected the limited opportunity for ventilation within the brood chamber. Percentage mortality of *L. arcana* embryos increased from background levels of 4.6–7.9% at salinities of 32–15‰ to 82.4% at 0‰. Brooded embryos of *L. saxatilis* suffered no mortality except at zero salinity, when the death rate was 57.5%.

All newly emerging young of *L. arcana* and *L. saxatilis* died by the third hour of desiccation and the mortality distributions over time were not significantly different between species (Survival Analysis, SPSS 1990, *P*>0.9). When exposed to low salinities mortality reached at least 80% by day 10. At salinities of 0, 10 and 15‰ distributions of mortality over time were not significantly different between species (Survival Analysis, *P*≥0.3), but at 5‰ *L. arcana* began dying

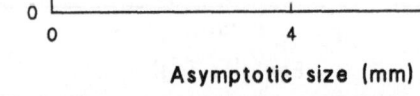

Fig. 7. Size on first attaining sexual maturity plotted as a function of final, adult size. Size at maturation was determined by dissection. Asymptotic size was estimated by the maximum size found in the population. Species are referred to by their trivial names. (original data from Hughes, 1980; Roberts, 1979; Roberts & Hughes, 1980).

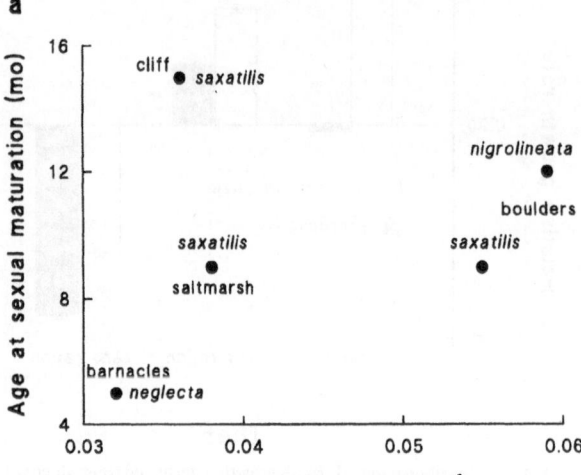

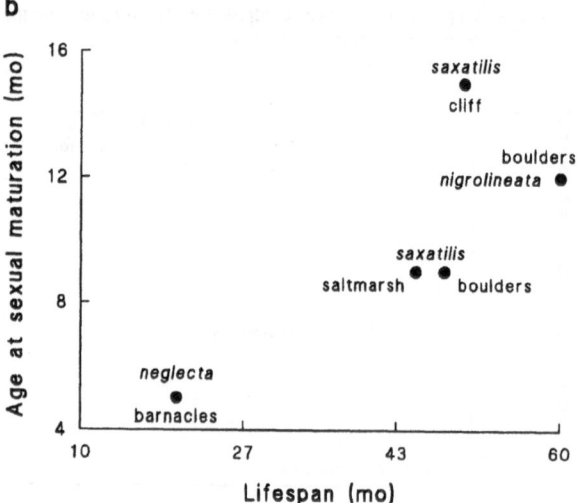

Fig. 8. (a) Age at first attaining sexual maturation plotted as a function of growth rate. Age at sexual maturation was determined by dissection. Growth rate is expressed coefficient, k, from the Bertalanffy equation. Species are referred to by their trivial names. (original data from Hughes & Roberts, 1980; Roberts & Hughes; 1980, Fish & Sharp, 1985). mo, months. (b) Age at sexual maturation plotted as a function of maximum longevity. Longevity was estimated by using growth curves to predict the age upon reaching the maximum size recorded in the population. Species are referred to by their trivial names. (original data from Roberts, 1979; Roberts & Hughes, 1980; Fish & Sharp, 1985). mo, months

earlier than *L. saxatilis* ($P = 0.04$), this departure from the general trend possibly being a batch, rather than a specific, effect.

Mortality of juveniles (young 2-3 mm in shell height) during 96 h of desiccation ranged was not significantly different between species (*L. arcana* mean $= 58.3\%$, *L. saxatilis* mean $= 56.7\%$, $t_4 = 0.45$, $P = 0.20$) (mortality distributions over time not recorded). Mortality caused by salinity stress reached 100% within 12 d. Mortality distributions over time were not significantly different between species (Survival Analysis, all salinities $P \geq 0.16$).

### (iii) Parental investment/adult survivorship

There was no mortality among adult *L. arcana* or *L. saxatilis* after desiccation for 96 h. There was no significant difference between species in resistance to high-temperature stress (probit analysis, $P > 0.05$).

## Discussion

### Experimental procedures

The resolution of the resource-allocation experiments was impaired by three factors. First, Palmer's method (1982) of partitioning total mass into flesh and shell mass has a characteristic error of 5-10%. Consistent patterns among taxa, however, lend confidence in the

ability of the technique to reveal differential responses of flesh growth to ration. Second, gestation time in *L. arcana* and *L. saxatilis* is of the order of 40 d (Roberts, 1979) and the lag between copulation and reproductive output may exceed 100 d (Warwick *et al.*, 1990). Consequences of any changes in the number of eggs fertilized after the first month of the experiment therefore might not have been detected. On the other hand, changes in gestation time or abortion rate of embryos established before, or early in, the experiment would have resulted in decreased output. Any significant effect of ration on reproductive allocation therefore should have been detectable. Third, only 32% of *L. arcana* laid egg masses. This may have been due to some unfavourable aspect of laboratory conditions, such as a lack of attractive oviposition sites, or to an intrinsically low rate of oviposition.

## Parts 1 and 2. The principle of allocation and associated trade-offs

Despite the above limitations, clear patterns emerged. There was a trade off in allocation of resources to somatic and reproductive functions: reproductive snails lost more flesh mass than non-reproductives when short of food. There was a priority in allocation: when resources became increasingly limited, flesh growth declined markedly in all species, whereas reproductive output was affected less consistently. Thus, reproductive output was unaffected by decreasing ration in *L. arcana* and sympatric *L. saxatilis*, slightly depressed in *L. saxatilis* sympatric with *L. neglecta* and more strongly depressed in *L. neglecta* itself. This graded effect of food deprivation on reproductive output corresponded to the order of decreasing body size (Methods). Only in smaller snails, with smaller energy reserves, did the decrease in allocation extend beyond flesh growth to reproductive output, a pattern that may be seen as a consequence of body size itself rather than of differences in allocation. Thus, priority in allocation was given to reproductive over somatic functions, but the degree of this priority remained the same among the different life histories.

Reproductive effort, expressed as the proportion of assimilated energy allocated to reproductive output, varied about a mean of approximately 10% without any clear relationship to ration. The principle of allocation therefore has no bearing upon the change from oviparity to ovoviviparity or from a perennial to an annual life cycle in these snails. Other allocation trade offs, also, seemed to be unimportant. Thus increased parental investment by brooding, while securing greater embryonic survivorship under severe desiccation and low-salinity stress, did not impair maternal resistance to physiological stressors.

## Part 3. Demography and the role of body size

Body size is an important variable among populations of rough periwinkles. The contrast between *L. saxatilis* and *L. neglecta* is an extreme example. There are lesser, but highly significant, differences in body size among populations of all species in the *L. saxatilis* complex and these differences are correlated with habitat (Raffaelli, 1982; Hart & Begon, 1982; Janson, 1983; Grahame *et al.*, 1990; Grahame & Mill, 1992). Body size reveals its importance at two stages in the life history: size of the newborn young and size of the adult.

Size of young emerging from the egg mass or brood chamber varies among populations inhabiting wave-beaten cliffs, moderately exposed boulder shores and sheltered, stony beaches or saltmarshes. Wide-scale surveys have shown that size of the newborn young of *L. saxatilis* is least on boulder shores, intermediate on exposed cliffs and greatest on sheltered shores (Raffaelli, 1976; Hughes & Roberts, 1980). Having failed to find any demographical characteristics that relate these differences to the predictions of life-history theory, Hughes & Roberts (1980, 1981) proposed that size of the newborn young reflects not so much the density-dependent or -independent nature of mortality agents, but the mechanical, physiological or biological nature of these agents. On boulder shores, mobility of the substratum during storms causes heavy mortality and, at the small juvenile stage, increased size would not secure significantly greater protection from mechanical impact. Moreover, damp, shaded microhabitats among boulders provide refuge from desiccation and heat stress. Size at birth, therefore, is not of paramount importance and smaller young may be produced at a gain in fecundity. On exposed cliffs, there is a significant chance of snails failing to find a crevice when the bare rock above the barnacle zone dries out on the receding tide (Raffaelli & Hughes, 1978). Larger size, through lessening the surface/volume ratio, potentially increases resistance to desiccation. Whereas adults are unlikely to be stressed long enough to be killed in this way (Cannon & Hughes, 1992), newborn young are more vulnerable (present study) and larger size may enhance their survivorship on dry rocks. Mean adult size on exposed cliffs is relatively small. The reason for this, whether related to crevice characteristics or feeding conditions, has not been determined unequivocally (Raffaelli & Hughes, 1978). Nevertheless, the smaller adult size limits brooding capacity and so the size of the newborn may represent a compromise between increasing resistance to physiological stressors and the retention of an adequate level of fecundity. It should be kept in mind, however, that differential resistance to stressors among different sized newborn has not been measured and could be small. On sheltered shores juvenile crabs, *Carcinus maenas* (L.), are a seasonally abundant predator on young periwinkles (personal observation). Larger newborn require a shorter time to grow to a size which offers a refuge from juvenile crab predation. On these shores, sheltering sites or feeding conditions evidently do not constrain adult size and so despite the large size of the newborn, fecundity is relatively high (Hughes & Roberts, 1980, 1981).

Other authors have interpreted such patterns somewhat differently. Hart & Begon (1982) and Naylor & Begon (1982) retained the concept of density-dependent and density-independent factors to explain the smaller newborn and larger adults of both *L. saxatilis* and *L. nigrolineata* on a boulder shore compared with a neighbouring exposed cliff. Larger size was thought to give the newborn a competitive advantage in the occupation of limited crevices on the cliff. This may be unlikely, however, for two reasons. First, newborn snails can fit into very small crevices and these are generally more abundant than larger crevices. Second, the presence of larger snails creates sheltered interstices in crevices that otherwise would be too wide for smaller snails (Raffaelli, 1976). Therefore it cannot safely be assumed that the limited availability of crevices for adults on cliffs (Raffaelli & Hughes, 1978) applies equally to the newborn. Both schools of thought agree that size of the newborn is adaptive, but ascribe this to different sources of mortality. Microhabitat requirements of the newborn in the field remain unstudied and further experimentation will be required to resolve the issue. The smallest newborn, however, are produced by *L. neglecta* and here size of the newborn probably reaches its functional minimum, reflecting the need to maximize clutch size when brooding capacity is severely restricted.

Adult size is greatest on boulder shores and on shores of various types that are sheltered from strong wave action. Sheltering sites are abundant and foraging is seldom curtailed by inclement weather, allowing body size to reach the maximum set by phylogeny. Greater body size, through greater clutch capacity and longer lifespan, increases lifetime reproductive output (Fig. 4). Adult size is intermediate at high levels on exposed cliffs, where crevices are of limited availability and where heavy wave action or hot dry spells interrupt foraging. It is least within the midshore barnacle zone on exposed shores, where small size is necessary for adults to fit within the interstices of empty barnacle tests.

## The growth curve and other correlates of body size

Population differences in size of the newborn and adults result from adjustments to egg size and to the ensuing growth curve. Among ovoviviparous populations, egg size is highly correlated with size of the newborn (Fig. 5). The outlying datum for the oviparous *L. nigrolineata* may be biased because egg-mass jelly was included in the estimation (Hughes & Roberts, 1980). Growth curves of rough periwinkles can be well described by the Bertalanffy equation (Roberts & Hughes, 1980). Larger asymptotic size is associated with faster growth, individuals on a boulder shore growing disproportionately faster than those on other types of shore (Fig. 6).

Excluding *L. neglecta* and presumably other barnacle-inhabiting ecotypes, body size at sexual maturation is only weakly correlated with asymptotic size (Fig. 7). Maturity is reached at a shell height of some 5.5-7.0 mm (Hughes & Roberts, 1980) and this may be the minimum size at which brooding becomes feasible. In barnacle ecotypes such as *L. neglecta*, allometric changes in morphology allow brooding to begin at a shell height of about 1.5 mm (Fish & Sharp, 1985) and their inclusion in the data set considerably strengthens the correlation between size at maturation and asymptotic size. In contrast to size, age at maturation varies considerably among populations, and although bearing no significant relationship to growth rate (Fig. 8a), it is correlated to life span (Fig. 8b).

In summary, putatively adaptive, intra- and interspecific differences occur among populations in size at birth and in adult size. Size at birth is determined by egg size and so is inversely related to fecundity. Other things being equal, therefore, there should be a tendency to keep eggs small. The smallest newborn, however, are produced by *L. neglecta* as a consequence of limited brooding capacity associated with the small adult size necessary to exploit the barnacle-zone habitat. Other brooders, less constrained volumetrically, produce larger young that presumably survive better after release. The size range of the newborn among these brooders overlaps that of newborn emerging from the egg capsules of non-brooders. In both cases, variation in size is linked to habitat, probably through the type of mortality factors threatening the newborn.

*A habitat-exploitation model*

The evolution of brooding, by protecting the embryonic stages, enables *L. saxatilis* to thrive in habitats such as saltmarshes, estuaries, mobile pebble beaches and silty shores that would be lethal to egg masses and therefore incompatible with oviparity, as retained from the ancestral condition by *L. arcana* and *L. nigrolineata* (Mill & Grahame, 1992). This evolutionary step has required little or no adjustment to the allocation programme and has compromized no attribute other than fecundity, which may be an order of magnitude less in brooders (Hughes & Roberts, 1981).

The evolution of an annual life history in *L. neglecta*, again, has required no adjustment to the allocation programme other than that brought about by compression of the physiological time scale. The suite of morphological and physiological allometric adjustments involved in this compression enable the life history to be completed at a body size smaller than the interstices of barnacle shells. This widespread, relatively stable and physiologically benign mid-shore habitat evidently generates strong selection pressure for small body size. Not only does *L. neglecta* persist as an ecotype in the face of potential introgression with neighbouring *L. saxatilis* (Johannesson and Johannesson 1990b), but equivalent barnacle-adapted forms have arisen independently from the oviparous *L. arcana* and *L. nigrolineata* (Reid, 1993).

*The persistence of oviparity*

Radiation of life histories in the rough periwinkles is seen to be linked to the exploitation of a progressively wider range of habitats, through the protection of embryos by brooding, through adjustments of body size and through allometric, morphological and physiological adjustments affecting life span. These coarsegrained evolutionary changes seem not to have involved conflicts in resource allocation. Life-history theory concerned with allocation therefore is not applicable in this context. Demographical life-history theory, placing emphasis on life tables, also seems inapplicable. Nevertheless, the above habitat-exploitation model fails to explain the frequent coexistence of oviparous and ovoviviparous forms. This problem is particularly acute in the case of *L. arcana* and *L. saxatilis*, whose populations often are intermingled. Competition between these sibling species for sheltering sites could be severe (Raffaelli & Hughes, 1978) and may even have resulted in character displacement

regarding foot area and associated shell shape (Grahame & Mill, 1989). Since oviparous and coexisting ovoviviparous forms have similar population dynamics and survivorship curves beyond the embryonic stage (Hughes & Roberts, 1981), it is difficult to explain why oviparity persists. Selective neutrality of reproductive mode in favourable habitats eventually should lead to random, local extinction. Greater fecundity presumably pays off where egg masses survive well. This could occur on a small spatial scale, the relative advantages of greater fecundity on the one hand and brood protection on the other differing both within and among shores. On boulder shores the oviparous *L. nigrolineata* predominates in the physiologically less demanding mid-shore region, whereas *L. saxatilis* predominates at higher levels (Hughes & Roberts, 1981). On these same shores, however, *L. arcana* appears to mingle with *L. saxatilis*, although there is some evidence of a downshore migration to lay egg masses (Hannaford Ellis, 1983). *L. arcana* is known to replace *L. saxatilis* on parts of shores or even on entire shores where exposure to wave action is severe (Mill & Grahame, 1990, 1992), suggesting that oviparity is more than a mere legacy of the past.

In conclusion, life-history theory based either on the principle of allocation or on demography cannot readily be applied to inter- or even intra-specific comparisons among rough periwinkles. Conservative allocation priority and the paucity of trade offs associated with allocation may reflect the ability of prosobranchs to draw heavily upon somatic tissues for energy without critically impairing survivorship (Hughes, 1986). The scarcity of demographical correlations with life-history traits may reflect the overwhelming influence of physiological factors on adaptive variation.

## Acknowledgements

Practical work was undertaken by research assistant, Jonathan Cannon, salaried under grant GR3/6981A generously awarded to RNH by the NERC.

## References

Brody, S. J., 1945. Bioenergetics and growth. Reinhold, New York, 1023 pp.

Cannon, J. P. & R. N. Hughes, 1992. Resistance to environmental stressors in *Littorina saxatilis* (Olivi) and *L. neglecta* Bean. In J. Grahame, P. J. Mill & D. G. Reid (eds), Proceedings of the 3rd

14

International Symposium on Littorinid Biology. The Malacological Society of London, London: 61–68.

Elliott, J. M. & W. Davidson, 1975. Energy equivalents of oxygen consumption in animal energetics. Oecologia (Berlin) 19: 195–201.

Ernsting, G. & J. A. Isaaks, 1991. Accelerated ageing: a cost of reproduction in the carabid beetle *Notiophilus biguttatus* F. Funct. Ecol. 5: 299–303.

Fish, J. D. & L. Sharp, 1985. The ecology of the periwinkle, *Littorina neglecta* Bean. In P. G. Moore & R. Seed (eds), The ecology of rocky coasts. Hodder & Stoughton, Lond.: 143–156.

Grahame, J. & P. J. Mill, 1989. Shell shape variation in *Littorina saxatilis* (Olivi) and *L. arcana* Hannaford Ellis: a case of character displacement? J. mar. biol. Ass. U.K. 69: 837–855.

Grahame, J. & P. J. Mill, 1992. Local and regional variation in shell shape of rough periwinkles in southern Britain. In J. Grahame, P. J. Mill & D. Reid (eds), Proceedings of the 3rd International Symposium on Littorinid Biology. The Malacological Society of London, London: 99–106.

Grahame, J., P. J. Mill & A. Brown, 1990. Adaptive and non-adaptive variation in two species of rough periwinkle (*Littorina*) on British shores. Hydrobiologia 193 (Dev. Hydrobiol. 56): 223–231.

Hart, A. & M. Begon, 1982. The status of general reproductive-strategy theories, illustrated in winkles. Oecologia (Berlin) 52: 37–42.

Hannaford Ellis, C. J., 1979. Morphology of the oviparous rough periwinkle, *Littorina arcana* Hannaford Ellis, 1978, with notes on the taxonomy of the *L. saxatilis* species complex (Prosobranchia: Littorinidae). J. Conch. 30: 43–56.

Hannaford Ellis, C. J., 1983. Patterns of reproduction in four *Littorina* species. J. moll. Stud. 49: 98–106.

Hart, A. & M. Begon, 1982. The status of general reproductive-strategy theories, illustrated in winkles. Oecologia (Berlin) 52: 37–42.

Holland, D. L., R. Tantanasiriwong & P. J. Hannant, 1975. Biochemical composition and energy reserves in the larvae and adults of the four British periwinkles *Littorina littorea*, *L. littoralis*, *L. saxatilis* and *L. neritoides*. Mar. Biol. 33: 235–239.

Hughes, R. N. 1980. Population dynamics, growth and reproductive rates of *Littorina nigrolineata* Gray from a moderately sheltered locality in North Wales. J. exp. mar. Biol. Ecol. 44: 211–228.

Hughes, R. N. 1986. A functional biology of marine gastropods. Croom-Helm, London and Sydney, 245 pp.

Hughes, R. N. & D. J. Roberts, 1980. Reproductive effort of winkles (*Littorina* spp.) with contrasted methods of reproduction. Oecologia (Berlin) 47: 130–136.

Hughes, R. N. & D. J. Roberts, 1981 Comparative demography of *Littorina rudis*, *L. nigrolineata* and *L. neritoides* on three contrasted shores in North Wales. J. anim. Ecol. 50: 251–268.

Johannesson, B. & K. Johannesson, 1990a. *Littorina neglecta* Bean, a morphological form within the variable species *Littorina saxatilis* (Olivi)? Hydrobiologia 193 (Dev. Hydrobiol. 56): 71–87.

Johannesson, K. & B. Johannesson, 1990b. Genetic variation within *Littorina saxatilis* (Olivi) and *Littorina neglecta* Bean: is *L. neglecta* a good species? Hydrobiologia 193 (Dev. Hydrobiol. 56): 89–97.

Mill, P. J. & J. Grahame, 1990. Distribution of the species of rough periwinkle (*Littorina*) in Great Britain. Hydrobiologia 193 (Dev. Hydrobiol. 56): 21–27.

Mill, P. J. & J. Grahame, 1992. Distribution of the rough periwinkles in Great Britain. In J. Grahame, P. J. Mill & D. G. Reid (eds), Proceedings of the 3rd International Symposium on Littorinid Biology. The Malacological Society of London, London: 305–307.

Naylor, R. & M. Begon, 1982. Variations within and between populations of *Littorina nigrolineata* Gray on Holy Island, Anglesey. J. Conch. 31: 17–30.

Palmer, A. R., 1982. Growth in marine gastropods: a nondestructive technique for independently measuring shell and body weight. Malacologia 23: 63–73.

Pianka, E. R., 1970. On 'r' and 'K' selection. Am. Nat. 104: 592–597.

Raffaelli, D. G., 1976. The Determinants of Zonation of *Littorina neritoides* and *Littorina saxatilis* species-complex. Ph.D. thesis, University of Wales, Bangor, 177 pp.

Raffaelli, D. G. 1978. Factors affecting the population structure of *Littorina neglecta* Bean. J. moll. Stud. 44: 223–230.

Raffaelli, D. G. 1982. Recent ecological research on some European species of *Littorina*. J. moll. Stud. 48: 342–241.

Raffaelli, D. G. & R. N. Hughes, 1978. The effects of crevice size and availability on populations of *Littorina rudis* and *Littorina neritoides*. J. anim. Ecol. 47: 71–83.

Reid, D. G. 1990. Trans-Arctic migration and speciation induced by climatic change: the biogeography of *Littorina* (Mollusca: Gastropoda). Bull. mar. Sci. 47: 35–49.

Reid, D. G. 1993. Barnacle-dwelling ecotypes of three British *Littorina* species and the status of *Littorina neglecta* Bean. J. moll. Stud. 59: 51–62.

Roberts, D. J. 1979. Reproductive strategies of *Littorina neritoides* and the *Littorina saxatilis* species complex. Ph.D. thesis, University of Wales, Bangor, pp. 160.

Roberts, D. J. & R. N. Hughes, 1980. Growth and reproductive rates of *Littorina rudis* from three contrasted shores in North Wales, UK. Mar. Biol. 58: 47–54.

Ryan, B. F., B. L. Joiner & T. Ryan, 1985. Minitab Handbook. Duxbury Press, Boston.

Sibly, R. M. & P. Calow, 1986. Physiological ecology of animals: an evolutionary approach. Blackwell Scientific Publications, Oxford, 179 pp.

Stearns, S. C., 1992. The evolution of life histories. Oxford University Press, Oxford, 179 pp.

Warwlck, T., 1983. A method of maintaining and breeding members of the *Littorina saxatilis* (Olivi) species complex. J. moll. Stud. 48: 368–370.

Warwick, T., A. J. Knight & R. D. Ward, 1990. Hybridisation in the *Littorina saxatilis* species complex (Prosobranchia: Mollusca). Hydrobiologia 193 (Dev. Hydrobiol. 56): 109–116.

*Hydrobiologia* **309**: 15–27, 1995.
*P. J. Mill & C. D. McQuaid (eds), Advances in Littorinid Biology.*
©1995 *Kluwer Academic Publishers.*

# TBT effects on the female genital system of *Littorina littorea*: a possible indicator of tributyltin pollution

Barbara Bauer[1], Pio Fioroni[1], Imke Ide[2], Stefanie Liebe[2], Jörg Oehlmann[3],
Eberhard Stroben[1] & Burkard Watermann[2]
[1]*Institut für Spezielle Zoologie und Vergleichende Embryologie, Hüfferstrasse 1, D-48149 Münster, Germany*
[2]*LimnoMar, Wulfsdorfer Weg 200, D-22926 Ahrensburg, Germany*
[3] *Internationales Hochschulinstitut, Markt 23, D-02763 Zittau, Germany*

*Key words: Littorina littorea*, reproductive failure, TBT contamination, biomonitoring, antifouling paints, histopathology

## Abstract

Specimens of the prosobranch *Littorina littorea* (L., 1758) collected along the East Frisian North Sea coast in summer 1993 exhibited alterations of the pallial oviduct termed as intersex in response to tributyltin (TBT) pollution. The range of TBT body burden was between 150.9 and 1289.5 $\mu$g as Sn kg$^{-1}$ (dry wt.). Five stages of intersex development (0–4) could be distinguished and are documented with scanning electron micrographs. In stages 2–4, which can be found in the direct vicinity of harbours and marinas, the morphological malformations of the oviduct inhibit successful copulation and capsule formation, resulting in sterilization.

The intersex index (ISI, calculated as the average intersex stage of a population) and the average prostate length of females were used as parameters for the determination of intersex intensities in the populations. Both indices show significant and positive correlations to the TBT body burden of *L. littorea* and are promising parameters for TBT biomonitoring. A comparison of TBT bioconcentration factors with populations from England and France indicates that the threshold concentration for intersex development is in the range of 15 ng TBT as Sn/l.

Morphometric analyses of the midgut gland revealed no significant differences between sampling stations. In the ovary a retardation and blockage of maturation (atresia) was observed in populations close to harbours. Lytic processes in ovary follicles were observed not only at TBT exposed sites but also at reference stations.

## Introduction

Tributyltin (TBT) compounds used as biocides in antifouling paints and in various other formulations are known to produce a variety of malformations in marine animals, with molluscs being one of the most TBT-sensitive groups of invertebrates (for review Bryan & Gibbs, 1991). TBT-induced detrimental effects in Great Britain and France, e.g. malformations of oyster shells and the imposex phenomenon of prosobranchs, became evident in the 1980s. Consequently, legislative restrictions were drawn up to reduce TBT contamination in coastal waters. The imposex phenomenon of prosobranchs has been successfully used as a biomonitoring system to determine the degree of environmental TBT pollution (e.g. Gibbs *et al.*, 1987; Oehlmann *et al.*, 1993).

The established European imposex species for TBT biomonitoring (e.g. *Nucella lapillus* (L.), *Hinia reticulata* (L.), *Ocenebra erinacea* (L.)) are absent on the German North Sea coast or can only be found in restricted areas. The periwinkle *Littorina littorea* (L.) is the only prosobranch which is very common and can be sampled in sufficient numbers. In the present study the potential of *L. littorea* was evaluated for TBT biomonitoring.

*Littorina littorea* is a shallow water species which lives on rocky and sandy shores. Though the bulk of the population occurs intertidally, some specimens can be found up to a depth of 15 m. Their geographical distribution ranges from Asturia to northern Norway in the east Atlantic and from New Jersey to Greenland in the western Atlantic (Nordsieck, 1968; Graham, 1988). The species feeds largely on epilithic algae and vegetable detritus (Taylor & Miller, 1989), although it may occasionally feed on dead animal matter (Matthiessen *et al.*, 1991). The sexes are separate and, after reaching maturity at shell heights of 10–12 mm and at an age of 12–18 months, females produce about 500 planktonic egg capsules each containing 1–5 eggs. After 5–6 days the animals hatch as free swimming veliger larvae drifting with the plankton. Metamorphosis occurs after 4–7 weeks (Linke, 1933; Thorson, 1946; Matthiessen *et al.*, 1991). Adults can live for more than 9 years (Heller, 1990) and can reach a shell height of 40 mm (Nordsieck, 1968; Fretter & Graham, 1980).

## Materials and methods

More than 500 *Littorina littorea* were collected at 11 stations in summer 1993 along the East Frisian North Sea coast between Emden and Cuxhaven (Fig. 1). Sample sites included marinas, ferry and fishing harbours as well as reference stations far away from shipping activities.

Snails were relaxed using 7% $MgCl_2$ in distilled water and shell and aperture height were measured to the nearest 0.1 mm. The shells were then crushed in a vice to remove the soft tissues and the specimens sexed. External dimensions of the genital tract, including pallial glands of females and males and penis length, were determined to 0.1 mm. *L. littorea* is often parasited by various trematodes such as *Cryptocotyle lingua*, *Himasthla elongata* and *Renicola roscovita* (Lauckner, 1980). Because parasitation causes suspicious modifications of the genital system, including gonad and midgut gland, infected animals were excluded from morphometric and histological analyses.

As a result of TBT exposure the winkle *L. littorea* exhibits malformations of the pallial oviduct which were termed as intersex in contrast to the imposex phenomenon of neogastropod species. Intersex is defined as any disturbance of the congruity between gonad and genital tract, while imposex is a superimposition of male sex organs (penis and/or vas deferens) on females (Smith, 1971). During intersex development no superimposition of male characters occurs but the organs of the pallial oviduct are modified towards a male morphological structure. For all populations two indices for the measurement of intersex intensity were employed: (1) Intersex index (ISI) is calculated

as the average intersex stage (according to Fig. 2) of a population; (2) Average prostate length of females.

For serial sections and scanning electron microscopy, specimens were fixed in Bouin's fluid and then preserved in 70% ethanol. After embedding in paraplast, serial sections (7 $\mu$m) were made and stained with haemalun-chromotrope (Romeis, 1989). Specimens for scanning electron microscopy were critical-point dried.

For histopathological analyses, 10 specimens from every site were fixed in Bouin's fluid and embedded in Technovit (hydroxymethylmethacrylate). 5 $\mu$m sections were stained with H & E and PAS (Romeis, 1989). The areas of midgut gland tubules, lumina, crypt- and digestive cells were measured using a graphic tablet as described by Vega *et al.* (1989). Only tubules in the holding phase were used for morphometric measurement (Langton, 1975).

The determination of TBT was based largely on Ward *et al.* (1981), modified according to Stroben *et al.* (1992b). Five to seven snails were homogenized in stoppered tubes, and 10 ml of concentrated HCl (Merck 'suprapur') were added. After shaking for 30 min, the homogenate was extracted with 10 ml of hexane (pesticide grade) on an automatic shaker for 30 min and then centrifuged. TBT as Sn (TBT-Sn) was determined in the hexane extract after shaking with 3 ml 1N NaOH for 3 min using a Perkin-Elmer HGA-500 attached to a Perkin-Elmer 5000 AAS (wave length 224.6 nm; slit 0.7 nm; injection volume 25 $\mu$l). Internal standardization (standard addition with spiked samples) was employed. Certified reference material (CRM: PACS-1, National Research Council of Canada) was also analysed. Our own results were within the standard deviation of the certified values for the CRM. Recovery factors were 78.0±12.6%. The detection limit ($3\sigma$) in a single sample was 6.7 ng TBT-Sn.

## Results

### Genital system of Littorina littorea

The anatomy and histology of the normal female and male genital tract of *Littorina littorea* are described in detail by Linke (1933), Fretter & Graham (1962), Fretter (1980) and Reid (1986a, 1989) In both sexes gonadial, renal and pallial sections of the genital tract can be distinguished. The testis and ovary are large and diffuse organs which lay in the upper parts of the visceral mass, branching between the tubules of the

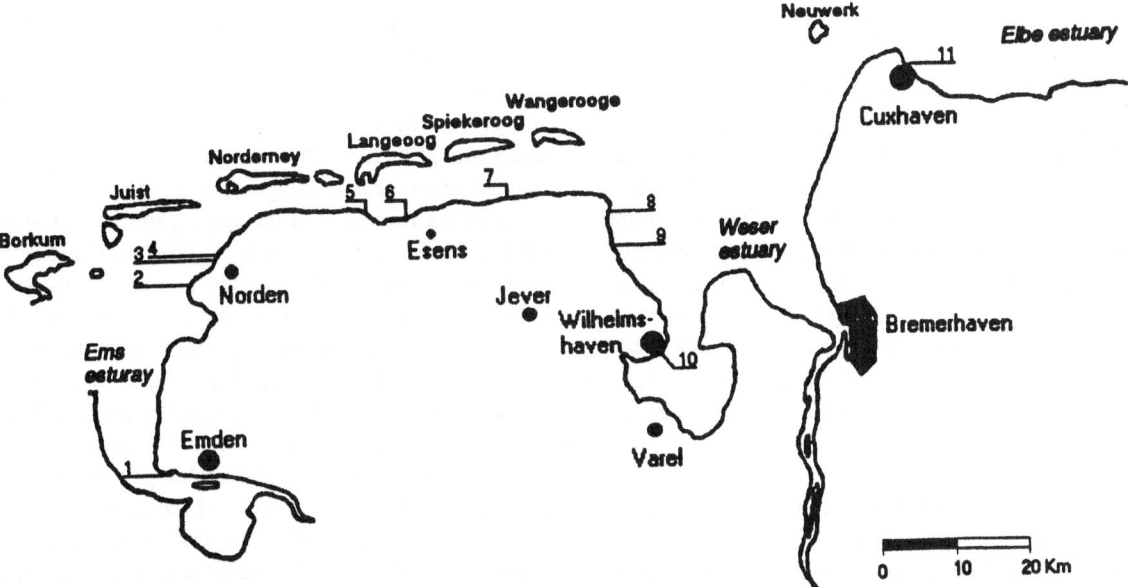

*Fig. 1.* Map of East Frisian coast showing stations sampled in summer 1993. (1) Knock, near Emden, (2) Westermarsch II, (3) Norddeich: outside harbour, (4) Norddeich: marina, (5) Dornumersiel: marina, (6) Bensersiel: marina, (7) Harlesiel: 800 m outside ferry harbour, (8) Horumersiel: entrance of harbour, (9) Hooksiel: entrance of outer harbour (the marina is separated from the sampling site by a sluice), (10) Wilhelmshaven: Nassauhafen, (11) Cuxhaven: ferry harbour.

midgut gland. The most proximal parts of the male and female genital tract are the gonadial sections of the vas deferens and oviduct. These run parallel to the posterior aorta at the columellar side of the visceral hump and continue in the rather short renal section.

The most elaborated part of the reproductive tract of *L. littorea* is the pallial section of males and females, which originates ontogenetically from an infolding of the mantle epithelium. The pallial section of males is an open sperm groove for its entire length. In the most proximal part the walls of the groove are tall and characterised by the development of folds of glandular tissue which form a well developed prostate gland. From the prostate the sperm groove runs forward on the floor of the mantle cavity to the base of the penis behind the right ocular tentacle. The copulatory organ is conical, flattened laterally and carries the sperm groove to its summit along the dorsal edge. On the ventral border of the penis numerous mamilliform penial glands can

be found. Their secretions hold the penis in position during copulation.

The pallial section of females is highly developed in order to store sperm, to provide extraembryonic nourishment for the embryos within the capsule and to produce the planktonic egg capsules. It consists of a receptaculum seminis, and albumen, capsule and jelly glands, and is completed by the bursa copulatrix in a ventral position and the vagina (Fig. 2, stage 0). In adult females the pallial oviduct is a closed tube originating from an infolding of the mantle epithelium with a consecutive fusion of its flaps (Fig. 3b, c). The ovoid-shaped vaginal opening is the only aperture towards the mantle cavity. Distally from the vaginal opening an egg channel extends over the floor of the mantle cavity and runs down the right side towards the foot. It ends in the so-called ovipositor (Fig. 3a). The ovipositor is a muscular and glandular organ with the function of launching the egg capsules on their pelagic life phase. Its topographical position, and histologi-

18

cal and ultrastructural properties are identical with the base of the male penis which remains after the periodic shedding of the copulatory organ during the sexual repose phase. Consequently, the male penis base and the ovipositor were deemed to be homologous by Fioroni *et al.* (1991).

*Intersex development and TBT*

Especially in direct proximity to harbours and marinas we found malformations of the female genital tract which were termed as intersex (see above). The female specimens affected by intersex were either characterised by the development of male features on the female pallial organs (inhibition of the ontogenetic closure of the pallial oviduct) or female sex organs were supplanted by the corresponding male formations. The intersex phenomenon of *L. littorea* is a gradual transformation of the female pallial tract, which can be described by an evolutive scheme with five stages (0 to 4) (Fig. 2).

*Stage 0.* Normal female without intersex characteristics (Fig. 3a). The entire pallial oviduct is a closed tube (Fig. 3b, c). This stage was found in 200 specimens (=77.4% of the 258 females analysed).

*Stage 1.* The bursa copulatrix is split ventrally, exposing its internal lobes (Fig. 3d). This malformation inhibits successful copulation because sperm can be spilled into the mantle cavity (Fig. 3e) (27 specimens = 10.5% of all analysed females).

*Stage 2.* The entire pallial oviduct (bursa copulatrix and jelly gland) is split ventrally (Fig. 3f, 4a). This open structure is a male characteristic because the prostate gland of males is also an open groove. (2 specimens = 0.8% of all analysed females).

*Stage 3.* The distal part of the pallial oviduct (capsule and jelly gland with bursa copulatrix and vagina) are supplanted by a prostate gland. In some specimens the free edges of the open prostate gland fuse to form a closed tube in order to establish a more female anatomical structure (Fig. 4b, c). The line of fusion of the prostate edges is detectable by its white coloration. The proximal female parts of the pallial oviduct (seminal receptacle and albumen gland) are conserved (28 specimens = 10.9% of all analysed females).

*Stage 4.* Additionally to the characteristics of stage 3, a penis with an open sperm groove is developed (Fig. 4d, e). This stage was found in a single specimen (=0.4% of all analysed females).

All analysed intersex females have an ovary with no signs of the onset of spermiogenesis but with signs of a disturbance of oogenesis (c.f. below). Intersex development causes restrictions of the reproductive capability of females. In stage 1 a loss of sperm during copulation is possible and consequently the reproductive success is reduced. Females in stages 2–4 are definitively sterile because the capsular material is spilled into the mantle cavity (stage 2) or the glands responsible for the formation of the egg capsule are missing (stages 3 and 4). Due to female sterility, populations of *L. littorea* can be in decline but are not likely to become extinct because of the planktonic veliger larvae of the species. Veligers produced by populations with lower intersex intensities can guarantee a minimum abundance of periwinkles even at sites suffering from high TBT contamination and reproductive failure.

Two of the 258 analysed females (one in each of stages 1 and 3) were characterised by a coiled renal and gonadial oviduct (Fig. 4f). This condition was interpreted by Smith (1980, 1981a–d) as a mimic seminal vesicle in *Ilyanassa obsoleta* and can also be found in many other imposex-affected species as a result of high TBT exposure (Fioroni *et al.*, 1991).

Contrary to the imposex phenomenon of muricids, sterilization caused by intersex development in *L. littorea* provokes poor recruitment of juveniles but not high female mortality. In *Nucella lapillus* (Gibbs *et al.*, 1987; Oehlmann *et al.*, 1991), *Nucella lima* (Short *et al.*, 1989), *Nucella lamellosa* (Bright & Ellis, 1990) the final stages of imposex development are sterilized due to a blockage of the female opening by proliferating tissues. This leads to an accumulation of abortive egg capsules, to a distension and finally a rupture of the pallial oviduct, causing the death of the female. In *L. littorea*, as in the imposex stage 5 of *Ocenebra erinacea* (Gibbs *et al.*, 1990; Oehlmann *et al.*, 1992) and *Urosalpinx cinerea* (Gibbs *et al.*, 1991), the pallial oviduct is split ventrally. Consequently, these malformations inhibit the formation of egg capsules and prevent their accumulation in the pallial tract. Thus the sex ratio, even in highly polluted periwinkle populations, does not change towards male dominance ($\chi^2$-Test; $p > 0.9$) (Table 1).

For the measurement of intersex intensities in the populations two indices were established. The intersex index (ISI) is the average intersex stage in a popula-

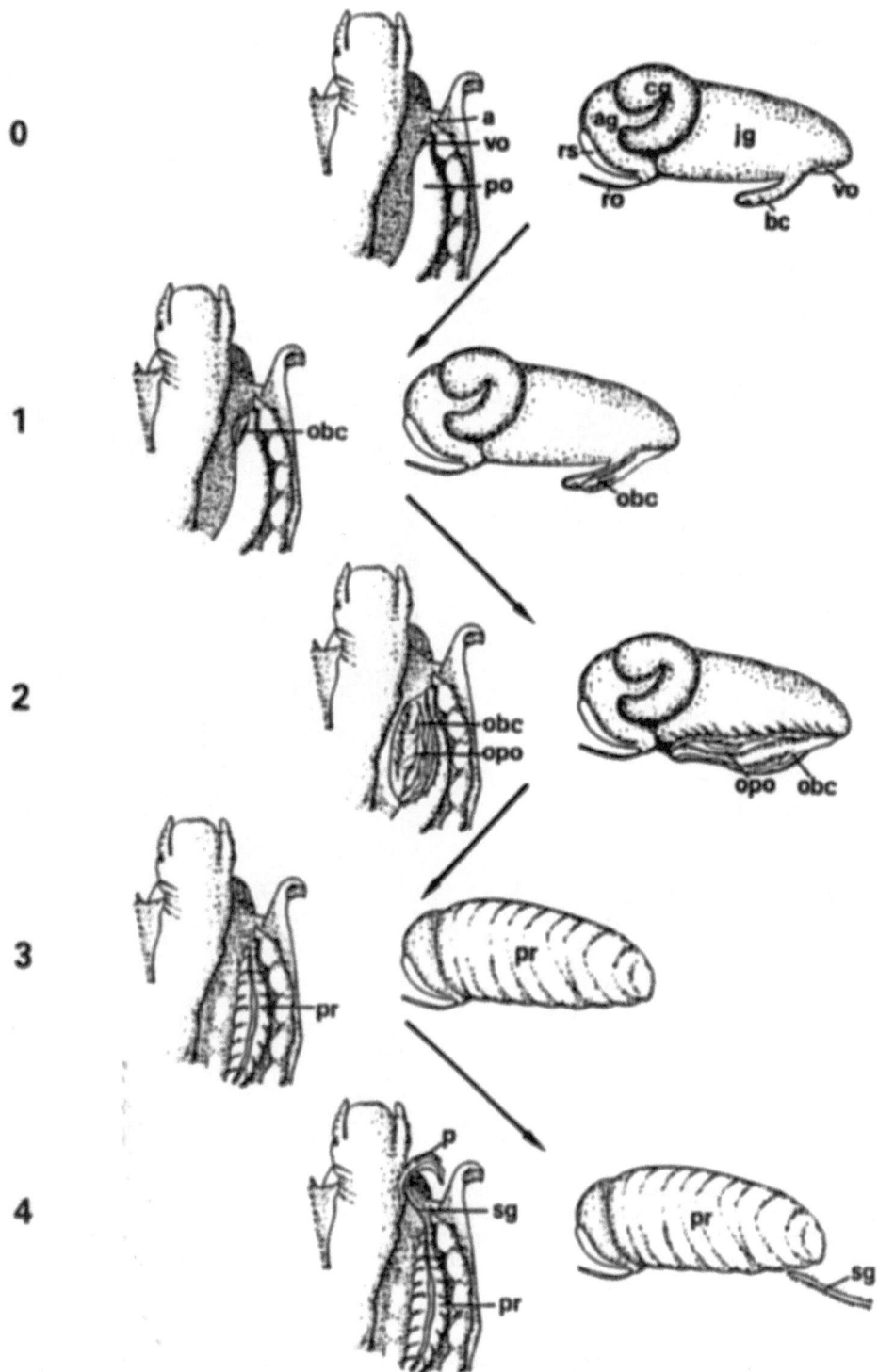

*Fig. 2.* *Littorina littorea.* Scheme of intersex development. Dorsal views with opened mantle cavity (left) and lateral views of pallial oviduct (right). Abbreviations: a, anus; ag, albumen gland; bc, bursa copulatrix; cg, capsule gland; jg, jelly gland; obc, open bursa copulatrix; opo, open pallial oviduct; p, penis; po, pallial oviduct; pr, prostate; ro, renal oviduct; rs, receptaculum seminis; sg, sperm groove; vo, vaginal opening.

20

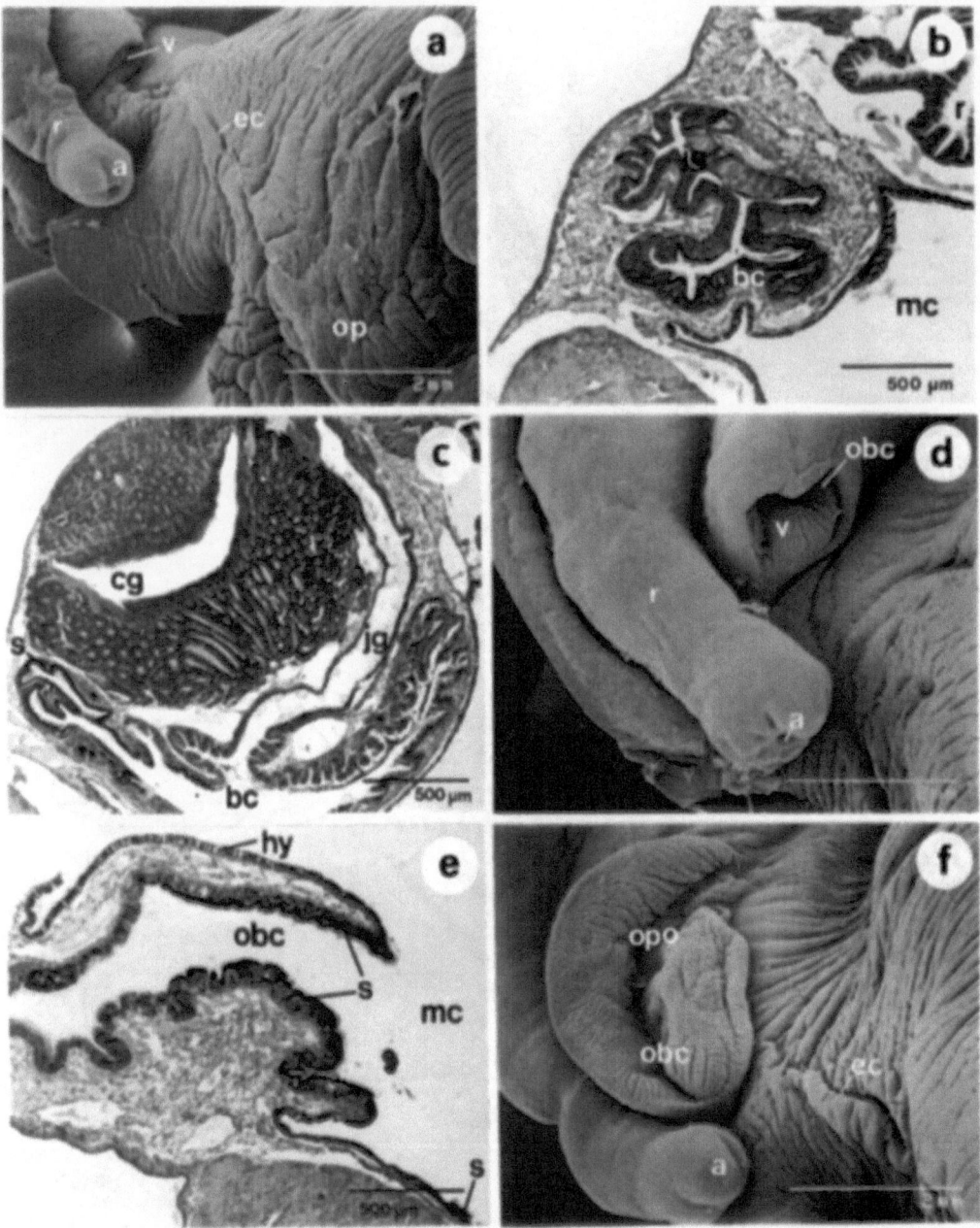

*Fig. 3.* *Littorina littorea.* Scanning electron micrographs of the mantle cavity and histological sections of intersex stages 0 to 2. (a) Stage 0 (normal female). (b) Stage 0: transverse section of bursa copulatrix and jelly gland. (c) Stage 0: transverse section of bursa copulatrix, jelly and capsule gland. (d) Stage 1. (e) Stage 1: transverse section of open bursa copulatrix. (f) Stage 2. Abbreviations: a, anus; bc, bursa copulatrix; cg, capsule gland; ec, egg channel; hy, hypobranchial gland; jg, jelly gland; mc, mantle cavity; obc, open bursa copulatrix; op, ovipositor; opo, open pallial oviduct; r, rectum; s, sperm; v, vaginal opening.

tion according to Fig. 2 (Σ intersex stages in a sample ÷ number of analysed females). A value of 0.0 indicates that only normal females (stage 0) occur and no restrictions of the reproductive capability have to be expected. ISI values above 0 show that intersex-affected females can be found and that reproductive success may be reduced. An ISI of 1.0 indicates that all females exhibit stage 1 and an ISI above 2.0 that most females in a population are sterilized due to intersex development. The second parameter is the average

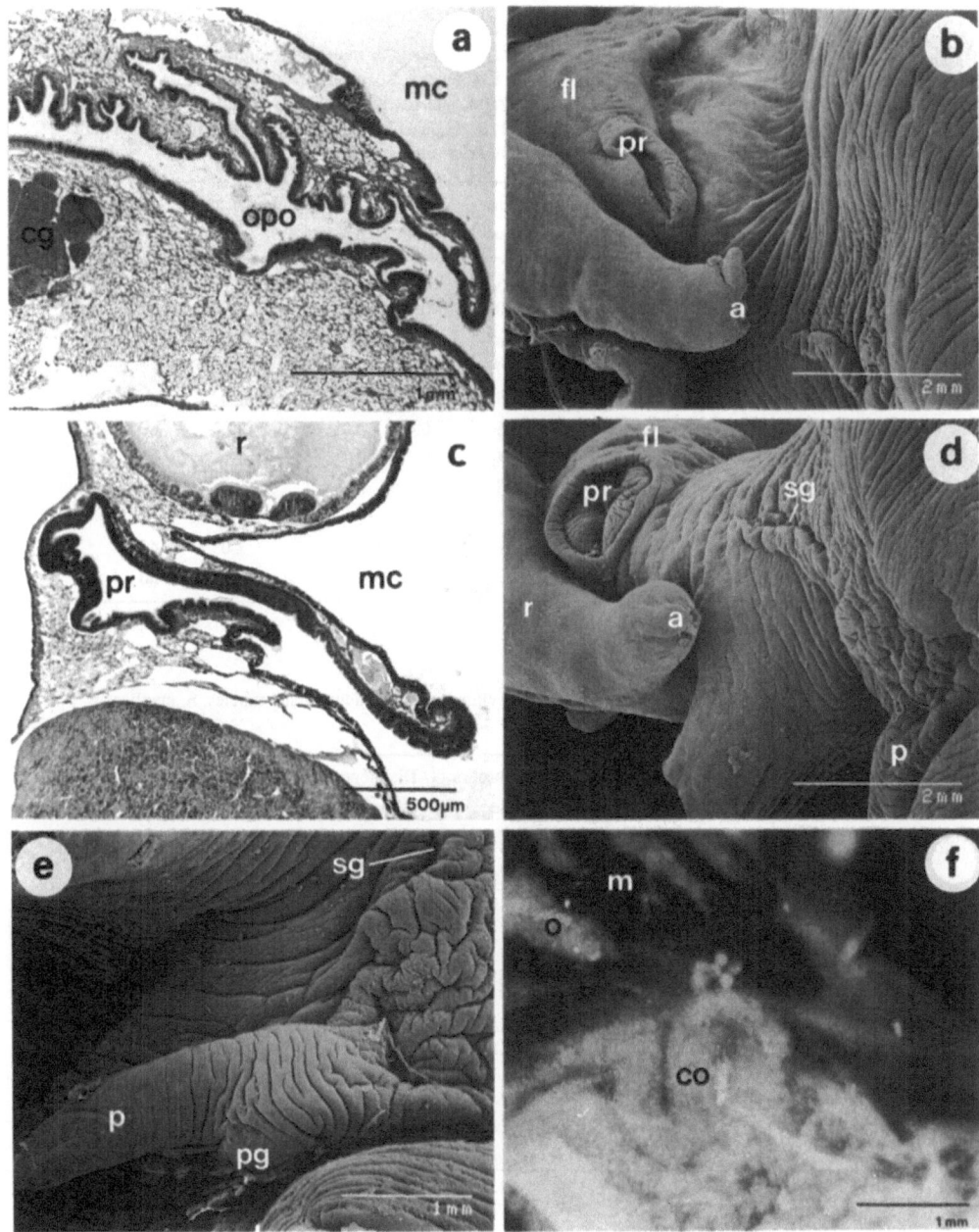

*Fig. 4. Littorina littorea.* Scanning electron micrographs of the mantle cavity and histological sections of intersex stages 2 to 4. (a) Stage 2: transverse section of open pallial oviduct. (b) Stage 3 with subsequent closed prostate gland. (c) Stage 3: transverse section of prostate gland. (d) Stage 4 with subsequent closed prostate gland. (e) Stage 4: detail of penis. (f) Detail of coiled oviduct. Abbreviations: a, anus; cg, capsule gland; co, coiled oviduct; fl, fusion line of free prostate edges; m, midgut gland; mc, mantle cavity; o, ovary; opo, open pallial oviduct; p, penis; pg, penial glands; pr, prostate gland; r, rectum; sg, sperm groove.

length of the prostate gland in females. This index has the disadvantage that it does not give authentic measurements of the proportion of sterile females in the population. Values above 0.0 are only attained if the sterile stage 3 occurs. But individuals already in stage 2 are not capable of producing egg capsules although females have no prostate, and consequently the length of the female prostate gland is 0.0 in such populations.

Table 1. *Littorina littorea.* Sampling stations (cf. Fig. 1; capitals indicate proximity to: D, dock yard; F, ferry harbour; M, marina; S, fishing harbour. R, reference, i.e. no apparent TBT sources) with TBT body burden (in $\mu$g as Sn kg$^{-1}$ dry wt.), sex ratio, intersex index (ISI), average length of prostate gland in females in mm, proportion of females (of 4–7 analysed specimens) affected by atresia and lytical processes in follicles of the ovary. nd, not determined.

| Sampling station | TBT body burden | Males ÷ females | ISI | Ø length of prostate | % of females with atresia | lysis |
|---|---|---|---|---|---|---|
| Knock near Emden (R) | 202.2 | 0.54 | 0.08 | 0 | 0 | 33 |
| Westermarsch II (R) | 244.9 | 0.40 | 0.10 | 0 | 20 | 20 |
| Norddeich (M) | 978.9 | 0.86 | 1.50 | 3.44 | 40 | 40 |
| Norddeich (F, S) | 482.4 | 1.50 | 0.25 | 0.39 | 0 | 17 |
| Dornumersiel (D, F, M) | 1289.5 | 1.05 | 2.75 | 7.02 | 20 | 60 |
| Bensersiel (F, M) | 666.2 | 0.95 | 0.43 | 0.78 | 40 | 60 |
| Harlesiel (R) | 150.9 | 0.64 | 0 | 0 | 29 | 71 |
| Horumersiel (M) | 564.3 | 1.05 | 0.30 | 0 | 0 | 75 |
| Hooksiel (R) | 156.6 | 0.71 | 0 | 0 | 0 | 17 |
| Wilhelmshaven (M) | 380.4 | 1.41 | 0.12 | 0 | 0 | 40 |
| Cuxhaven (F, M) | 247.6 | 0.73 | 0.08 | 0 | nd | nd |

Table 2. *Littorina littorea.* Sampling stations with phases of midgut gland tubules. *n*, number of analysed specimens.

| Sampling station | Holding phase | Absorption phase | Disintegration phase | Reconstitution phase | *n* |
|---|---|---|---|---|---|
| Knock near Emden (R) | 7 | 36 | 58 | 0 | 9 |
| Westermarsch II (R) | 92 | 8 | 0 | 0 | 10 |
| Norddeich (M) | 24 | 23 | 43 | 8 | 9 |
| Norddeich (F, S) | 16 | 14 | 67 | 0 | 9 |
| Dornumersiel (D, F, M) | 57 | 0 | 41 | 2 | 10 |
| Bensersiel (F, M) | 80 | 12 | 2 | 2 | 10 |
| Harlesiel (R) | 16 | 46 | 39 | 0 | 9 |
| Horumersiel (M) | 58 | 33 | 10 | 0 | 9 |
| Hooksiel (R) | 51 | 34 | 12 | 0 | 11 |
| Wilhelmshaven (M) | 78 | 21 | 0 | 1 | 9 |
| Cuxhaven (F, M) | 65 | 17 | 18 | 0 | 6 |

In this study TBT concentrations in the water column were not determined because the investigation period was too short to give reliable information of TBT pollution on the background of high season and tide related changes in coastal waters (Oehlmann *et al.*, 1993). However, stations were chosen on the background of former investigations on TBT concentrations in water and sediments (Kalbfus *et al.*, 1991). The range of TBT body burdens was between 150.9 (Harlesiel) and 1289.5 $\mu$g TBT as Sn kg$^{-1}$ dry wt. (Dornumersiel) (Table 1). In a reference population

of *L. littorea* from Roscoff harbour (Brittany, France; mean TBT concentration in sea water between 1989 and 1992: 15.4±5.84 ng TBT as Sn l$^{-1}$, *n* = 19) a TBT body burden of 406.6 to 534.5 $\mu$g TBT as Sn kg$^{-1}$ dry wt. was determined. This shows, that at least at some sites of the German North Sea coast, aquatic TBT concentrations exceed the values measured at Roscoff harbour considerably.

There are highly significant sigmoid correlations between the TBT body burden and the ISI or the average female prostate length (Fig. 5a, b). This shows that both parameters give not only an assessment of the intersex intensity in the populations but allow a determination of the TBT pollution in different populations. The ISI is characterized by a higher sensitivity, *i.e.*, this index increases at lower TBT body burdens compared to female prostate length. The threshold concentrations are 200 $\mu$g TBT as Sn kg$^{-1}$ dry wt. for the ISI and 500 $\mu$g TBT as Sn kg$^{-1}$ dry wt. for the prostate length.

*Histopathological results*

Parallel histological investigations were performed to elucidate if the observed morphological changes were accompanied by alterations at the level of cells and tissues. Morphometric measurements of the midgut gland revealed no significant differences between the sampling stations with regard to the area of the digestive epithelium and the luminal area in relation to the whole tubule (Fig. 6).

At Norddeich marina and Dornumersiel, tubules with a diminished area of the digestive epithelium were observed. These occurred at very low frequencies and did not affect the mean values. The synchrony in the phases of the tubules was pronounced at Westermarsch, Bensersiel and Wilhelmshaven with a dominating holding phase. At all other stations a distinct asynchrony was found (Table 2).

Tubules in disintegration and reconstitution phases occurred at the reference stations (Knock, Harlesiel) and in the harbours of Norddeich and Dornumersiel. Consequently, this characteristic was not useful for assessing pathological responses to TBT.

Most of the pathological disorders were observed in the female gonad. Retardation and interruption in the maturation of the oocytes (atresia) occurred mostly at stations in the vicinity of harbours. Lytic processes in follicles were observed at all stations and were frequent in harbours as well as at reference stations (Table 1). With increasing intensity of atresia in the follicles, glycogen was found to be accumulated in the connective tissue cells around the follicles, instead of stored in the yolk masses of the eggs. Additionally, diffuse haemocytic infiltrations were present in and around the gonads of female specimens. At Knock, near Emden one case of a granulocytoma was recorded.

**Discussion**

Despite the widespread distribution of *Littorina littorea* only a few TBT effects have been reported in this species. Matthiessen *et al.* (1991) reported a TBT-caused decline of periwinkle populations in southwest England and also a decrease in the abundance of veligers in the plankton until 1987. The situation has improved since the partial TBT ban in England in 1987. In the literature we have found no indication for the phenomenon described in this study as intersex. Only Fioroni *et al.* (1992) found a single penis-bearing female on the rocky shores of Helgoland but unfortunately the pallial oviduct of this female was not analysed in detail. Generally, littorinids are gonochoristic with the exception of the protandric species *Mainwaringia rhizophila* (Reid, 1986b). Penis-bearing females were also reported in *Bembicium auratum* and *Nodilittorina acutispira* (Muggeridge, 1979) but according to Reid (1986b) a trematode infection can be responsible for this malformation. In the present study parasited specimens were excluded from morphometric and histological analyses.

Only stage 4 of intersex development reported here, with an additional female penis, shows parallels to the imposex situation of other prosobranch species. Imposex is a superimposition of male sex organs (penis and/or vas deferens) on females (Smith, 1971). During intersex development no superimposition of male characters occurs but the organs of the pallial oviduct are modified towards a male morphological structure. Only in the final stages are female organs supplanted by the corresponding male formations. We want to stress that the main indices for intersex (ISI) and imposex measurement (vas deferens sequence (VDS) index according to Gibbs *et al.* (1987) and Fioroni *et al.* (1991)) are not comparable.

The imposex phenomenon has proven to be a reliable and widely used biomonitoring system for the detection and estimation of environmental TBT contamination (e.g. Gibbs *et al.*, 1987; Oehlmann *et al.*, 1992; Stroben *et al.*, 1992a, b). However, the established European imposex species for TBT biomonitoring are absent on the German North Sea coast or can only be found in restricted areas. For the TBT survey of the North Sea Task Force (Harding *et al.*, 1992) *Nucella lapillus* was transplanted in cages for three months to numerous sites on the Dutch, German and Danish North Sea coast. Because of the unnatural environmental conditions at these locations most of the

24

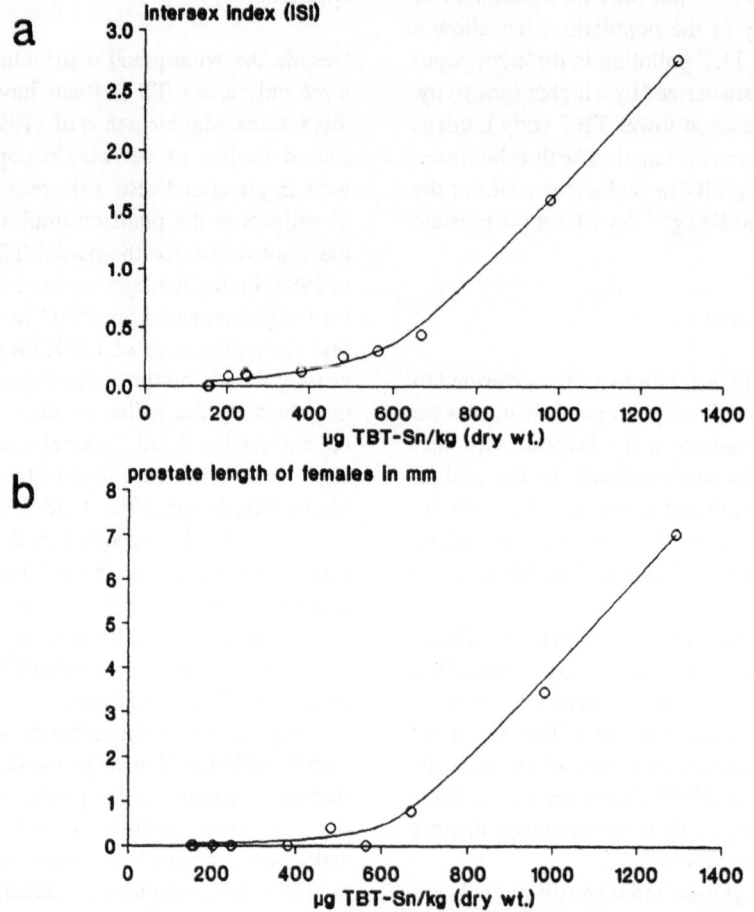

**Fig. 5.** *Littorina littorea*. Relationship between TBT body burden and intersex parameters ($n = 11$ populations with $\geq 40$ specimens analysed from each). (a) Intersex index (ISI), with calculated sigmoid correlation: $y = -3.28 \div (1 + e^{0.0055(x-995)}) + 3.29$; $r = 0.997$; $p < 0.0005$. (b) Average prostate length of females, with calculated sigmoid correlation: $y = -7.76 \div (1 + e^{0.0075(x-995)}) + 7.79$; $r = 0.990$; $p < 0.0005$.

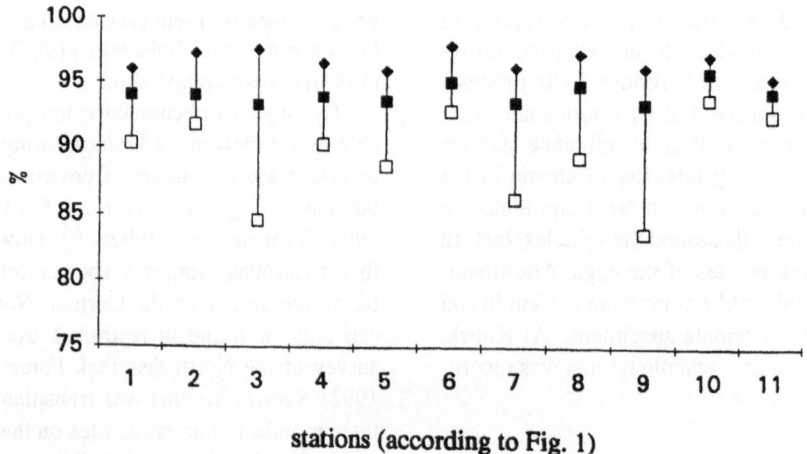

**Fig. 6.** *Littorina littorea*. Proportion of epithelial area from whole tubulus area. ■ mean value, □ minimum value, ◆ maximum value.

animals died and the results obtained are probably not reliable.

Because of intersex development *L. littorea* is a promising candidate for TBT biomonitoring at those sites where imposex-affected species are missing due to high TBT contamination or inappropriate habitats. The development scheme (Fig. 2) gives a very simple description of easily detectable alterations in the pallial oviduct. Furthermore, it is the basis of the intersex index (ISI). The present (and still preliminary) results indicate that the ISI is more suited for the description of intersex intensities and for the determination of TBT exposure than the average female prostate length. ISI values increase with the first occurrence of stage 1, whereas prostate length increases are detectable only after stage 3 is reached.

Although the correlation between TBT body burden and ISI or female prostate length (Fig. 5) is a strong indication that intersex is caused by TBT, this needs to be confirmed by laboratory experiments under controlled conditions. These investigations will have to answer the question whether or not these malformation can be induced by TBT in adult and sexually mature females or only during a earlier specific sensitive phase in the life cycle of *L. littorea*.

Bryan & Gibbs (1991) found TBT body burdens of $403.6 \pm 171.2$ $\mu$g TBT as Sn kg$^{-1}$ dry wt. in *L. littorea* from Northam Bridge (Itchen estuary; average aqueous concentration: $27.3 \pm 16.2$ ng TBT as Sn l$^{-1}$; $n = 8$). The calculated bioconcentration factor (bcf) is $1.48 \times 10^4$. Langston *et al.* (1987, 1990) reported values between 100–1120 $\mu$g TBT as Sn kg$^{-1}$ dry wt. in periwinkles from Poole Harbour (south England) at ambient TBT concentrations of 2–139 ng as Sn l$^{-1}$ (bcf: $8.1 \times 10^3$–$5.0 \times 10^4$). The values determined by Langston *et al.* (1987, 1990) are in good accordance to the body burdens of periwinkles analysed here. Based on the reported bcf from England, the present TBT contamination in our investigation area has to be expected to be in the range of 2–150 ng TBT as Sn l$^{-1}$. This indicates that, in spite of the ban of TBT containing antifouling paints on boats <25 m overall length, there is still a considerable degree of TBT pollution in German coastal waters. The calculated bcf values at Roscoff harbour are $2.64 \times 10^4$–$3.46 \times 10^4$. This result from France confirms the findings of Langston *et al.* (1987, 1990) and Bryan & Gibbs (1991) in England.

The analysed *Littorina* populations from Roscoff harbour allow an estimation of the threshold concentrations for intersex development. Calculated ISI indices in these samples were in the range of 0.0–0.06, *i.e.* only up to 6.1% of the females in the populations exhibited intersex. Consequently, the average aqueous TBT concentration of 15.4 ng as Sn l$^{-1}$ at this station seems to reflect the most probable threshold concentration for intersex development.

Histological observations of the midgut gland did not reveal any severe lesions. Normally, the digestive tubules show a synchrony of the different phases in each specimen. This means that one phase is clearly dominant (Langton, 1975). Several stressors have been shown to upset the digestive rhythm. These include spawning activity, starvation, temperature and pollutants. If all biological factors can be separated, the mean height of digestive cells and the loss of synchrony are estimated as good indicators for the influence of anthropogenic stressors (Bayne *et al.*, 1984; Couch, 1985).

During our investigations, no severe pathological deviations could be found in the midgut gland. This might indicate that the general pollution load along the East Frisian coast (Koopmann *et al.*, 1993) is insufficient to induce persistent histopathological changes in *L. littorea*. This has to be proved in more detailed studies with seasonal sampling.

The pathological disorders in the female gonad, such as atresia and lysis, coincide only partially with TBT body burdens and intersex intensities. Probably additional stressing factors have to be taken into account. Beside TBT copper and other heavy metals can also induce a retardation in gonadal maturation (Myint & Tylor, 1982; Sindermann, 1985).

The glycogen accumulation around the follicles, in combination with lytic and atretic processes inside the follicles, indicates changes in the energy metabolism and the incorporation of glycogen into the yolk. This phenomenon is well-known to occur following starvation and under the influence of heavy metals (Thompson *et al.*, 1974; Bayne *et al.*, 1978). The combination of histopathological techniques with morphological parameters revealed dominant alterations in the female gonad at histological and morphological levels. TBT body burdens of *L. littorea* were comparable with those measured in periwinkles from other European coasts with shipping activities (Langston *et al.*, 1987, 1990). On the other hand, the accumulation of other pollutants, such as heavy metals, from the East Frisian cast revealed relatively low levels for periwinkles (Thiel *et al.*, 1992). The effects reported here are therefore likely to be mainly associated with TBT.

26

## Acknowledgments

The financial support of 'Aktion Seeklar e.V.' (Hamburg) and the German section of 'World Wild Fund for Nature' (Frankfurt/M.) is greatly acknowledged.

## References

Bayne, B. L., D. A. Brown, K. Burns, D. R. Dixon, A. Ivanovitci, D. R. Livingstone, D. M. Lowe, M. N. Moore, A. R. D. Stebbing & J. Widdows, 1984. The effects of stress and pollution on marine animals. Praeger, New York, 385 pp.

Bright, D. A. & D. V. Ellis, 1990. A comparative survey of imposex in northeast Pacific neogastropods (Prosobranchia) related to tributyltin contamination, and a choice of a suitable bioindicator. Can. J. Zool. 68: 1915–1924.

Bryan, G. B. & P. Gibbs, 1991. Impact of low concentrations of tributyltin (TBT) on marine organisms: a review. In M. C. Newman & A. W. McIntosh (eds), Metal ecotoxicology: concepts and applications. Lewis Publisher, Ann Arbor: 323–361.

Couch, J. A., 1985. Prospective study of infectious and noninfectious diseases in oysters and fishes in three Gulf of Mexico estuaries. Dis. Aquat. Org. 1: 59–82.

Fioroni, P., U. Deutsch, E. Stroben & J. Oehlmann, 1992. Artificially induced pseudohermaphroditism in prosobranchs and its absence in littorinids. In J. Grahame, P. J. Mill & D. Reid (eds), Proceedings of the 3rd International Symposium on Littorinid Biology. The Malacological Society of London, London: 313–315.

Fioroni, P., J. Oehlmann & E. Stroben, 1991. The pseudohermaphroditism of prosobranchs; morphological aspects. Zool. Anz. 226: 1–26.

Fretter, V., 1980. Gross anatomy of the female genital duct of British, UK Littorina spp. J. moll. Stud. 46: 148–153.

Fretter, V. & A. Graham, 1962. British prosobranch molluscs. Their functional anatomy and ecology. Ray Society, London, 755 pp.

Fretter, V. & A. Graham, 1980. The prosobranch molluscs of Britain and Denmark. Part 5 – Marine Littorinacea. J. moll. Stud., Suppl. 7: 243–284.

Gibbs, P. E., G. W. Bryan, P. L. Pascoe & G. R. Burt, 1987. The use of the dog-whelk, Nucella lapillus, as an indicator of tributyltin (TBT) contamination. J. mar. biol. Ass. U.K. 67: 507–523.

Gibbs, P. E., G. W. Bryan, P. L. Pascoe & G. R. Burt, 1990. Reproductive abnormalities in female Ocenebra erinacea (Gastropoda) resulting from tributyltin-induced imposex. J. mar. biol. Ass. U.K. 70: 639–656.

Gibbs, P. E., B. E. Spencer & P. L. Pascoe, 1991. The American oyster drill, Urosalpinx cinerea (Gastropoda): evidence of decline in an imposex-affected population (R. Blackwater, Essex). J. mar. biol. Ass. U.K. 71: 827–838.

Graham, A., 1988. Molluscs: prosobranch and pyramidellid gastropods. Academic Press, London, New York, 112 pp.

Harding, M. J. C., S. K. Bailey & I. M. Davies, 1992. TBT imposex survey of the North Sea. Annex 4: Germany. UK Department of the Environment, Contract PECD 7/8/214. 20 pp.

Heller, J., 1990. Longevity in molluscs. Malacologia 31: 259–295.

Kalbfus, W., A. Zellner, S. Frey & E. Stanner, 1991. Gewässergefährdung durch organozinnhaltige Antifouling-Anstriche. Final report R + D project UBA, Berlin, 169 pp.

Koopmann, C., J. Faller, K.-H. van Bernem, A. Prange & A. Müller, 1993. Schadstoffkartierung in Sedimenten des deutschen Wat-

tenmeeres Juni 1989–Juni 1992. Final report R + D project UBA, Berlin, 156 pp.

Langston, W. J., G. W. Bryan, G. R. Burt & P. E. Gibbs, 1990. Assessing the impact of tin and TBT in estuaries and coastal regions. Functional Ecology 4: 433–443.

Langston, W. J., G. R. Burt & Z. Mingjiang, 1987. Tin and organotin in water, sediments, and benthic organisms from Poole Harbour. Mar. Pollut. Bull. 18: 634–639.

Langton, R. W., 1975. Synchrony in the digestive diverticula of Mytilus edulis L. J. mar. biol. Ass. U.K. 55: 221–229.

Lauckner, G., 1980. Diseases of Mollusca: Gastropoda. In O. Kinne (ed.), Diseases of marine animals. Wiley & Sons, Chichester: 311–424.

Linke, O.,1933. Morphologie und Physiologie des Genitalapparates der Nordseelittorinen. Wiss. Meeresunters. Abt. Helgoland 19: 1–60.

Matthiessen, P., R. Waldock, J. E. Thain, S. Milton & S. Scorpe-Howe, 1991. Changes in periwinkle (Littorina littorea) population following the ban on TBT-based antifoulings on small boats. Int. Council for Exploration of the Sea. Marine Environmental Quality Committee CM 1991/E: 5.

Muggeridge, P. L., 1979. The reproductive biology of the mangrove littorinids Bembicium auratum (QUOY & GAIMARD) and Littorina scabra scabra (LINNE) (Gastropoda, Prosobranchiata), with observations in the reproductive cycles of rocky shore littorinids of New South Wales. PhD Thesis, University of Sydney.

Myint, U. M. & P. A. Tyler, 1982. Effects of temperature, nutritive and metal stressors on the reproductive biology of Mytilus edulis. Mar. Biol. 67: 209–223.

Nordsieck, F., 1968. Die europäischen Meeres-Gehäuseschnecken (Prosobranchia). Vom Eismeer bis Kapverden und Mittelmeer. Fischer, Stuttgart, 273 pp.

Oehlmann, J., E. Stroben & P. Fioroni, 1991. The morphological expression of imposex in Nucella lapillus (LINNAEUS) (Gastropoda: Muricidae). J. moll. Stud. 57: 375–390.

Oehlmann, J., E. Stroben & P. Fioroni, 1992. The rough tingle Ocenebra erinacea (Gastropoda: Muricidae): an exhibitor of imposex in comparison to Nucella lapillus. Helgoländer Meeresunters. 46: 311–328.

Oehlmann, J., E. Stroben & P. Fioroni, 1993. Fréquence et degré d'expression du pseudohermaphrodisme chez quelques Prosobranches Sténoglosses des côtes françaises (surtout de la baie de Morlaix et de la Manche). 2. Situation jusqu'au printemps de 1992. Cah. Biol. mar. 34: 343–362.

Reid, D. G., 1986a. The littorinid molluscs of mangrove forests in the Indo-Pacific region. The genus Littoraria. British Museum (Natural History), London, 227 pp.

Reid, D. G., 1986b. Mainwaringia NEVILL, 1885, a littorinid genus from Asiatic mangrove forests and a case of protandrous hermaphroditism. J. moll. Stud. 52: 225–242.

Reid, D. G., 1989. The comparative morphology, phylogeny and evolution of the gastropod family Littorinidae. Phil. Trans. r. Soc., Lond. B. 324: 1–110.

Romeis, B., 1989. Mikroskopische Technik. Urban & Schwarzenberg, München, Wien, Baltimore, 697 pp.

Short, J. W., S. D. Rice, C. C. Brodersen & W. B. Stickle, 1989. Occurrence of tri-$n$-butyltin-caused imposex in the North Pacific marine snail Nucella lima in Auke Bay, Alaska. Mar. Biol. 102: 291–297.

Sindermann, C., 1985. Notes of a pollution watcher. In F. J. Vernberg, F. P. Thurberg, A. Calabrese & W. Vernberg (eds), Marine Pollution and Physiolgy. Recent Advances. University of South Carolina Press, Durham: 11–30.

Smith, B. S., 1971. Sexuality in the American mud snail, *Nassarius obsoletus* SAY. Proc. malac. Soc. Lond. 39: 377–378.

Smith, B. S., 1980. The estuarine mud snail, *Nassarius obsoletus*: abnormalities in the reproductive system. J. moll. Stud. 46: 247–256.

Smith, B. S., 1981a. Male characteristics in the female *Nassarius obsoletus*: variations related to locality, season and year. Veliger 23: 212–216.

Smith, B.S., 1981b. Reproductive anomalies in stenoglossan snails related to pollution from marinas. J. appl. Toxicol. 1: 15–21.

Smith, B. S., 1981c. Male characteristics on female mud snails caused by antifouling bottom paints. J. appl. Toxicol. 1: 22–25.

Smith, B. S., 1981d. Tributyltin compounds induce male characteristics on female mud snails *Nassarius obsoletus* = *Ilyanassa obsoleta*. J. appl. Toxicol. 1: 141–144.

Stroben, E., C. Brömmel, J. Oehlmann & P. Fioroni, 1992a. The genital systems of *Trivia arctica* and *Trivia monacha* (Prosobranchia: Mesogastropoda) and tributyltin induced imposex. Zool. Beitr. N.F. 34: 349–374.

Stroben, E., J. Oehlmann & P. Fioroni, 1992b. The morphological expression of imposex in *Hinia reticulata* (Gastropoda: Buccinidae): a potential biological indicator of tributyltin pollution. Mar. Biol. 113: 625–636.

Taylor, J. D. & J. A. Miller, 1989. The morphology of the osphradium in relation to feeding habits in meso- and neogastropods. J. moll. Stud. 55: 227–237.

Thiel, H., M. Kaiser, J. Lade, H. Marencic & D. Lorch, 1992. Vergleichende Untersuchungen über die Eignung von Wattorganismen unterschiedlicher Trophiestufen zum Trendmonitoring ausgewählter Schwermetalle und polychlorierter Biphenyle. UBA, Berlin, 140 pp.

Thompson, R. J., N. A. Ratcliffe & B. L. Bayne, 1974. Effects of starvation on the structure and function in the digestive gland of the mussel (*Mytilus edulis*). J. mar. biol. Ass. U.K. 54: 699–712.

Thorson, G., 1946. Reproductive and larval development of Danish marine bottom invertebrates. Medd. Komm. og Havundersøg. Kbh. Ser. Plankton 4: 1–523.

Vega, M. M., J. A. Marigomez & E. Angulo, 1989. Quantitative alterations in the structure of the digestive cell of *Littorina littorea* on exposure to cadmium. Mar. Biol. 103: 547–553.

Ward, G. S, G. C. Cramm, P. R. Parrish, H. Trachmann & A. Slesinger, 1981. Bioaccumulation and chronic toxicity of bis(tributyltin)oxide (TBTO): tests with a saltwater fish. In D. R. Branson & K. L. Dickson (eds), Aquatic toxicity and hazard assessment. Associate Committee on Scientific Criteria for Environmental Quality, Philadelphia: 183–200.

*Hydrobiologia* **309**: 29–35, 1995.
*P. J. Mill & C. D. McQuaid (eds), Advances in Littorinid Biology.*
©1995 *Kluwer Academic Publishers.*

# Comparison of imposex response in three Prosobranch species

M. Huet[1], P. Fioroni[2], J. Oehlmann[2] & E. Stroben[2]
[1]*Laboratoire de Biologie Marine U.R.A. C.N.R.S. D 1513, Université de Bretagne Occidentale, F-29 275 Brest Cedex, France*
[2]*Institut für Spezielle Zoologie und Vergleichende Embryologie der Universität Münster, D-48 149 Münster, Germany*

*Key words:* imposex, gastropod, TBT, *Nucella, Ocenebra, Hinia*

## Abstract

A comparative study of the three gastropod species *Nucella lapillus* (L.), *Ocenebra erinacea* (L.) and *Hinia (Nassarius) reticulata* (L.) reveals that *Nucella* is the most TBT sensitive species while *Hinia* is the least sensitive. Of the two imposex indices VDSI and RPSI, good interspecies correlations were obtained only for VDSI. The three species can be considered as complementary not only in terms of their ecology but also for their levels of sensitivity. Indeed, *Nucella* and *Ocenebra* are useful test species at TBT concentrations below 2 ng Sn $l^{-1}$ while *Hinia* is the more appropriate species at higher TBT levels. For the first time, two sterilised *Hinia* females are recorded. This sterilization does not seem to be due to proliferation of vas deferens tissue in the vaginal opening and further investigation is needed to find intermediary VDS stages between stage $4^+$ and sterilization. Studies using *Nucella* show that the use of narcotization in imposex analysis leads to an underestimation of RPSI compared with non-narcotization methods. Indeed, narcotization straightens the penis and increases its length when compared with non-narcotized animals, but this increase is proportionally higher in males than in females.

## Introduction

Imposex in gastropods is a useful response for assessing TBT pollution. This pseudohermaphroditism was discovered by Blaber (1970), termed imposex by Smith (1971) and consists of the addition of male sexual characters in females (notably a penis and a spermduct or vas deferens). The species which has been the most studied is *Nucella lapillus* (L.) because it is common on European rocky shores, is sensitive to TBT, has direct development and is relatively immobile. *Nucella*, however, is not present along all of the European coasts because of inappropriate habitats, and some populations have disappeared following female sterility in highly polluted areas. Thus it would be interesting to examine other gastropods which exhibit imposex. The purpose of the present paper is to compare the imposex responses of three Prosobranch species: *Nucella lapillus* (L.), *Ocenebra erinacea* (L.) and *Hinia (Nassarius) reticulata* (L.).

## Materials and methods

Sampling stations were located along the West French coast, from Luc-sur-Mer in the north to Arcachon in the south west (Fig. 1). Sampling took place from March 1988 to September 1993. Where possible, 30 adult individuals of each species (*Nucella lapillus, Ocenebra erinacea* and *Hinia reticulata*) were collected in each of 110 samples. (if present).

In the laboratory, individuals were narcotized in 70‰ $MgCl_2$ in distilled water. The shells were crushed in a vice and the animals removed and observed under a binocular microscope. Sex determination was made based on the presence of the typical sex glands, i.e. prostate and testis in males, capsule gland, sperm ingesting gland, albumen gland and ovary in females.

Two imposex indices were calculated: Relative Penis Size Index (RPSI) and Vas Deferens Sequence

*Fig. 1.* Map of France indicating the position of three sites: Luc-sur-Mer (A), Brest (B) and Arcachon (C)

Index (VDSI) (Bryan *et al.*, 1986; Gibbs *et al.*, 1987):

$$RPSI = [(\text{mean female penis length})^3/$$

$$(\text{mean male penis length})^3] \times 100$$

$$VDSI = \text{average of VDS stages in females}$$

Penis lengths were measured to the nearest 0.1 mm. VDS stages extend from 0 to 6 on the scale of Gibbs *et al.* (1987) for *Nucella lapillus*, this scale was expanded by Fioroni *et al.* (1991) and Stroben *et al.* (1992) in order to obtain a VDS scale suitable for all gastropod species (Fig. 2). VDS stage 7 was assigned to females whose gonad was partially or totally transformed into a testis, producing spermatozoids.

The above procedure has not been used by all authors. Thus Gibbs *et al.* (1987) did not narcotize animals before observation. In order to compare the two methods, 40 supplementary *Nucella* were collected from each of 58 stations and analyzed without narcotization. TBT analysis in sea water was performed using the method described by Stroben *et al.* (1992). Some individuals possessing peculiar characteristics were observed histologically. Interesting parts of these animals were embedded in paraplast, and 7 μm thick serial sections were cut and stained before observation.

All statistical analyses were performed using 'Statgraphics' software.

## Results

Some aphallic *Nucella* (males and females) were sampled at polluted sites from the Brest region (Fig. 1). Such a phenomenon has been recorded by Gibbs (1993) for a polluted English site and seems to have a genetic basis. Thus aphallic animals were considered to be non-imposex and therefore discarded from the calculation of indices.

*Comparison of imposex response of the three species*

This comparison was made using two indices, keeping in mind that RPSI is based on penis length, which is a continuous variable while VDSI is based on a discontinuous variable.

*RSPI* Figure 3 shows RSPI distributions for each station for each pair of species. There is a clear positive correlation in the relationship between *Hinia* and *Nucella* ($r^2 = 0.77$) but for the other two pairings, *Ocenebra* and *Nucella* ($r^2 = 0.29$) and *Hinia* and *Ocenebra* ($r^2 = 0.20$) there is no obvious correlation.

*VDSI* Comparisons of VDS indices were performed after ranking VDS indices of each pair of species at each site, using a sign test. Analyses revealed that there is a significance difference in species sensitivity: *Nucella*>*Hinia* ($P = 1$), *Nucella*>*Ocenebra* ($P>0.99$), *Ocenebra*>*Hinia* ($P>0.99$). Thus, we can assume that *Nucella* is the most sensitive species while *Hinia* is the least.

In the sampled populations of *Nucella*, maximal VDS indices are above 5, due to sterilised females with abortive egg-capsules within capsule-glands (VDS stage 6a) and sex changes, i.e. females producing spermatozoids (VDS stage 7). *Ocenebra* showed only one VDSI above 5, at a station where one sex change was observed and histological examination revealed a suppression of oogenesis in another female but no spermiogenesis. Such results have been reported by Gibbs *et al.* (1990). In stations where at least two species were collected, the maximum VDS stage in *Hinia* was 4[+] (considered as 4 in the calculation of the index). However, two sterilised *Hinia* females were found at a highly polluted site where both *Ocenebra* and *Nucella* were absent. Sterilization of these two *Hinia* females was not proven to be due to a vas deferens tissue proliferation blocking the vulva. The first sterilised female had excrescences of tissues in the bursa copulatrix, causing a narrowness of the vaginal

31

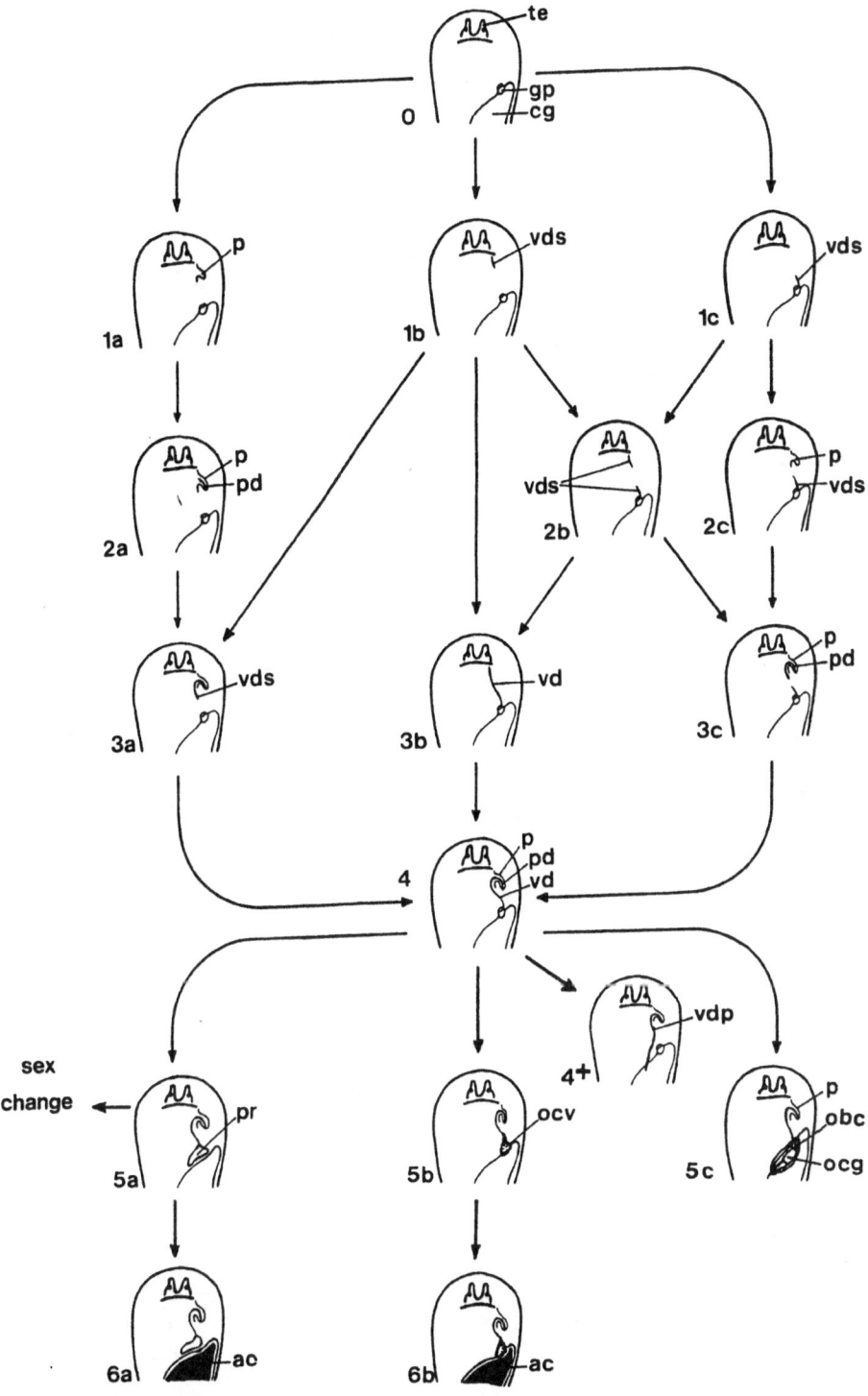

*Fig. 2.* Scheme of VDS evolution. The numbers refer to stages. ac, aborted capsules; cg, capsule gland; gp, genital papilla; obc, open bursa copulatrix; ocg, open capsule gland; ocv, occlusion of the vulva; p, penis; pd, penis duct; pr, prostate; te, tentacle; vd, vas deferens; vdp, vas deferens passage into capsule gland; vds, vas deferens section

32

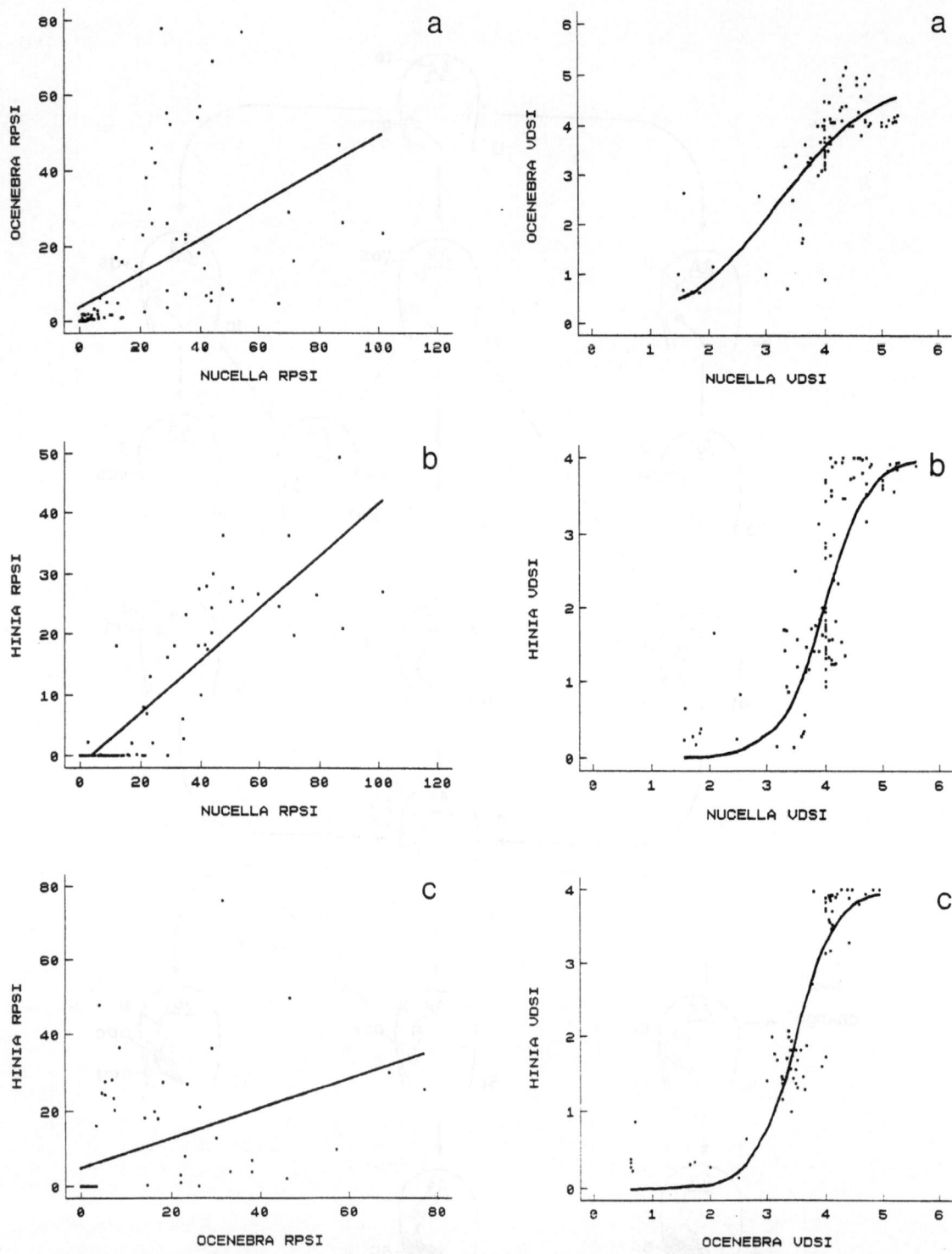

*Fig. 3.* Relationships between RPS indices of paired species. Each point corresponds to one sample. (a) *Ocenebra* and *Nucella* $\text{RPSI}_{Ocenebra} = 3.8 + 0.5 \text{RPSI}_{Nucella}$, $n = 82$, $r^2 \approx 0.29$; (b) *Hinia* and *Nucella* $\text{RPSI}_{Hinia} = -1.6 + 0.4 \text{RPSI}_{Nucella}$, $n = 110$, $r^2 \approx 0.77$; (c) *Hinia* and *Ocenebra*, $\text{RPSI}_{Hinia} = 4.8 + 0.4 \text{RPSI}_{Ocenebra}$, $n = 69$, $r^2 \approx 0.20$

*Fig. 4.* Relationships between VDS indices of paired species. Each point corresponds to one sample. (a) *Ocenebra* and *Nucella*, $\text{VDSI}_{Ocenebra} = 4.9/(1+59 \exp[-1.3 \text{ VDSI}_{Nucella}])$, $n = 89$, $r^2 \approx 0.65$ (b) *Hinia* and *Nucella*, $\text{VDSI}_{Hinia} = 4/(1+5 \cdot 10^5 \exp[-2.7 \text{ VDSI}_{Nucella}])$, $n = 115$, $r^2 \approx 0.63$ (c) *Hinia* and *Ocenebra*, $\text{VDSI}_{Hinia} = 4/(1+2.5 \cdot 10^5 \exp[-2.9 \text{ VDSI}_{Ocenebra}])$, $n = 70$, $r^2 \approx 0.84$

channel; an abortive capsule was found in the capsule gland. These excrescences seem to be constituted by vaginal neoplasic tissues and not by a vas deferens tissue proliferation. The second sterilised *Hinia* female had a brown mass made up of abortive egg-capsules fused in the shell. The capsule gland had disappeared, probably joined with the abortive egg-capsules.

The relationship between VDSIs for each station for pairs of species is shown in Fig. 4. The best fits are obtained with sigmoid curves. For *Ocenebra* and *Nucella* the relationship is almost linear (Fig. 4a), indicating a similar development of VDSI in the two species. However, *Hinia* reacts differently, as is shown by the markedly sigmoid curves (Fig. 4b,c). First, imposex starts more slowly in *Hinia* and we observe a plateau at low indices. Secondly, a very marked increase in VDSI is seen around a value of 4 for *Nucella* and 3.5 for *Ocenebra*. Thirdly, this increase stops and a plateau is reached at a VDSI value of 4, the maximum VDSI value in this species.

*Water TBT concentration and Imposex.* In comparisons between the RSPIs of the three species, no clear correlation can be observed for RSPIs and sea water TBT levels. However, a relatively clear correlation exists between VDSIs and sea water TBT levels (Fig. 5). Unfortunately, data points corresponding to low TBT levels cannot be included in the analysis as the method employed for TBT determinations in sea water has a detection limit of 1.5 ng Sn $l^{-1}$. Nevertheless we can see that *Nucella* exhibits the highest indices whilst *Hinia* exhibits the lowest. The relationship curves for *Nucella* and *Ocenebra* seem parallel and show only a small increase with increasing TBT levels, whereas *Hinia* VDSI values increase sharply between 1.5 and 10 ng Sn $l^{-1}$.

*Comparison of indices calculated with narcotized and non-narcotized Nucella*

The comparison between narcotized and non-narcotized *Nucella* indicates that narcotization did not affect the calculated VDSIs significantly ($P>0.95$; Kolmogorov-Smirnov Test). A comparison of RPSIs obtained with the two methods at the same sites is shown in Fig. 6. The slope of the regression line is significantly different from 1. However, the intercept is not significantly different from 0 ($P>0.95$). From this it is clear that narcotization leads to an underestimation of the RSPI when compared with non-narcotization. Thus the size of the male penis must show a relatively greater increase in length than that of the female penis as a

result of narcotization although, since mean male penis length is greater than mean female penis length in all of the sites sampled for this comparative study, this may simply be an effect of initial penis length rather than any sexual difference. The change in penis length at each site (calculated separately for males and females as mean narcotized penis length divided by mean non-narcotized penis length) was plotted against narcotized penis length (Fig. 7). The relationship is: length change = 0.92 × (narcotized penis length)$^{0.25}$ ($n=116$, $r^2=0.49$). This indicates that, for the vast majority of sites sampled, the mean penis length of narcotized animals was greater than that of non-narcotized ones. Furthermore, the percentage increase in penis length in narcotized animals increases with increase in penis length.

## Discussion

### Comparison of the narcotized and non-narcotized methods

RPS indices obtained with narcotized *Nucella* are lower than indices calculated with non-narcotized *Nucella*. Thus it is not possible to directly compare ecological TBT surveys using RPS indices obtained with the two different methods. Nevertheless, a recalculation can be made using the relationship shown in Fig. 6. For example, an RSPI of 40 calculated from non-narcotized animals corresponds to 49 for the non-narcotization method. In *Hinia*, analysis of imposex seems impossible without narcotization because this species is very active and the penis tends to have a corkscrew shape which is straightened by $MgCl_2$.

### Comparison of the three species

No clear correlation is observed between paired species RPSIs when considering *Ocenebra* (Fig. 3) but no explanation can be given at the moment. VDSIs, however, seem to be useful in comparing imposex responses between the three species (Fig. 4). Increase in the indices show an almost linear relationship for *Nucella* and *Ocenebra*, a similar result to that found by Oehlmann *et al.* (1992), but the relationship between *Hinia* and the other two species is different. *Hinia* is less sensitive to TBT; female sterility occurs only at very high TBT levels, at which *Nucella* and *Ocenebra* are unable to survive. Also, imposex seems to start at a higher TBT level than for the other two species.

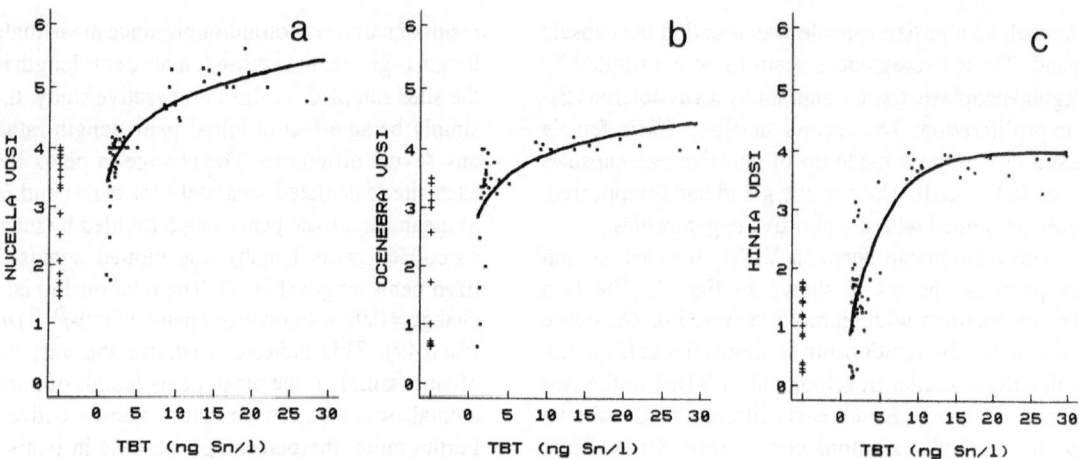

*Fig. 5.* Relationships between sea water TBT concentrations and species VDS indices. Each point corresponds to one sample. $^+$, samples correspond to sea water TBT concentrations under the detection limit. (a) *Nucella*, $\text{VDSI}_{Nucella} = 3.4+0.6 \text{ LOG(TBT)}$, $n = 68$, $r^2 \approx 0.62$ (b) *Ocenebra*, $\text{VDSI}_{Ocenebra} = 2.8+0.5 \text{ LOG(TBT)}$, $n = 44$, $r^2 \approx 0.41$ (c) *Hinia*, $\text{VDSI}_{Hinia} = 4(1-e^{-0.2TBT})$, $n = 68$, $r^2 \approx 0.71$

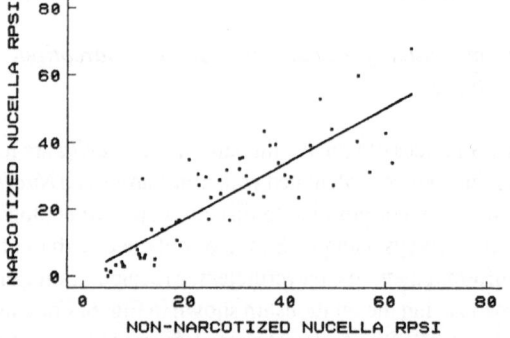

*Fig. 6.* Relationship between RPS indices calculated with narcotized and non-narcotized *Nucella*. Each point corresponds to one sample. $\text{RPSI}_{narcotizedNucella} = 0.5 + 0.8 \text{ RPSI}_{non-narcotizedNucella}$, $n = 58$, $r^2 \approx 0.74$

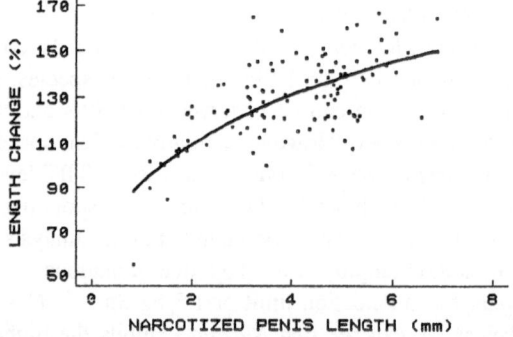

*Fig. 7.* Relationship between R and narcotized *Nucella* penis length (R = mean narcotized *Nucella* penis length/mean non- narcotized *Nucella* penis length, calculated separately for males and females). Each point corresponds to one sample. $R = 0.92 \times \text{(penis length)}^{0.25}$.

Thus VDSIs below 1 are observed in *Hinia* when corresponding VDSIs for *Nucella* and *Ocenebra* are 3 and 2.5 respectively. These results agree with those of Bryan & Gibbs (1991), who found an initiation of penis development in *Nucella* at TBT sea water concentrations below 0.4 ng Sn $l^{-1}$, and Bryan *et al.* (1993), who estimated this initiation in *Hinia* at about 1 ng Sn $l^{-1}$. However, these authors did not observe any sterilised *Hinia* females, which were found in the present study. Indeed, two sterilised *Hinia* females (one with a testis) were collected at a location where a few *Nucella* had been sampled one year before. These animals were suspected of having been transplanted accidentally by oyster culturing; no *Ocenebra* were found despite the suitable habitat.

The plotting of VDSIs against sea water TBT levels (Fig. 5) reveals that there is a marked increase in *Hinia* VDSI between 2 and 5 ng Sn $l^{-1}$ whereas *Nucella* and *Ocenebra* VDSIs increase slowly (see also Fig. 4b,c). Since TBT determination in sea water is an immediate measurement whereas, on the other hand, the imposex evaluation in adult animals is a function of exposure to TBT over several years, this comparison must be viewed with caution. However, Fig. 4 is of interest because comparable data are presented. Indeed, we can assume populations from the same site have received a similar TBT exposure and their imposex can therefore be compared. Thus we can assume that *Hinia* reacts more slowly at lower TBT levels and that it has a more gradual reaction to intermediate levels of exposure.

Consequently, *Nucella* and *Ocenebra* are the more useful test species at lower TBT levels while *Hinia* is the most appropriate species to estimate intermediate TBT concentrations. At the higher levels of pollution, none of the species examined in the present study is clearly appropriate since imposex reachs its maximum. These three species can be considered as complementary in ecological TBT surveys, *Hinia* being the more appropriate in very polluted sites since this species can survive where populations of *Ocenebra* and *Nucella* cannot because of female sterilization. Indeed, higher TBT concentrations are required to obtain sterilization in *Hinia*. Unlike *Ocenebra* and *Nucella*, this species has a planktonic phase and therefore populations in polluted areas may also be sustained by larval settlement. Nevertheless, the two sterilised *Hinia* females which were found indicate that there is a limit to the resistance of this species to TBT-induced sterilization and further investigation is needed to assess imposex response of this species in very polluted sites.

## Acknowledgments

We thank Prof. Lasserre (Station d'océanologie et de biologie marine, Roscoff) and Prof. Déniel (Laboratoire de biologie animale, Brest) for excellent working conditions in their laboratories. Many thanks to Dr Baron for his help in statistical analyses. We are very grateful to Barbara Hasert, Christel Mehlis, Gaby Nierster and Robert Marc for technical assistance. This work was partly supported by the Brest community, within the 'Rade de Brest' contract.

Contribution n° 95-005 URA CNRS D1513.

## References

Blaber, S. J. M., 1970. The occurrence of a penis-like structure outgrowth behind the right tentacule in spent females of *Nucella lapillus* (L.). Proc. malac. soc. Lond. 39: 231–233.

Bryan, G. W. & P. E. Gibbs, 1991. Impact at low concentrations of tributyltin (TBT) on marine organisms: a review. In: M. C. Newman & A. W. McIntosh (eds), Metal Ecotoxicology: Concepts and Applications. Lewis, Ann Arbor: 323–361.

Bryan, G. W., G. R. Burt, P. E. Gibbs & P. L. Pascoe, 1993. *Nassarius reticulatus* (Nassariidae: Gastropoda) as an indicator of tributyltin pollution before and after TBT restrictions. J. mar. biol. Ass. U.K. 73: 913–929.

Bryan, G. W., P. E. Gibbs, L. G. Hummerstone & G. R. Burt, 1986. The decline of the gastropod *Nucella lapillus* around South-West England: evidence for the effect on tributyltin from antifouling paints. J. mar. biol. Ass. U.K. 66: 611–640.

Fioroni, P., J. Oehlmann & E. Stroben, 1991. The pseudo-hermaphroditism of prosobranchs; morphological aspects. Zool. Anz. 226: 1–26.

Gibbs, P. E., 1993. A male genital deffect in the dog-whelk, *Nucella lapillus* (Neogastropoda), favouring survival in a TBT-polluted area. J. mar. biol. Ass. U.K. 73: 667–678.

Gibbs, P. E., G. W. Bryan, P. L. Pascoe & G. R. Burt, 1987. The use of the dog-whelk, *Nucella lapillus*, as an indicator of tributyltin (TBT) contamination. J. mar. biol. Ass. U.K. 67: 507–523.

Gibbs, P. E., G. W. Bryan, P. L. Pascoe & G. R. Burt, 1990. Reproductive abnormalities in female *Ocenebra erinacea* (Gastropoda) resulting from tributyltin-induced imposex. J. mar. biol. Ass. U.K. 70: 639–656.

Smith, B. S., 1971. Sexuality in the American mud snail, Nassarius obsoletus: abnormalities in the reproductive system. J. moll. Stud. 46: 247–256.

Oehlmann, J., E. Stroben & P. Fioroni, 1992. The rough tingle *Ocenebra erinacea* (Neogastropoda: Muricidae): an exhibitor of imposex in comparison to *Nucella lapillus*. Helgoländer wiss. Meeresunters 46: 311–328.

Stroben, E., J. Oehlmann & P. Fioroni, 1992. The morphological expression of imposex in *Hinia reticulata* (Gastropoda: Buccinidae): a potential indicator of tributyltin pollution. Mar. Biol. 113: 625–636.

*Hydrobiologia* **309**: 37–44, 1995.
*P. J. Mill & C. D. McQuaid (eds), Advances in Littorinid Biology.*
©1995 *Kluwer Academic Publishers.*

# Validation of a planimetric procedure to quantify stress in *Littorina littorea* (Gastropoda: Mollusca): is it independent of the reproductive cycle?

G. Calvo-Ugarteburu[1,2], V. Saez[1], C. D. McQuaid[2] & E. Angulo[1]
[1]*Zitologi eta Histologi Laborategia, Biologia Zelularra eta Zientzia Morfologikoen Saila, Zientzi Fakultatea, Euskal Herriko Unibertsitatea, 644 P.K., 48080 Bilbo, Spain*
[2]*Department of Zoology and Entomology, Rhodes University, Grahamstown 6140, South Africa*

*Key words: Littorina littorea*, planimetry, digestive gland, reproductive state

## Abstract

This study forms part of a larger project in which five planimetric parameters have been used to study changes in the digestive epithelium of *Littorina littorea* under different environmental and physiological conditions. Our aim was to examine the effect of the reproductive cycle on these parameters in order to assess their usefulness as indicators of stress.

A one way anova shows that the absolute parameters of mean epithelial thickness (MET), mean diverticulum radius (MDR) and mean luminal radius (MLR) vary significantly depending on the time of the year. This variation is highly correlated to the amount of digestive tissue present and hence negatively correlated to the reproductive state because gonad and digestive tissue volume are inversely related. Consequently, these parameters are not good indicators of stress. However, whereas the absolute size of the digestive acinus varies with the reproductive state of the animal, the MET, MDR and MLR retain the same proportions and therefore the ratios MET/MDR and MLR/MET remain constant. This makes them useful indicators of stress because they are independent of intrinsic variables such as the reproductive cycle.

## Introduction

Pollution of the marine ecosystem is becoming a serious problem for the exploitation of aquatic resources. Marine organisms have the capacity to accumulate environmental toxins in their tissues and to concentrate them above sea water levels (Fossato & Siviero, 1974; Phillips, 1977; Harrinson & Berger, 1982; Simkiss *et al.*, 1982; Widdows *et al.*, 1982, 1983; Simkiss & Mason, 1984; Etxeberria, 1990), which makes them ideal for monitoring levels of contaminants. Over the past few years the use of marine molluscs as sentinel organisms in this way has become world-wide (Goldberg, 1975, 1986; Goldberg *et al.*, 1978) and special attention has been paid to cellular responses (Simkiss & Mason, 1984; Moore, 1985, 1986). The importance of cellular responses is that they take place before the effects of contaminants become evident at other levels, and therefore they can be used to provide an ear-

ly indication of environmental stress (Moore, 1985; Underwood & Peterson, 1988; Vega *et al.*, 1989).

In particular, attempts have been made to link stress to changes in the digestive epithelium because of the importance of this organ in the total physiology of the animal. The digestive gland of molluscs has been reported to be an important storage site for accumulated hydrocarbons (Widdows *et al.*, 1983) and heavy metals (Janssen & Scholz, 1979; Scholz, 1980). Simkiss & Mason (1984) suggested that the basophilic cells of the digestive gland of gastropods may take part in detoxification mechanisms and this may have direct morphological consequences under stress.

Several studies describe a thinning of the digestive epithelium of molluscs in response to different stressors. For example, oil in *Mytilus* sp., (Lowe *et al*, 1981; Widdows *et al.*, 1982; Cajaraville *et al.*, 1992); 17 $\beta$ oestradiol and anthracene in *Mytilus edulis* (Moore *et al.*, 1978a, b); nutritional, thermal and exposure stressors in *M. edulis* and *M. californianus* (Thompson *et al.*, 1974, 1978; Moore *et al.*, 1979); salinity in *Littorina littorea* (Marigomez *et al.*, 1991); oil in *Venus verrucosa* (Axiak *et al.*, 1988); copper in *Arion ater* (Marigomez *et al.*, 1986a, b); organic pollutants in *Crassostrea virginica* (Couch, 1984); and inorganic pollutants in *Mercenaria mercenaria* (Tripp *et al.*, 1984). This thinning is due to the induction of cellular lysis in the epithelium as a result of the detoxification processes and the general response to stress (Soto, 1988, Saez *et al.*, 1992), and is easily quantified using planimetry (Agirregoikoa, 1988; Recio *et al.*, 1988).

Many studies have been done on cellular responses from a biochemical and metabolic perspective (Moore, 1980; Moore *et al.*, 1980; Harrinson & Berger, 1982), as well as from a morphological point of view (Lowe *et al.*, 1981; Marigomez *et al.*, 1986b; Recio *et al.*, 1988; Cajaraville *et al.*, 1989). However, Mason *et al.* (1984) concluded that the analysis of invertebrates for environmental monitoring should involve a range of physiological processes which are independent of each other. For example, changes in the digestive cell structure may occur as a response to natural phenomena such as tidal cycles (Morton, 1971; Bernard, 1973; Langton & Gabbot, 1974; Langton, 1975, 1977) and feeding cycles (Thompson *et al.*, 1974; Langton, 1975; Robinson, 1983).

This study is part of a larger project in which planimetric parameters have been used to interpret changes in the digestive epithelium of several molluscs under different environmental and physiological conditions (Castillero *et al.*, 1987; Recio *et al.*, 1988; Soto, 1988; Marigomez, 1989; Vega *et al.*, 1989; Etxeberria, 1990; Marigomez *et al.*, 1990, 1991, 1992; Saez *et al.*, 1992).

One of the natural variables which may affect these parameters is the reproductive state of the animal because the relative proportions of gonad, digestive gland and connective tissue vary during the reproductive cycle (Martel *et al.*, 1986; Morvan & Answell, 1988; Soto, 1988; Margiomez *et al.*, 1992). The visceral hump of *Littorina littorea* is formed mainly by the digestive gland and the gonad, together with some connective tissue (Fretter & Graham, 1962; Soto, 1988). Since there is a finite space in the visceral hump, changes in the volume of one of the organs would be expected to affect that of the other. In this part of the study we were concerned with the effects of changes during the reproductive cycle on the planimetric parameters of the digestive gland of *L. littorea* in order to assess their value as indicators of envi-

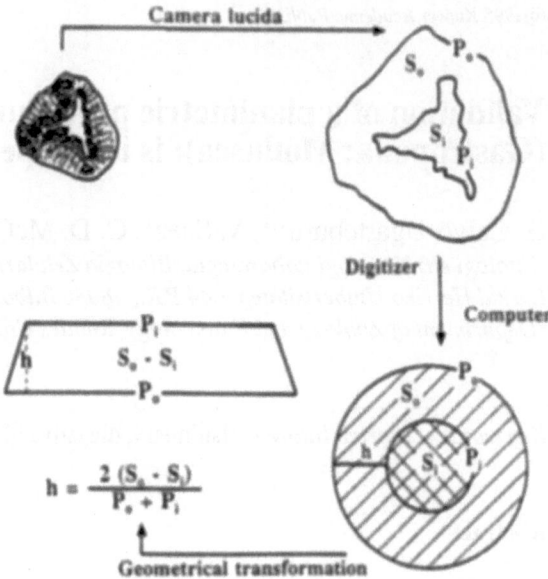

*Fig. 1.* Method of geometrical transformation for the calculation of the planimetric parameters (after Marigomez *et al.*, 1990) ($S_0$, area of epidermal layer; $S_i$, area of lumen; $P_0$, perimeter of the tubule; $P_i$, perimeter of the lumen; h, mean epithelial thickness (MET)).

ronmental stress. No attempt was made to examine mechanisms.

## Materials and methods

Monthly samples of *Littorina littorea* from the Bay of Clew (Ireland) were purchased from a commercial dealer between March 1989 and May 1990. Five animals of each sex with a columellar height between 21 and 24 mm were chosen and the total wet weight was measured, with and without the shell.

The digestive gland was carefully removed and weighed and the posterior end was placed in Bouin's fixative for histological examination. Six 8 $\mu$m sections were cut for each animal and stained with haematoxylin-eosin.

A total of 30 digestive acini per slide were randomly selected and five planimetric parameters were calculated by means of geometrical transformation (Recio *et al.*, 1988; Marigomez, 1989; Vega *et al.*, 1989). This method is given in detail in Marigomez *et al.* (1990). It is based on the geometric transformation of the irregular outline of the digestive acini into a regular figure of the same area (a ring). This figure is then transformed into a trapezium, which has, as the short base, the internal perimeter of the ring, and as the long base

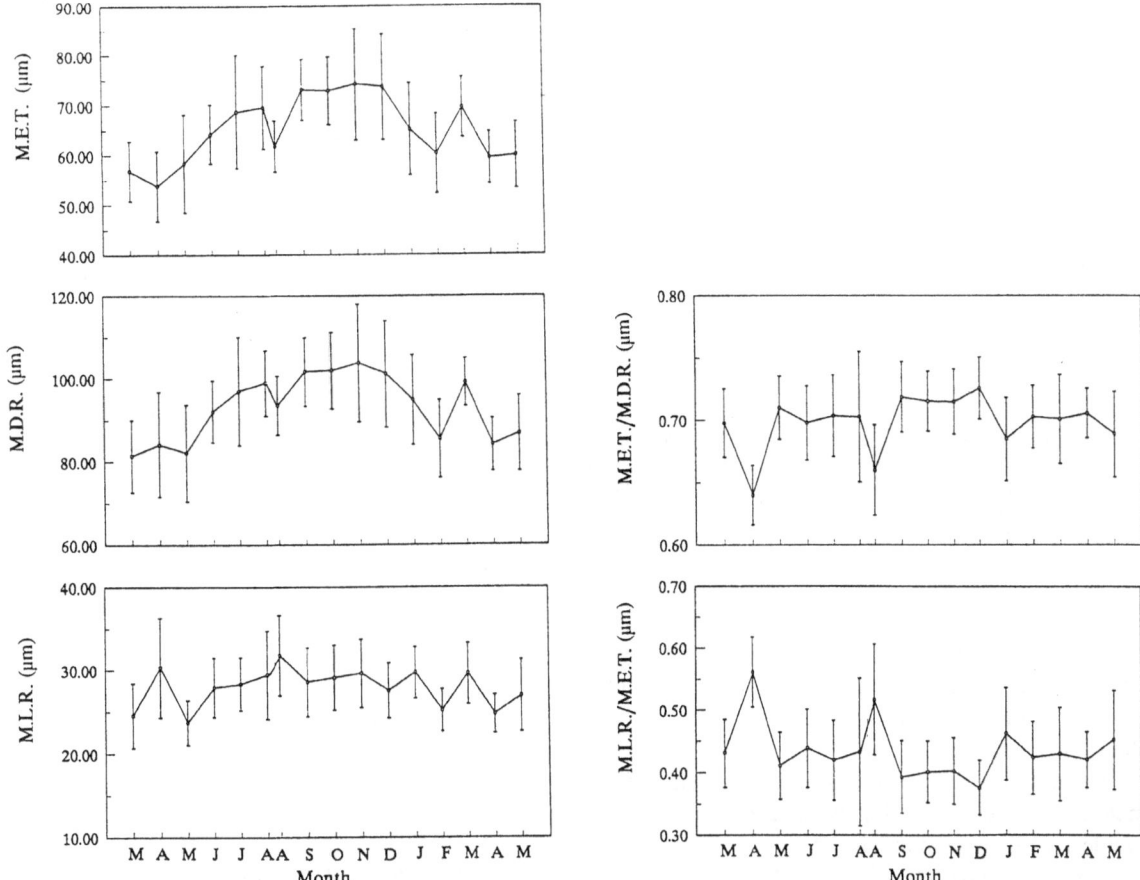

*Fig. 2.* Changes in the planimetric parameters of the digestive gland of *L. littorea* during the sampling period (from March 1989 to May 1990 with two samples in August 1989).

the external perimeter, while the height (h) is the mean epithelial thickness (MET) (Fig. 1). The mean diverticulum (i.e. tubule) radius (MDR) is the radius of the circle defined by the circumference $P_0$ and the mean lumen radius (MLR) is the radius of the circle defined by $P_i$ (Fig. 1). The ratios MLR/MET and MET/MDR were calculated to indicate relative changes between luminal size and digestive cell volume, and between digestive cell volume and tubule size respectively.

The results were analyzed using two- and one-way analysis of variance to detect the effects of month and sex on the planimetric parameters. In cases where there were significant differences Duncan's multiple range test (Zar, 1984) was applied to locate where these differences lay. The reproductive state of the animals was calculated by a planimetric procedure in which two sections per slide were projected onto a digitizer with

the aid of a camera lucida adapted to a dissecting microscope. The total area of the section was measured, as well as the relative areas of digestive gland, gonad and connective tissue.

Variations in the relative proportions of the different tissues during the year were tested using a one-way analysis of variance after arcsin transformation of the data. Correlation analysis was used to examine the relationships between the proportion of the total area occupied by the different tissues and the planimetric parameters.

## Results

The changes in the measured parameters of the digestive gland during the sixteen month study period are

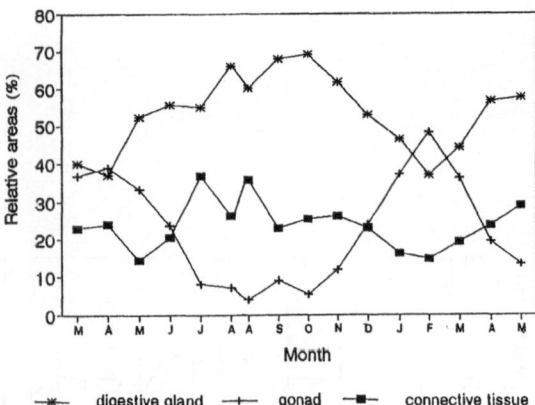

*Fig. 3.* Percentage of the area of cross section of the visceral hump occupied by the digestive gland, the gonad and the connective tissue during the sixteen months of sampling.

*Table 1.* One-way ANOVA for the five planimetric parameters used to quantify monthly changes in the digestive epithelium of *L. littorea* between March 1989 and May 1990 (* = $P<0.05$).

| Source | d.f. | Sum of squares | Mean squares | F ratio |
|---|---|---|---|---|
| **(MET)** | | | | |
| Between groups | 15 | 6553 | 436.89 | 7.00 * |
| Within groups | 140 | 8743 | 62.45 | |
| Total | 155 | 15297 | | |
| **(MDR)** | | | | |
| Between groups | 15 | 9167 | 611.14 | 6.25 * |
| Within groups | 140 | 13696 | 97.83 | |
| Total | 155 | 22864 | | |
| **(MLR)** | | | | |
| Between groups | 15 | 714 | 47.61 | 3.13 * |
| Within groups | 140 | 2127 | 15.19 | |
| Total | 155 | 2841 | | |
| **(MET/MDR)** | | | | |
| Between groups | 15 | 0.064 | 0.004 | 4.45 * |
| Within groups | 140 | 0.134 | 0.001 | |
| Total | 155 | 0.197 | | |
| **(MLR/MET)** | | | | |
| Between groups | 15 | 0.287 | 0.019 | 4.30 * |
| Within groups | 140 | 0.623 | 0.004 | |
| Total | 155 | 0.909 | | |

shown in Fig. 2. A two-way ANOVA for month and sex showed that month had a highly significant effect on the variation of the parameters, but that sex did not. Hence we treated both sexes together for the rest of the statistical analyses.

A one-way ANOVA showed that the absolute planimetric parameters were highly dependant on month (Table 1). Likewise month had a significant effect on the two ratios (one-way ANOVA, $P<0.01$) (Table 1), but Duncan's multiple range test showed that the difference resides primarily in two months (April 1989 and the second sample of August) in both cases. If these two months are taken out of the analysis month has no significant effect on any of the ratios ($P>0.01$). Re-examining the raw data for these two samples we found that, in April, when sexing the animals by external anatomy, six out of ten animals were misclassified, and that two of the other remaining four were infected by trematodes. In the second sample of August we found unusually high rates of trematode infection. This suggests that the animals were already under some sort of stress other than the reproductive cycle during these two months, resulting in variation in the MET/MDR and MLR/MET.

From Fig. 3 it is clear that there is a negative correlation between the areas occupied by the gonad and the digestive gland during the whole study period. The proportion of visceral mass occupied by digestive gland increased steadily during the spring and early summer, reaching a maximum between August and October, during which period the gonad was very small and the sex could hardly be determined. It declined again after

*Table 2.* Correlation analyses between the five planimetric parameters and the percentage of total area occupied by digestive tissue (Dig.). The levels of significance are given in parentheses.

| | MET | MDR | MLR | MET/MDR | MLR/MET |
|---|---|---|---|---|---|
| Dig. | 0.64 | 0.64 | 0.33 | 0.39 | −0.35 |
| | (0.008) | (0.008) | (0.21) | (0.14) | (0.18) |

October, reaching a minimum in February, by which time the gonad size was at a maximum. This was true for both males and females (Fig. 4). Before spawning the area occupied by the gonad was extensive (Fig. 4a, b) and the digestive gland tended to be somewhat compressed, the digestive cells being much thinner than after spawning (Fig. 4c, d).

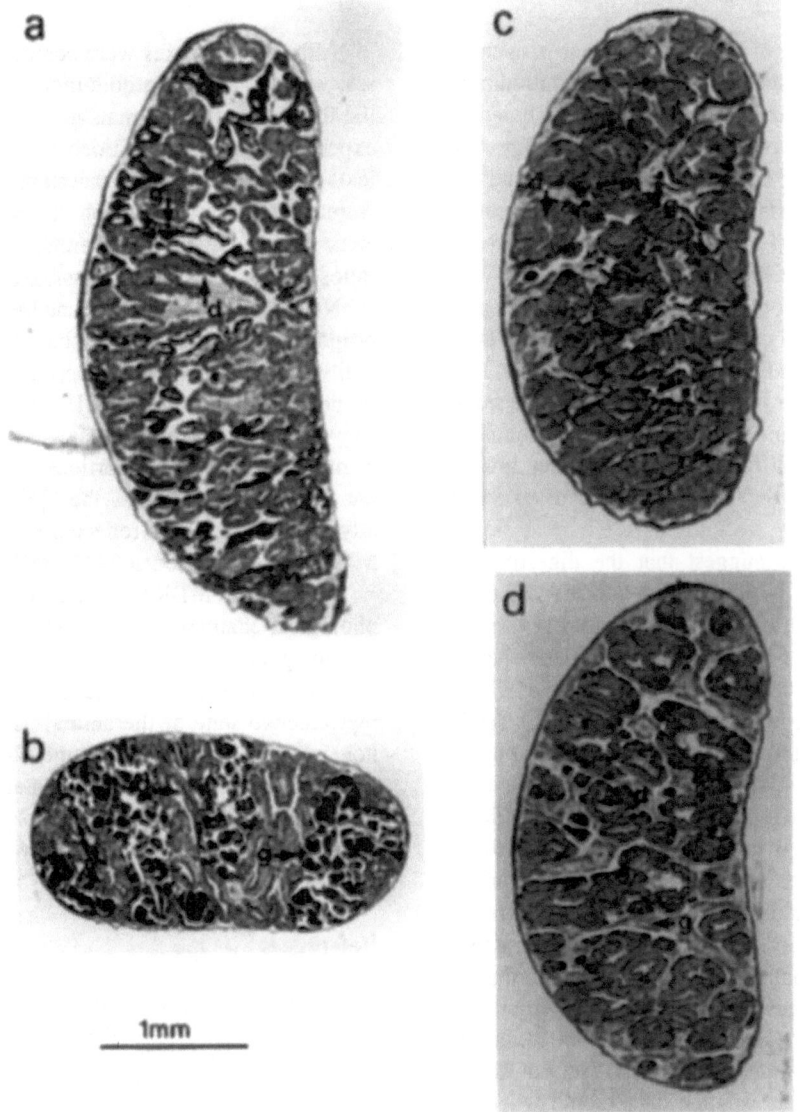

*Fig. 4.* Cross sections of the digestive gland-gonad complex of male and female *L. littorea* before and after spawning. (a) male before spawning, (b) female before spawning, (c) male after spawning, (d) female after spawning. (d, digestive gland; g, gonad).

A one-way ANOVA (of the arcsin transformed data) showed that the relative proportions of the different tissues varied significantly depending on the time of year ($P<0.01$). The proportion of digestive tissue was significantly correlated with both MET and MDR ($r = 0.638$, $P<0.01$ and $r = 0.635$, $P<0.01$ respectively). However, the MLR and the two ratios were not correlated with the amount of digestive tissue (Table 2).

## Discussion

Fretter & Graham (1962), in their description of the anatomy of *Littorina littorea*, said that greater or lesser areas of gonad will lie over the digestive gland depending on the reproductive state of the animals. This corresponds with the present findings, which clearly show a negative correlation between the areas occupied by the gonad and the digestive gland. Furthermore, the results of the current study show that there is a high variability in both the MET and the MDR depending on the time of the year, and that both parameters are highly correlated

42

with the proportion of the total area that is occupied by the digestive tissue. This appears to contradict the results of Marigomez *et al.* (1992), who concluded that only minor changes occurred in the morphology of the digestive tubules of *L. littorea* over the year and that, in any case, the variability recorded did not seem to be related to changes in the digestive gland. However, they recognised the limitations of their study, which was based on only four months (May, July, September and December), and pointed out the necessity of doing a more extensive study of the reproductive cycle. If we consider our data for just those four months the results are basically the same. May was the only month that showed significant differences from the other three for the MET and from September and December for the MDR.

Previous studies suggest that the digestive cell height on its own is not a good indicator of pollution, since it may be affected by some natural variables (Robinson, 1983; Soto, 1988), and indeed there is a strong negative correlation between the MET and MDR and the reproductive cycle. Vega *et al.* (1989) stated that, whilst the absolute parameters are not sufficient to indicate a pathological state in the digestive gland, the use of the ratios MLR/MET and MET/MDR is an appropriate way to interpret digestive gland morphology. From our results we can see that these ratios remain constant despite the influence of the reproductive cycle. The only two samples that were significantly different from the rest (April 1989 and the second sample of August) were unusual. The number of animals assigned to the wrong sex when sexing the animals by external anatomy in April (six out of ten), and the fact that this was the only month when this occurred, suggests that this was not simply a mistake in the procedure, but a reflection of a real state of the animals. Several studies describe the induction of pseudohermaphroditism in prosobranchs by exposure to pollution (e.g., Oehlmann *et al.*, 1991; Fioroni *et al.*, 1992) and this could have been the case for the April sample. Although there is a clear danger of circularity of argument, it does seem likely that these animals may have already been stressed by some factor other than the reproductive cycle, resulting in an abnormal thinning of the digestive tubules. In the second sample for August we found unusually high rates of trematode infection. Soto (1988) found that higher variability of the MET occurs in animals infested by trematodes. This could be because the parasites are not uniformly distributed within the digestive gland, but variability of the MET may also be a sign of stress.

Although the sexes were combined for our analyses, we can say that in both months the mean epithelial thickness of the animals was lower than would be expected in normal conditions, which is usually a typical response to stress. Furthermore, these two months were the only ones in which the two relative parameters were significantly different, suggesting that the ratios are also sensitive to possible stress.

Normally the MET, MDR and MLR retain the same proportions, although the absolute size of the acinus will vary depending on the reproductive state. In consequence the ratios MET/MDR and MLR/MET also stay more or less constant, as noted above. In our two aberrant samples the proportions between the absolute parameters changed, i.e., the MET was proportionally thinner. This was reflected in the ratios, so that MET/MDR decreased and MLR/MET increased.

While MET, MDR and MLR are indicative of morphological changes of the digestive epithelium, they are not good indicators of stress on their own, due to their variation with natural factors, such as the reproductive state of the animal. On the other hand, not only do the relative parameters MET/MDR and MLR/MET allow more reliable interpretation of morphological changes in the digestive gland, but they are also independent of the reproductive cycle.

## References

Agirregoikoa, M. G., 1988. *Mytilus edulis* L. bibalbioaren liseri-epitelioaren batezbesteko lodieraren aldakuntz espazio-tenporala Bizkaiko kostaldean. Tesis de Licenciatura, Universidad de Pais Vasco - Euskal Herriko Unibertsitatea, 76 pp.

Axiak, V., J. J. George & M. N. Moore, 1988. Petroleum hydrocarbons in the marine bivalve *Venus verrucosa*: accumulation and cellular responses. Mar. Biol. 97: 225–230.

Bernard, F. R., 1973. Crystalline style formation and function in the oyster *Crassostrea gigas* (Thunberg, 1795). Ophelia 12: 159–170.

Cajaraville, M. P., J. A. Marigomez & E. Angulo, 1989. A stereological survey of lysosomal structure alterations in *Littorina littorea* exposed to 1-naphthol. Comp. Biochem. Physiol. 93C: 231–237.

Cajaraville, M. P., J. A. Marigomez, G. Diez & E. Angulo, 1992. Comparative effects of the water accommodated fraction of three oils on mussels. 2- Quantitative alterations in the structure of the digestive tubules. Comp. Biochem. Physiol. Ser. C 102: 113–123.

Castillero, J., J. A. Marigomez & J. Moya, 1987. Un nuevo metodo planimetrico para el calculo de la altura media de epitelios. Cuad. Invest. Biol. 10: 89–95.

Couch, J. A., 1984. Atrophy of diverticular epithelium as an indicator of environmental irritants in the oyster, *Crassostrea virginica*. Mar. envir. Res. 14: 525–526.

Etxeberria, M., 1990. Alteraciones en la estructura cuantitativa de los tubulos digestivos y de la actividad fasica de la glandula digestiva en mejillones expuestos a cadmio (Cd), cobre (Cu) y cinc (Zn). Tesis de Licenciatura, Universidad del Pais Vasco - Euskal Herriko Unibertsitatea, 94 pp.

Fioroni, P., U. Deutsch, E. Stroben & J. Oehlmann, 1992. Artificially induced pseudohermaphroditism in prosobranchs and its absence in littorinids. In J. Grahame, P. J. Mill & D. G. Reid (eds), Proceedings of the 3rd International Symposium on Littorinid Biology. The Malacological Society of London, London: 313–315.

Fossato, V. U. & E. Siviero, 1974. Oil pollution monitoring in the Lagoon of Venice using the mussel *Mytilus galloprovincialis*. Mar. Biol. 25: 1–6.

Fretter, V. & A. Graham, 1962. British prosobranch mollusca. Ray Society, London, 755 pp.

Goldberg, E. D., 1975. The mussel watch - A first step in global marine monitoring. Mar. Pollut. Bull. 6: 111.

Goldberg, E. D., 1986. The mussel watch concept. Envir. Monit. Ass. 7: 91–103.

Goldberg, E. D., V. T. Bowen, J. W. Farrington, G. Harvey, J. H. Martin, P. L. Parker, R. W. Risebrough, W. Robertson, E. D. Schneider & E. Gamble, 1978. The mussel watch. Envir. Conserv. 5: 101–125.

Harrinson, F. L. & R. Berger, 1982. Effects of copper on the latency of lysosomal hexosaminidase in the digestive cells of *Mytilus edulis*. Mar. Biol. 68: 109–116.

Janssen, H. H. & N. Scholz, 1979. Uptake and cellular distribution of Cadmium in *Mytilus edulis*. Mar. Biol. 55: 133–141.

Langton, R. W., 1975. Synchrony in the digestive diverticula of *Mytilus edulis* L. J. mar. biol. Ass. U.K. 55: 221–230.

Langton, R. W., 1977. Digestive rhythms in the mussel *Mytilus edulis*. Mar. Biol. 41: 53–58.

Langton, R. W. & P. A. Gabbot, 1974. The tidal rhythm of extracellular digestion and the response to feeding in *Ostrea edulis*. Mar. Biol. 24: 181–187.

Lowe, D. M., M. N. Moore & K. R. Clarke, 1981. Effects of oil on digestive cells in mussels: quantitative alterations in cellular and lysosomal structure. Aquat. Toxicol. 1: 213–226.

Marigomez, J. A., 1989. Aportaciones cito-histologicas a la evaluacion ecotoxicologica de niveles subletales de cadmio en el medio marino: estudios de laboratorio en el gasteropodo prosobranquio *Littorina littorea* (L.). Tesis doctoral, Universidad del Pais Vasco - Euskal Herriko Unibertsitatea, 430 pp.

Marigomez, J. A., E. Angulo & J. Moya, 1986a. Copper treatment of the digestive gland of the slug *Arion ater* L. 1. Bioassay conduction and histochemical analysis. Bull. envir. Contam. Toxic. 36: 600–607.

Marigomez, J. A., E. Angulo & J. Moya, 1986b. Copper treatment of the digestive gland of the slug *Arion ater* L. 2. Morphometrics and histophysiology. Bull. envir. Contam. Toxic. 36: 608–615.

Marigomez, J. A., V. Saez, M. P. Cajaraville & E. Angulo, 1990. A planimetric study of the mean epithelial thickness of the molluscan digestive gland over the tidal cycle and under environmental stress conditions. Helgoländer wiss Meeresunters. 44: 81–94.

Marigomez, J. A., M. Soto & E. Angulo, 1991. Responses of winkles digestive cells and their lysosomal system to environmental salinity changes. Cell. Mollec. Biol. 37: 29–39.

Marigomez, I., M. Soto & E. Angulo, 1992. Seasonal variability in the quantitative structure of the digestive tubules of *Littorina littorea*. Aquat. Living Resour. 5: 299–305.

Martel, A., D. H. Larrive, R. R. Klein & J. H. Himmelman, 1986. Reproductive cycle and seasonal feeding activity of the neogastropod *Buccinum undatum*. Mar. Biol. 92: 211–221.

Mason, A. Z., K. Simkiss & K. P. Ryan, 1984. The ultrastructural localization of metals in specimens of *Littorina littorea* collected from clean and polluted sites. J. mar. biol. Ass. U.K. 64: 699–720.

Moore, M. N., 1980. Cytochemical determination of cellular responses to environmental stressors in marine organisms. Rapp. P.-v. Rean. Cons. perm. int. Explor. Mer. 179: 7–15.

Moore, M. N., 1985. Cellular responses to pollutants. Mar. Pollut. Bull. 16: 134–139.

Moore, M. N., 1986. Molecular and cellular indices of pollution. In: Giam, C. S. and M. J. M. Dow (eds), NATO ASI series. Vol 69. Strategies and advanced techniques in marine pollution studies. Mediterranean sea. NATO, Berlin.

Moore, M. N., D. M. Lowe & P. E. M. Fieth, 1978a. Responses of lysosomes in the digestive cells of the common mussel, *Mytilus edulis*, to sex steroids and cortisol. Cell. Tissue Res. 188: 1–9.

Moore, M. N., D. M. Lowe & P. E. M. Fieth, 1978b. Lysosomal responses to experimentally injected anthracene in digestive cells of *Mytilus edulis*. Mar. Biol. 48: 297–302.

Moore, M. N., D. M. Lowe & S. L. Moore, 1979. Induction of lysosomal destabilisation in marine bivalve mollusca exposed to air. Mar. Biol. Lett. 1: 47–57.

Moore, M. N., A. Bubel & D. M. Lowe, 1980. Cytology and Cytochemistry of the pericardial gland cells of *Mytilus edulis* and their lysosomal responses to horseradish peroxidase and anthracine. J. mar. biol. Ass. U.K. 60: 135–149.

Morton, B., 1971. The diurnal rhythm and tidal rhythm of feeding and digestion in *Ostrea edulis*. Biol. J. linn. Soc. 3: 329–342.

Morvan, C. & A. D. Ansell, 1988. Stereological methods applied to reproductive cycle of *Tapes rhomboides*. Mar. Biol. 97: 355–364.

Oehlmann, J., E. Stroben & P. Fioroni, 1991. The morphological expression of imposex in *Nucella lapillus* (Linnaeus) (Gastropoda: Muricidae). J. moll. Stud. 57: 375–390.

Phillips, D. J. H., 1977. The use of biological indicator organisms to monitor trace metal pollution in marine and estuarine environments: a review. Envir. Pollut. 13: 281–317.

Recio, A., J. A. Marigomez, E. Angulo & J. Moya, 1988. Zinc treatment of the digestive gland of the slug *Arion ater* L. 2. Sublethal effects at the histological level. Bull. Environ. Contam. Toxicol. 41: 865–871.

Robinson, W. E., 1983. Assessment of bivalve intracellular digestion based on direct measurements. J. moll. Stud. 49: 1–8.

Saez, V., G. Calvo-Ugarteburu, L. A. Aldonza & E. Angulo, 1992. Effects of oil-derived hydrocarbons on the digestive gland of *Littorina littorea*: a planimetric study. In J. Grahame, P. J. Mill & D. G. Reid (eds), Proceedings of the 3rd International Symposium on Littorinid Biology. The Malacological Society of London, London: 317–319.

Scholz, N., 1980. Accumulation, loss and molecular distribution of Cadmium in *Mytilus edulis*. Helgolander wiss. Meeresunters. 33: 68–78.

Simkiss, K. & A. Z. Mason, 1984. Cellular responses of molluscan tissues to environmental metals. Mar. envir. Res. 14: 103–118.

Simkiss, K., M. Taylor & A. Z. Mason, 1982. Metal detoxification and bioaccumulation in mollusca (review). Mar. Biol. Lett. 3: 187–201.

Soto, M., 1988. Planimetria del grosor medio del epitelio de la glandula digestiva de *Littorina littorea* en relacion con la variabilidad intrapoblacional e intraindividual. Tesis de Licenciatura, Universidad del Pais Vasco - Euskal Herriko Unibertsitatea, Bilbo, 91 pp.

Thompson, R. J., C. J. Bayne, M. N. Moore & T. J. Carefoot, 1978. Haemolymph volume, changes in the biochemical composition of the blood and cytological responses of the digestive cells in *Mytilus californianus* Conrad induced by nutritional, thermal and exposure stress. J. comp. Physiol. 127: 287–298.

44

Thompson, R. J., N. A. Ratcliffe & B. L. Bayne, 1974. Effects of starvation on structure and function in the digestive gland of the mussel (*Mytilus edulis* L.). J. mar. biol. Ass. U.K. 54: 699–712.

Tripp, M. R., C. R. Fries, M. A. Craven & C. E. Grier, 1984. Histopathology of *Mercenaria mercenaria* as an indicator of pollutant stress. Mar. envir. Res. 14: 521–524.

Underwood, A. J. & C. H. Peterson, 1988. Towards an ecological framework for investigating pollution. Mar. Ecol. Prog. Ser. 46: 227–234.

Vega, M. M., J. A. Marigomez & E. Angulo, 1989. Quantitative alterations in the structure of the digestive cell of *Littorina littorea* on exposure to cadmium. Mar. Biol. 103: 547–553.

Widdows, J., T. Bakke, B. L. Bayne, P. Donkin, D. R. Livingstone, D. M. Lowe, M. N. Moore, S. V. Evans & S. L. Moore, 1982. Responses of *Mytilus edulis* on exposure to the water accommodated fraction of North Sea oil. Mar. Biol. 67: 15–31.

Widdows, J., S. L. Moore, R. K. Clarke & P. Donkin, 1983. Uptake, tissue distribution and elimination of (1-14C) Naphthalene in the mussels *Mytilus edulis*. Mar. Biol. 76: 109–114.

Zar, J., 1984. Biostatistical analysis. 2nd edn. Prentice Hall, New Jersey.

*Hydrobiologia* **309**: 45–52, 1995.
*P. J. Mill & C. D. McQuaid (eds), Advances in Littorinid Biology.*
©1995 *Kluwer Academic Publishers.*

# Fine structure of the cephalic sensory organ in veliger larvae of *Littorina littorea*, (L.) (Mesogastropoda, Littorinidae)

Daniela Uthe
*Institut für Spezielle Zoologie & Vergleichende Embryologie der Universität Münster, Hüfferstrasse 1, D-48149 Münster, Germany*

*Key words: Littorina littorea*, veliger, ultrastructure, cephalic sensory organ, ciliated sensory cells

## Abstract

The cephalic sensory organ (CSO) in planktonic veliger larvae of *Littorina littorea* is situated dorsally between the velar lobes at the level of the shell aperture. It consists of ciliated primary sensory cells, adjacent accessory cells and supporting epithelial cells. Cell bodies of the ciliated cells originate in the cerebral commissure and their dendrites pass to the epidermis. The flask-shaped sensory cells are characterized by a deep invaginated lumen with modified cilia arising from the cell surface in the lumen. These cilia are presumed to be non-motile because they lack striated rootlets and show a modified microtubular pattern (6 + 2, 7 + 2 and 8 + 2). The adjacent accessory cells never possess an invaginated lumen; occasionally cilia and branched microvilli arise from the apical surface. These cells may be sensory, but there is no obvious direct connection with the nervous system. The supporting epithelial cells are part of the epidermis and flank the apical necks of the sensory and accessory cells. Morphological evidence suggests that the CSO may function in chemoreception related to substrate selection at settlement, feeding or other behaviour.

## Introduction

Superficial sensory organs like 'apical sense organs' or 'apical tufts' have been demonstrated using light microscopy in the larvae of several gastropod species (Patten, 1886; Conklin, 1897; Pelseneer, 1991; Werner, 1955).

Preliminary fine structural observations of the larvae of nudibranchs (Bonar, 1978a; Chia & Koss, 1984) confirm the existence of a cephalic sensory organ (CSO), which was first described by Bonar (1978a) in larvae of *Phestilla sibogae* (Bergh). This CSO is a superficial sensory receptor located dorsally between the velar lobes. It consists of three cell types: the flask-shaped sensory cells bearing numerous cilia, the adjacent supporting cells with cilia and often branched microvilli arising from the surface, and vacuolated cells which occupy the centre of the area. Bonar (1978a) suggested that the CSO might be involved in the perception of chemical cues emitted by the food species of the adults of *Phestilla*, which in turn induce

settlement and metamorphosis. Further fine structural studies in the CSO were made by Chia & Koss (1984). They showed that the veliger larvae of *Rostanga pulchra* (MacFarland) possess a CSO situated dorsally between the rhinophores. The organ consists of at least three types of receptor cells, which may indicate a greater degree of specialization. The ampullary cells can be compared with the flask-shaped ciliated cells of *Phestilla sibogae*. Due to their deeply invaginated lumen they are tightly packed with cilia originating from the cell surface. The parampullary cells resemble the ampullary cells in general morphology, but they lack a ciliated lumen. Ciliary tuft cells are characterized by 20 or more cilia arising from the apical cell surface, forming a tuft of cilia in the intervelar area. According to Chia & Koss (1984) it is possible that the CSO may function in more than one modality, i.e. for chemoreception and mechanoreception. The present study deals with the fine structure of the CSO in larvae of *Littorina littorea*. Its structure is compared with

46

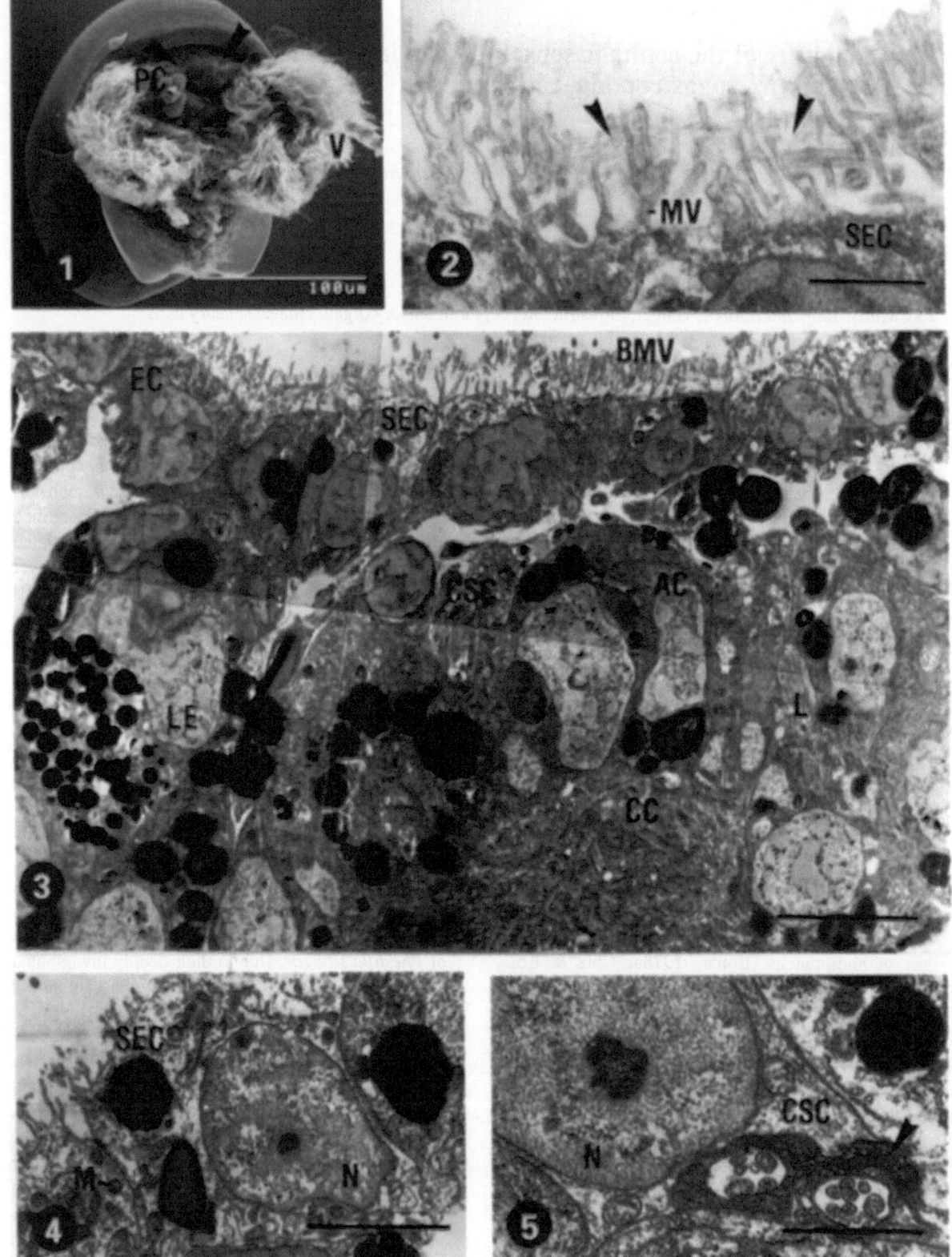

*Fig. 1–5.*

those of the opisthobranch species and also with similar cells in cephalopod olfactory organs.

## Material and methods

Egg capsules of *Littorina littorea* were selected from plankton hauls at Roscoff, France in April 1992. The veligers were pipetted out of the capsules into fresh seawater. Free swimming larvae were narcotized by adding 7% $MgCl_2$ to the seawater. Subsequently the veliger larvae were fixed for 1–11/2 h at 4 °C in 2% $OsO_4$ dissolved in 5% $K_2Cr_2O_7$ and filtered seawater (1:9, pH 7,2–7,4). All specimens were dehydrated via a graded ethanol series. For transmission electron microscopy (TEM) the larvae were embedded in Spurr's medium (Robinson *et al.*, 1985) and stained with lead-citrate (Reynolds, 1963; Nagl, 1981). The sections were made using a Reichert Ultracut and investigated with a Siemens TEM Elmiskop 101. For scanning electron microscopy (SEM) the specimens were critical-point dried, coated with gold and examined with a Hitachi SEM H–330.

## Results

The CSO in veliger larvae of *Littorina littorea* is located dorsally on the centre of the head between the two velar lobes (Fig. 1). It is composed of three cell types: ciliated cells, which are primary sensory cells; adjacent accessory cells and supporting epithelial cells. These cell types are arranged in one single epithelial layer. Cell bodies of the ciliated sensory cells and the accessory cells are subepithelial in origin. Their bases lie in the dorsal part of the cerebral commissure; the cells pass through a sinus and terminate in an area of the epidermis formed by the supporting epithelial cells (Figs 3, 12). Processes from the sensory cells and the accessory cells ascend to the epidermis in two separate tracts (Fig. 12). The surface of all three cell types is covered with a dense network of elongate microvilli forming a distinct brushborder. Those microvilli are often branched or bear short, lateral projections. An electron-opaque material forms a fine network that interconnects the microvilli (Fig. 2). Furthermore, all cells of the CSO are interconnected by septate junctions which are situated in the apical regions of the cell membranes (Figs 11, 12). Scattered pigment granules of large size (about 1 $\mu$m in diameter) occur in all three cell types (see Figs 3, 12).

### Ciliated sensory cells

Five or six cells bearing numerous cilia are present in the sensory area. A prominent feature of these flask-shaped sensory cells is the invagination of the apical surface to form a deep, ciliated lumen which extends almost to the base of the cell (Fig. 7). In just a few cases electron-dense material was observed in the cytoplasm surrounding this lumen. This material differed completely from the remaining cytoplasm surrounding the ciliated lumen (Fig. 5) and no organelles were observed within it (Fig. 5). This matrix may be artifactual. The expanded basal part of the cell of about 10 $\mu$m in diameter is located beneath the epidermis. It contains the nucleus and is embedded in the nervous tissue of the cerebral commissure. Axons of the ciliated cells run into the mass of neurites comprising the commissure (Fig. 11). Synaptic junctions between the sensory cells and the adjacent nervous tissue have not been detected. Such connections may exist but could not be found up to now. The apical portion of the ciliated cell protrudes with a slender neck to the epidermal surface, which is covered by a brushborder of branched microvilli. In some cases 2–4 cilia which originate in the apical part of the cell rise above the lumen (Fig. 7). A characteristic feature of the cilia of the sensory cells is a modified 6 + 2, 7 + 2, 8 + 2 or, sometimes, the normal 9 + 2 microtubular pattern (Fig. 10). In contrast to the long striated rootlets of motile cilia, the basal rootlets of cilia in the ciliated cells are either very short or missing, but common anchoring structures like basal bodies are present (see Figs 8, 11). The cytoplasm appears to be

*Figs 1–5.* Scanning electron micrograph of a planktonic veliger of *Littorina littorea* showing the bilobed velum (V) with protruding pigmented cells (PC). Note the cephalic sensory area (arrow).
Branched microvilli (MV) of supporting epithelial cells (SEC) are interconnected by an electron-opaque material forming a fine network (arrows). Scale bar = 1 $\mu$m.
Longitudinal section showing the cephalic sensory organ (CSO) and the right larval eye (LE). Note cells contributing to the CSO are in contact with the cerebral commissure (CC). Supporting epithelial cells (SEC) overlie the ciliated sensory cells (CSC) and the accessory cells (AC). BMV = brushborder of microvilli, L = ciliated lumen of sensory cell, EC = epidermal cell. Scale bar = 40 $\mu$m.
Section through a supporting epithelial cell (SEC). N = nucleus, M = mitochondion. Scale bar = 2 $\mu$m.
Transverse section of ciliated sensory cells (CSC). Note the electron-dense matrix surrounding the ciliated lumina (arrow). N = nucleus. Scale bar = 1 $\mu$m.

48

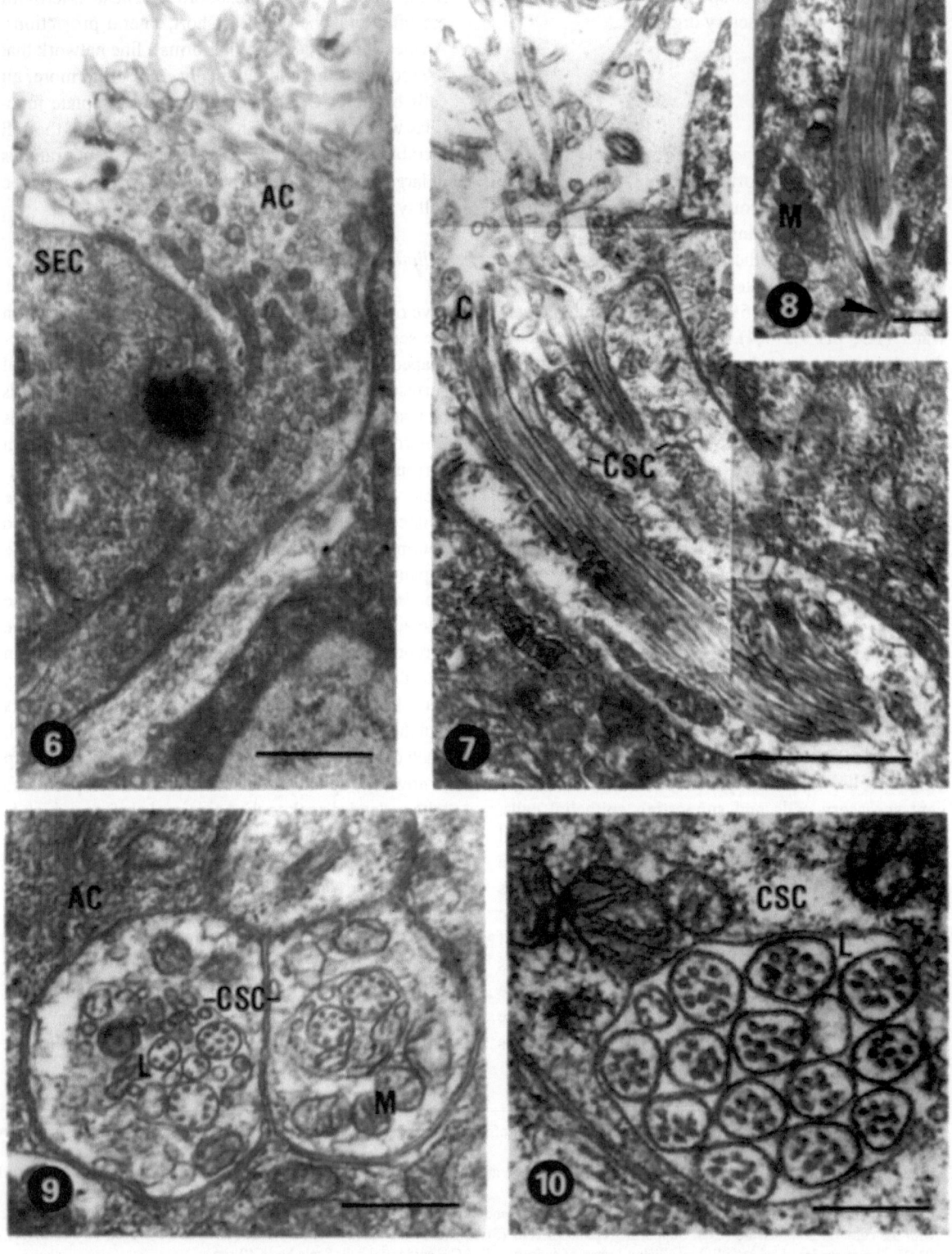

electron-lucent and cytoplasmic organelles are sparse (Fig. 9). Mitochondria of the tubular type are mainly located near the ciliated lumen and in the proximal part of the cell. Small amounts of granular and agranular endoplasmic reticulum of relatively homogenous distribution are found throughout the cytoplasm. Profiles of Golgi apparatus are rarely seen and free ribosomes occur just as dictyosomes in moderate numbers. An axon projects from the proximal part of the cell (Fig. 11). The cytoplasm of the adjacent nervous tissue resembles that of the axons. Both are filled with granules, vesicles and mitochondria; neurotubules are abundant (Fig. 11).

*Accessory cells*

Five to six accessory cells are located adjacent to the ciliated cells. In comparison to the latter the accessory cells never possess an invaginated lumen with the characteristic bundles of cilia. The apical surface of the accessory cell is endowed with branched microvilli (Fig. 6). The main part of the cell body is situated beneath the epidermis. The slender apical necks pass through a sinus to the epidermal surface (Figs 6, 12) accompanied by the apical portions of the ciliated cells (Fig. 12). Due to abundant vesicles and granules of ribosomal size the cytoplasm appears to be electron-dense. The organelles are more numerous than in the cytoplasm of the ciliated cells (Fig. 9). This may indicate a higher metabolic activity of the accessory cells. The elliptical nucleus occupies the base of the cell. Dictyosomes, mitochondria, granular and agranular endoplasmic reticulum occur mainly in the perinuclear region and in the distal part of the cell. The accessory cells may be sensory cells but they do not have the same intimate contact with the cerebral commissure as the ciliated sensory cells. Axons could not be detected.

*Supporting epithelial cells*

The six to seven supporting epithelial cells associated with the CSO are entirely epidermal and do not extend beneath the epidermal layer (Fig. 3). They surround the apical regions of the two other cell types (Figs 6, 12). These cuboidal cells are approximately 4–4.3 $\mu$m high and 3.5–4 $\mu$m across. The cytoplasm contains a large number of organelles; mitochondria are numerous and often closely packed (Fig. 4). A large nucleus occupies much of the central or the basal part of the cell (Fig. 4). The surface of the cell is endowed with a dense array of elongate, branched microvilli forming a brushborder 1–1.5 $\mu$m in thickness (Fig. 2). Several (motile) cilia arise from the surface. Normal epithelial cells show simple erect microvilli and no apical cilia.

## Discussion

The CSO of *Littorina littorea* resembles that described for *Phestilla sibogae* (Bonar, 1978a). The ciliated sensory cells in *Littorina* are similar to the flask-shaped cells in *Phestilla* and to the ampullary cells in *Rostanga pulchra* (Chia & Koss, 1984). In *Littorina littorea* these cells bear modified cilia in which the microtubules are arranged in 6 + 2, 7 + 2, 8 + 2 and only sometimes in the normal 9 + 2 pattern. No such modification of the cilia has been described for the opisthobranch species (Bonar, 1978a; Chia & Koss, 1984). In the sensory cilia of the flask-shaped cells of *Phestilla sibogae* the microtubules appear to lack the outer dynein arms of the A-tubules. This feature cannot be detected in either *Littorina* or in *Rostanga pulchra*. Basal rootlets cannot be demonstrated in the cilia of the sensory cells of *Littorina* or of the two opisthobranch species. In cephalopod olfactory organs an electron-dense matrix which contains a basophilic, osmiophilic material is present in the ciliated cavity of the sensory cells (Emery, 1975, 1976; Wildenburg & Fioroni, 1989). However, a similar electron-dense matrix cannot be demonstrated in *Littorina*, *Phestilla* or *Rostanga*. The CSO of *Rostanga pulchra* is more complex than the CSOs of *Littorina* and *Phestilla*, due to the presence of at least three types of receptor cells in *Rostanga*. A cell type corre-

*Figs 6–10.* Longitudinal section of an accessory cell (AC) and supporting epidermal cells (SEC). The main cell body of the accessory cell is located beneath the epidermis, only a slender apical neck passing through the epidermis to the surface. Scale bar = 1 $\mu$m.
Longitudinal section of ciliated sensory cells (CSC). Note the cell lumina opening at the epidermal surface. C = cilia. Scale bar = 10 $\mu$m.
Detail of a ciliated sensory cell. The cilia arising from the invaginated lumen lack striated rootlets (arrow). Mitochondria (M) occur in large numbers, mainly in the proximal part of the cell. Scale bar = 1 $\mu$m.
Transverse section through ciliated (CSC) and accessory (AC) cells. Note the electron-lucent cytoplasm of the ciliated sensory cells in comparison to the electron-dense cytoplasm of the accessory cells. Numerous mitochondria (M) are scattered around the invaginated ciliated lumina (L). Scale bar = 0.5 $\mu$m.
Detail of the modified 6 + 2, 7 + 2 and 8 + 2 axonemal structure in cilia of a ciliated sensory cell (CSC). L = ciliated lumen. Scale bar = 0.25 $\mu$m.

50

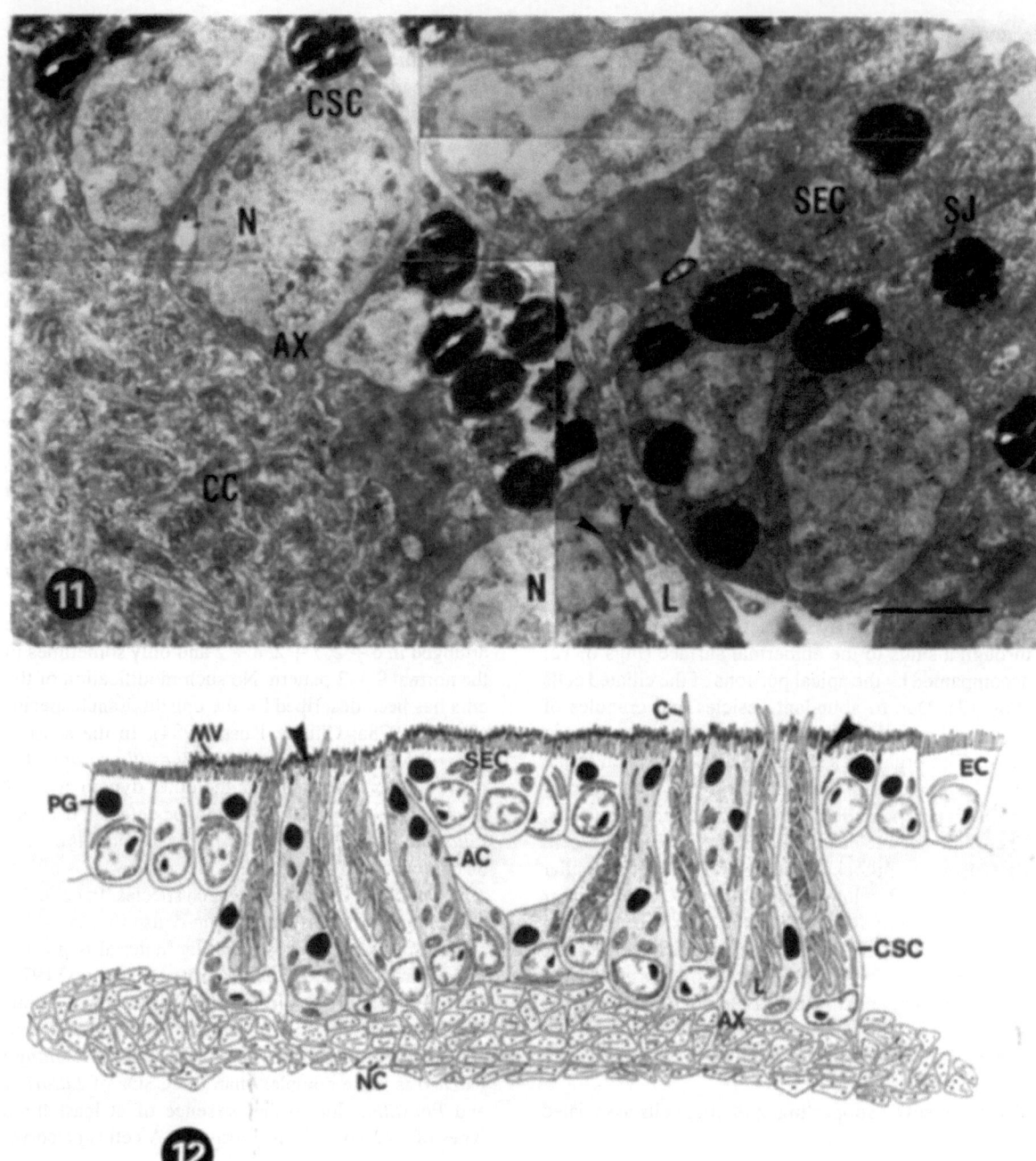

*Figs 11–12.* Transverse section through the bases of ciliated sensory cells (CSC) showing an axon (AX) entering the cerebral commissure (CC). The large nucleus (N) occupies most of the base of the cell. Note the cilia of the ciliated lumen lack striated rootlets but show normal basal feet (arrows). L = lumen of ciliated sensory cell, SEC = supporting epithelial cell, SJ = septate junction. Scale bar = 2 μm.

Schematic drawing of the cephalic sensory organ (CSO) illustrating the relationships of the contributing cell types: ciliated sensory cells (CSC), accessory cells (AC), supporting epithelial cells (SEC) and the nerve cells of the cerebral commissure (NC). The bases of the ciliated sensory cells and the accessory cells pass through a sinus and ascend to the epidermis in two separate tracts. All cells of the CSO are interconnected by septate junctions (arrows). AX = axon, C = sensory cilium, EC = epidermal cell, L = ciliated lumen, MV = microvilli, PG = pigment granule.

sponding to the ciliary tuft cells in *Rostanga* does not exist in *Littorina* and *Phestilla*. Furthermore, although the accessory cells in the CSO of *Littorina* resemble the parampullary cells in *Rostanga*, no axons can be detected in the accessory cells, whereas in the parampullary cells a proximally located axon passes into the cerebral commissure.

Ciliated sensory cells almost identical to those described for *Littorina littorea* have been reported from the olfactory organ of juvenile and adult cephalopods (Barber & Wright, 1969; Woodhams & Messenger, 1974; Emery, 1975, 1976; Wildenburg & Fioroni, 1989). However, the cephalopod organs are more complex than those described for *Littorina*, due to the larger number of different cell types in the former. The sensory cells of both organs are in contact with the nervous system, either by direct axonal contact to the cerebral commissure as in *Littorina* or by synaptic junctions with the olfactory nerve as described for the olfactory organ in cephalopods. All types of receptor cells in the olfactory organ possess cilia showing the normal $9 + 2$ configuration. These cilia are presumed to be non-motile because of their reduced basal feet and the lack, or reduced size, of the ciliary rootlets (Barber & Wright, 1969; Emery, 1975, 1976; Wildenburg & Fioroni, 1989). It should be emphasized that no morphological criteria have been established to distinguish chemo- and mechanoreceptors (Crisp, 1973; Barber, 1974; Laverack, 1988). Often sensory cilia lack one or more of the structural elements of a motile cilium. The axonemal tubules may be organized in other than the typical $9 + 2$ pattern (Horridge, 1965; Barber, 1974; Haszprunar, 1985; Laverack, 1988). Ciliated receptor cells with a suspected chemosensory function have been described in several gastopods (Welsch & Storch, 1969; Storch, 1972; Crisp, 1973; Bonar, 1978a, Phillips, 1979, Davis & Matera, 1981; Chia & Koss, 1982; Haszprunar, 1985) and in cephalopod species (Barber & Wright, 1969; Woodhams & Messenger, 1974; Emery, 1975, 1976). Furthermore, some ciliated receptor cells have been associated with mechanoreception or both chemo- and mechanoreception (Chia & Koss, 1984; Wildenburg & Fioroni, 1989).

Veliger larvae of *Littorina littorea* are too small for behavioural and electrophysiological studies to determine the exact receptor modality. Bonar (1978a) suggested that the CSO of *Phestilla sibogae* might be involved in the perception of chemical stimuli emitted by the adult's food, which may be a necessary condition to induce settlement and metamorphosis. The receptors of the CSO are in close proximity to the substrate, which can be explored for its appropriateness as a settling site. Chia & Koss (1984) agreed with Bonar (1978a) that the CSO may function in chemo- and mechanoreception related to substrate selection at settlement. Furthermore, they suggested a multifunctionality and multimodality for the CSO of *Rostanga pulchra* based on the presence of three different receptor cell types. Due to its position in the anterior region of the head between the velar lobes, the organ is exposed to the external environment and also to the water currents created by the velar cilia. This placement may support other possible functions of the CSO. For example it may be involved in feeding by selecting food items on the bases of taste, size, etc., which would include chemo- and mechanoreception.

In *Littorina littorea* the majority of cilia of the ciliated cells occur in the ciliated lumen located beneath the epidermis. Therefore most of the sensory surface is not in direct contact with the environment. This may imply that the larvae have to be in contact with the adult's food for a certain period. Too short a period of contact would not induce the presumed resulting induction of settlement and metamorphosis.

The mentioned differences in the structure of the CSO demonstrate that the question of homology is not yet solved: different types of cephalic sensory organs occur in opisthobranchia (Bonar, 1978a; Chia & Koss, 1984) as well as in prosobranchia (Uthe, unpublished). The origin of the cells comprising the CSO is not known exactly because embryological stages of *Littorina littorea* were not examined. According to Bonar (1978a) it is possible that the ciliated cells of the CSO were originally ciliated apical plate cells which invaginated and are finally displaced beneath the epithelium. Studies are currently in progress to describe the ontogeny of the cephalic sensory area by investigating the apical plate morphology. Such investigations are necessary to determine if fine structural features of the CSO could be useful in establishing phylogenetic relationships. Cells of the CSO seem to be lost after metamorphosis (Bonar, 1978b). Additional investigations are required to determine the fate of the CSO after metamorphosis.

## Acknowledgement

I would like to thank Prof. Dr P. Fioroni for his support and advice and Dr G. Sundermann for the donation of larvae of *Littorina littorea*.

52

# References

Barber, V. C., 1974. Cilia and sense organs. In M. A. Sleight (ed.), Cilia and flagella. Academic Press, London: 403–430.

Barber, V. C. & D. E. Wright, 1969. The fine structure of the sense organs of the cephalopod mollusc *Nautilus*. Z. Zellforsch. 102: 293–312.

Bonar, D. B., 1978a. Ultrastructure of the cephalic sensory organ in larvae of the gastropod *Phestilla sibogae*, (Aeolidacea, Nudibranchia). Tissue Cell 10: 153–165.

Bonar, D. B., 1978b. Morphogenesis at metamorphosis in opisthobranch molluscs. In F. S. Chia & E. Rice (eds), Settlement and metamorphosis of marine invertebrate larvae, Elsevier, New York: 177–196.

Chia, F. S. & R. Koss, 1982. Fine structure of the larval rhinophores of the nudibranch, *Rostanga pulchra*, with emphasis on the sensory receptor cells. Cell Tiss. Res. 225: 235–248.

Chia, F. S. & R. Koss, 1984. Fine structure of the cephalic sensory organ in the larva of the nudibranch *Rostanga pulchra* (Mollusca, Opisthobranchia, Nudibranchia). Zoomorph. 104: 131–139.

Crisp, M., 1973. Fine structure of some prosobranch osphradia. Mar. Biol. 22: 231–240.

Conklin, E. J., 1897. The embryology of *Crepidula*. J. Morph. 13: 1–227.

Davis, W. J. & E. M. Matera, 1981. Chemoreception in gastropod molluscs: Electron microscopy of putative receptor cells. J. Neurobiol. 13: 79–84.

Emery, D. G., 1975. The histology and fine structure of the olfactory organ of the squid *Lolliguncula brevis* Blainville. Tiss. Cell 7: 357–367.

Emery, D. G., 1976. Observations on the olfactory organ of adult and juvenile *Octopus joubini*. Tiss. Cell 8: 33–46.

Haszprunar, G., 1985. The fine morphology of the osphradial sense organs of the mollusca. 1. Gastropoda, Prosobranchia. Phil. Trans. r. Soc., Lond. B 307: 457–496.

Horridge, G. A., 1965. Non-motile sensory cilia and neuromuscular junctions in Ctenophores. Proc. r. Soc., Lond. B 162: 333–350.

Laverack, M. S., 1988. The diversity of chemoreceptors. In J. Atema *et al.* (eds), Sensory biology of aquatic animals, Springer Verlag 11: 287–312.

Nagl, W., 1981. Elektronenmikroskopische Laborpraxis. Springer Verlag, Berlin.

Patten, W., 1886. The embryology of *Patella*. Arb. zool. Inst. Wien, Triest 6: 149–174.

Pelseneer, P., 1911. Recherches sur l'embryologie des gastéropodes. Mém. acad. r. Belg. Sci. Sér. 2: 1–163.

Phillips, D. W., 1979. Ultrastructure of sensory cells on the mantle tentacles of the gastropod *Notoacmea scutum*. Tiss. Cell 11: 623–632.

Reynolds, F. S., 1963. The use of lead citrate at high pH as an electron opaque stain in electron microscopy. J. Cell Biol. 17: 208–212.

Robinson, D. G., U. Ehlers, T. Herken, B. Herrmann, F. Mayer & F.-W. Schürmann, 1985. Präparationsmethodik in der Elektronenmikroskopie. Springer Verlag, Berlin, Heidelberg, New York.

Storch, V., 1972. Elektronenmikroskopische und histochemische Untersuchungen über Rezeptoren von Gastropoden (Prosobranchia, Opisthobranchia). Z. wiss. Zool. 184: 1–26.

Welsch, U. & V. Storch, 1969. Über das Osphradium der prosobranchen Schnecken *Buccinum undatum* L. und *Neptunea antiqua* (L.). Z. Zellforsch. 95: 317–330.

Werner, B., 1955. Über die Anatomie, die Entwicklung und Biologie des Veligers und der Veliconcha von *Crepidula fornicata* L. (Gastropoda Prosobranchia). Helgoländer wiss. Meeresunters 5: 196–217.

Wildenburg, G. & P. Fioroni, 1989. Ultrastructure of the olfactory organ during embryonic development and at the hatching stage of *Loligo vulgaris* Lam. (Cephalopoda). J. Ceph. Biol. 1: 56–70.

Woodhams, P. L. & J. B. Messenger, 1974. A note on the ultrastructure of the *Octopus* olfactory organ. Cell Tiss. Res. 152: 253–258.

*Hydrobiologia* **309**: 53–59, 1995.
*P. J. Mill & C. D. McQuaid (eds), Advances in Littorinid Biology.*
©1995 *Kluwer Academic Publishers.*

# Egg capsule morphology of five Hong Kong rocky shore littorinids

Y. M.Mak
*The Swire Institute of Marine Science, Department of Ecology and Biodiversity, The University of Hong Kong, Hong Kong*

*Key words: Nodilittorina trochoides, Nodilittorina radiata, Nodilittorina vidua, Peasiella roepstorffiana, Littoraria articulata,* spawn

## Abstract

The morphology of the egg capsules of five Hong Kong rocky shore littorinids, *Nodilittorina trochoides, N. radiata, N. vidua, Peasiella roepstorffiana* and *Littoraria articulata* are illustrated. The egg capsules of the three *Nodilittorina* are cupola-shaped and never exceed 250 $\mu$m in diameter, whilst those of *P. roepstorffiana* and *L. articulata* are biconvex lens-shaped with a capsule diameter of 300-400 $\mu$m. Ova are always singly encapsulated. The ovum diameter of the three *Nodilittorina* species is 65 $\mu$m and approximately 70 $\mu$m for the other two littorinids. There are species-specific differences in the elaborate morphology of the egg capsules, which can be used as supplementary characters for littorinid identification. A new description of the egg capsule of *Nodilittorina vidua* is given based on evidence from SEM studies.

## Introduction

Planktonic egg capsules only occur in the prosobranch family, Littorinidae (Pilkington, 1971). A wide range of spawn and developmental types is now well known in the Littorinidae, and has been extensively reviewed (Bandel, 1974; Bandel & Kadolsky, 1982; Reid, 1986, 1989, 1990, 1992). Almost all tropical and warm temperate littorinid species show planktotrophic development, and most of these produce pelagic egg capsules (Reid, 1989).

Winckworth (1922) stated that members of the family Littorinidae were sufficiently distinct in their mode of reproduction to merit separation into distinct genera and classified the British Littorinidae based mainly on their reproductive type. This has since been revised based upon detailed anatomical studies which have revealed that in several groups of littorinids the form of the egg capsules is a useful diagnostic character (Murray, 1979; Bandel & Kadolsky, 1982). Other workers, such as Tokioka & Habe (1953) and Abbott (1954) have also noted the importance of the morphology of pelagic egg capsules as a taxonomic character for the identification of Littorinidae. The morphology of the pelagic egg capsules can be used as a taxonomic character. Firstly, at the generic level, genera usually

show a characteristic capsule form, e.g. cupola-shaped in *Nodilittorina*, biconvex lens-shaped in *Littoraria* (Bandel & Kadolsky, 1982; Reid, 1986, 1989). Secondly, different species in the same genus show distinct forms, especially in the genera *Nodilittorina* (Abott, 1954; Borkowski, 1971; Bandel & Kadolsky, 1982) and *Littorina* (Murray, 1979; Schmitt, 1979; Kojima, 1985a, b; reviewed by Reid, 1990).

The general morphology of these egg capsules has been classified by Bandel (1974) and three principal shapes can be distinguished:

(i) pill-box type – simple, almost symmetrical, flat and cylindrical in shape;

(ii) biconvex disc type – lens-shaped capsule with a wide circumferential flange;

(iii) cupola type – asymmetrically convex, with the more domed upper side sculptured with 1–10 concentric rings and an overhanging skirt below.

In Hong Kong, 13 species of the family Littorinidae have been recorded, inhabiting both rocky shores and mangals (Reid, 1992). In this study the egg capsules of the five most common rocky shore littorinids (*Nodilittorina trochoides* (Gray, 1839), *Nodilittorina radiata* (Eydoux & Souleyet, 1852), *Nodilittorina vidua* (Gould, 1859), *Peasiella roepstorffiana* (Nevill, 1885) and *Littoraria articulata* (Philippi, 1846)) were inves-

54

tigated. Ohgaki (1985) has described the distribution
of the Littorinidae on eight rocky shores of varying
exposure in Hong Kong. *Nodilittorina trochoides* (as
*N. pyramidalis*) and *N. radiata* (as *N. exigua*) were
commonly found at all eight sites with *Littoraria articulata* (as *Littorina scabra*) and *Peasiella roepstorffiana*
(as *Peasiella* sp.) only common on sheltered shores,
whilst *Nodilittorina vidua* (as *N. millegrana*) was only
common on exposed shores. The vertical zonation pattern on a typical shore was *N. trochoides*, *N. radiata*,
*N. vidua*, *L. articulata* and *P. roepstorffiana*, from high
to low shore. Although these littorinids are relatively common in Hong Kong, little is known about the
breeding and spawning of these species.

The aim of this work was to identify and describe
the egg capsules of these five littorinids and to use,
for the first time, the Scanning Electron Microscopy to
investigate the fine structure of the capsules.

## Materials and methods

### Description of sites

Samples of littorinids were collected at three sites,
Big Wave Bay, Cape d'Aguilar and South Bay (in
decreasing degree of exposure) on Hong Kong Island
(22 °10′N 114 °10′E). *Littoraria articulata* was only
found at South Bay and in the sheltered inshore areas
of Big Wave Bay. The other four species were commonly found year-round at all sites, although *Peasiella
roepstorffiana* occurred mainly in the more sheltered
areas.

To obtain a sufficient number (approximately 20)
of females, approximately 50 snails of each species
were collected from the three sites between April and
July 1993. Individuals were sexed by looking for a
penis behind the right cephalic tentacles. For laboratory spawning studies, females of each species were
placed individually in plastic vials (diameter: 20 mm;
height: 47 mm), half-filled (4 ml) with filtered sea
water (filter paper pore size 0.45 $\mu$m) and covered.
Vials were kept at room temperature (about 24–26 °C)
with no aeration and water was renewed daily. Adults
survived well under these conditions. Vials were examined each day for eggs,which were usually found in
clusters on the vial bottom. As in some cases the eggs
were dispersed when clusters were not visible, the vials
were checked under a dissecting microscope and occasionally a few floating eggs were detected.

Twenty egg capsules from different individuals
were identified and the capsule diameter and ovum
diameter measured ($\pm 1\ \mu$m) under a compound microscope. Photos of the lateral view were taken, from
which the capsule height was estimated. Egg capsules
were prepared by critical-point drying and observations were made by Scanning Electron Microscopy to
investigate fine structure.

## Results

In general, the pelagic egg capsules of the five littorinids always contain a single small ovum (never
multiple ova) surrounded by a thin layer of albumen.
The capsule itself is colourless and transparent while
the ovum is pinkish grey in colour. The egg capsules
of the studied species can be classified into two types,
the cupola and the biconvex shape.

### Nodilittorina trochoides (Fig. 1a)
This species is common on rocks in the littoral fringe,
ranging from exposed to sheltered shores. *N. trochoides* can be found high up the shore at about 2 m
above Chart Datum (+ 2.0 m C.D.). The egg capsule is
cupola-shaped (Figs 2a, 3a). There are a minimum
of 4 and a maximum of 5 concentric rings on the
convex side without any sculpturing on the overhanging circumferential skirt. The mean capsule diameter
(mean$\pm$S.D.) is 195$\pm$12 $\mu$m, ovum diameter 65$\pm$1 $\mu$m
and the capsule height about 110 $\mu$m (Table 1).

### Nodilittorina radiata (Fig. 1b)
*Nodilittorina radiata* is common on both exposed
and sheltered rocky shores, usually distributed below
*N. trochoides* at about + 1.7 m C.D. Their cupola-
shaped egg capsules have a minimum of 6 and a
maximum of 7 concentric rings on one side and the
overhanging peripheral edge is thinner than that of
*N. trochoides* (Figs 2b, 3b). The capsule diameter is
237$\pm$7 $\mu$m, ovum diameter 65$\pm$1 $\mu$m and the capsule
height is about 120 $\mu$m (Table 1).

### Nodilittorina vidua (Fig. 1c)
This species can be found at + 1.5 m C.D. on moderately exposed to sheltered rocky shores. The cupola-
shaped capsules have a conspicuous circumferential
skirt with 15–23 vertical ridges (Figs 2c, 3c, 3d). There
are a minimum of 4 and a maximum of 5 concentric
rings on the convex side of the egg capsule. The capsule

*Table 1.* Comparison of published measurements of Littorinid egg capsules. (Abbreviations: d, with diagram; p, with photo; ed, size estimated from diagram).

| Species | Capsule diameter/$\mu$m | Capsule height/$\mu$m | Number of rings | Ovum diameter/$\mu$m | References | Remarks |
|---------|--------|--------|--------|--------|------------|---------|
| *N. trochoides* | 183–207 | 110 | 4–5 | 64–66 | Mak, this study | d & p |
| | 160–220 | 95–110 | 10 | 75 | Kojima, 1958a | d |
| | 220 | / | 6 | / | Habe, 1956 | p |
| | 154 | 82 | 5 | 57 | Berry, 1986 | ed |
| *N. radiata* | 230–244 | 120 | 6–7 | 64–66 | Mak, this study | d & p |
| | 160 | / | / | 50 | Habe, 1955 | p |
| | 210–230 | / | 3 | / | Habe, 1956 | p |
| | i) 195 | 100 | 3 | 73 | Kojima, 1960 | ed; two types |
| | ii) 200 | 102 | 5–6 | 78 | | of capsules |
| *N. vidua* | 200–220 | 140 | 4–5 | 64–66 | Mak, this study | d & p; with 15-23 ridges |
| | 170–190 | / | / | / | Habe, 1956 | d, with 18-26 ridges |
| | 236 | 79 | | 61 | Berry, 1986 | ed |
| *P. roepstorffiana* | 360–390 | 100 | / | 69–71 | Mak, this study | d & p |
| | 170 | 80 | / | 60 | Amio, 1963 | ed |
| | / | / | / | / | Habe, 1956 | p |
| *L. articulata* | 300–340 | 90 | / | 69–71 | Mak, this study | d & p |
| | 350 | 80–110 | / | 70 | Kojima, 1958b | d |
| | 248–268 | / | / | 76–82 | Reid, 1986 | d |
| | i) 194 | 71 | / | 54 | Berry, 1986 | ed; two types of |
| | ii) 268 | 93 | | 68 | | capsules |

is $210 \pm 10\mu$m in diameter, ovum diameter $65\pm1$ $\mu$m and the capsule height 140 $\mu$m (Table 1).

### *Peasiella roepstorffiana* (Fig. 1d)

These small gastropods are common on sheltered shores at the same vertical level as the cyanobacterium *Kyrtuthrix maculans* Umezaki (+ 1.3 m C.D.). The egg capsule is biconvex lens shaped, with a wide circumferential flange composed of two layers (Figs 2d, 3e). The capsule is not symmetrical when viewed from the side. The capsule diameter is $375\pm15$ $\mu$m, ovum diameter $70\pm1$ $\mu$m and the height of the capsule about 100 $\mu$m (Table 1).

### *Littoraria articulata* (Fig. 1e)

This littorinid species is common only on sheltered rocky shores and is also found on leaves and trunks of mangroves. The pelagic egg capsule has a symmetrical biconvex disc shape, approximately $320\pm20$ $\mu$m in diameter (Figs 2e, 3f). The circumferential edge is

thickened into a flotation collar. The ovum diameter is $70\pm1$ $\mu$m and the height is about 90 $\mu$m (Table 1).

### Discussion

The planktonic egg capsules of these five species have been described previously (Table 1): *N. trochoides* (as *N. pyramidalis*) by Habe (1956); Kojima (1958a) and Berry (1986); *N. radiata* (as N. exigua) by Habe (1955, 1956) and Kojima (1960); *N. vidua* (as *N. millegrana*) by Habe (1956 as *N. picta*) and Berry (1986); *P. roepstorffiana* by Habe (1956) and Amio (1963); *L. articulata* by Kojima (1958b as *L. strigata*); Berry (1986 as *L. strigata*) and Reid (1986). However, most of these descriptions include either a simple drawing or a light microscope photograph of the top-view of the egg capsule, which makes identification difficult as the lateral view is more diagnostic. Specifically, as the size of the egg capsules is variable (Table 1), the shape or

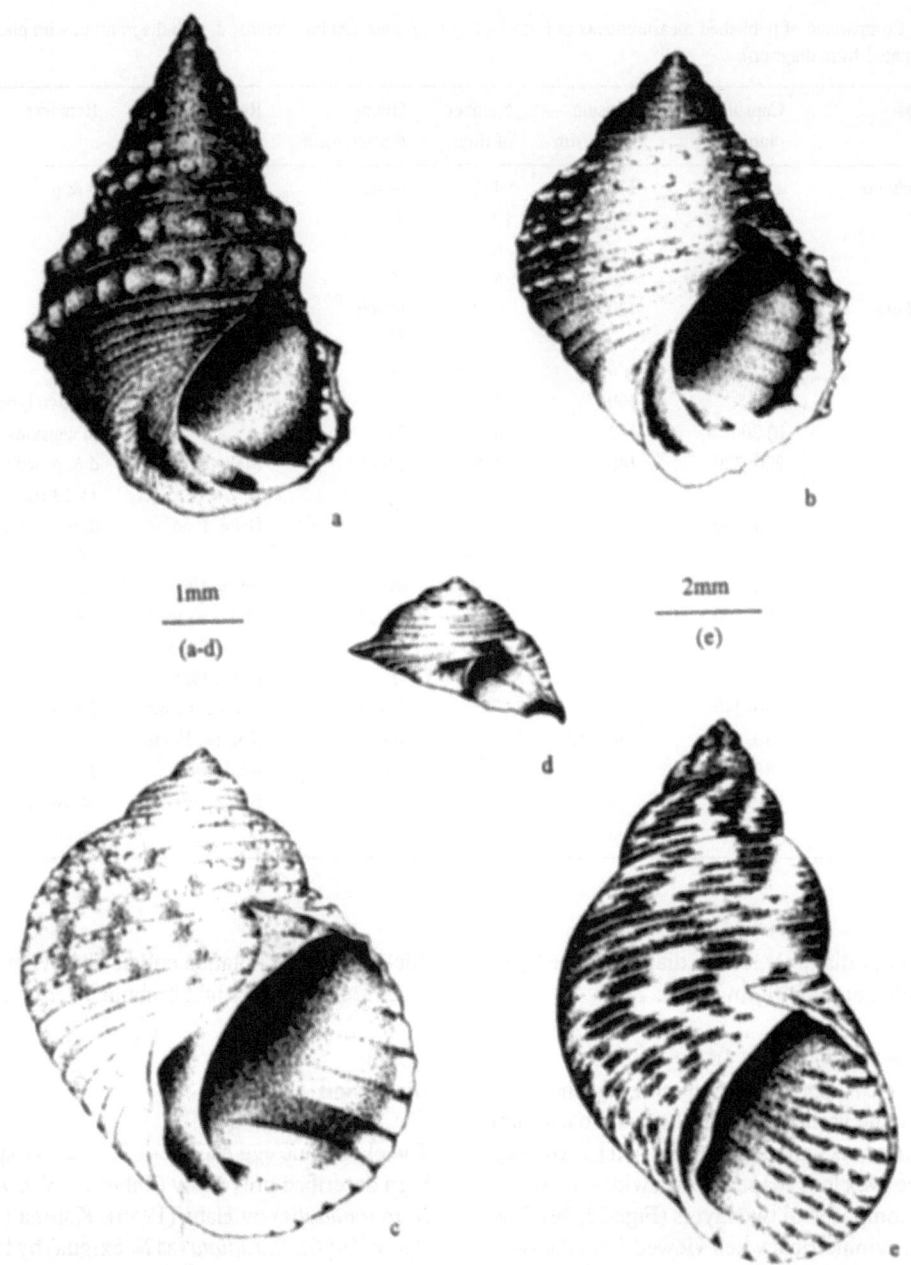

1mm
―――――
(a-d)

2mm
―――――
(e)

*Fig. 1.* The shells of the five rocky shore littorinids (a) *Nodilittorina trochoides*, (b) *N. radiata*, (c) *N. vidua*, (d) *Peasiella roepstorffiana* and (e) *Littoraria articulata*.

fine sculpturing of the egg capsules is more reliable for species identification.

The egg capsules of each of the 5 species are distinctive. They can be distinguished first by shape, whether they are cupola-shaped (the three *Nodilittorina* species) or lens-shaped. *Nodilittorina vidua* can

easily be distinguished by the undulating rim, even from the top view. Results from this study show the egg capsule of *Nodilittorina vidua* as a cupola form with vertical sculpturing on the overhanging skirt as opposed to a biconvex lens-shape with a distinctive spiral sculpture of the capsule rim as described by

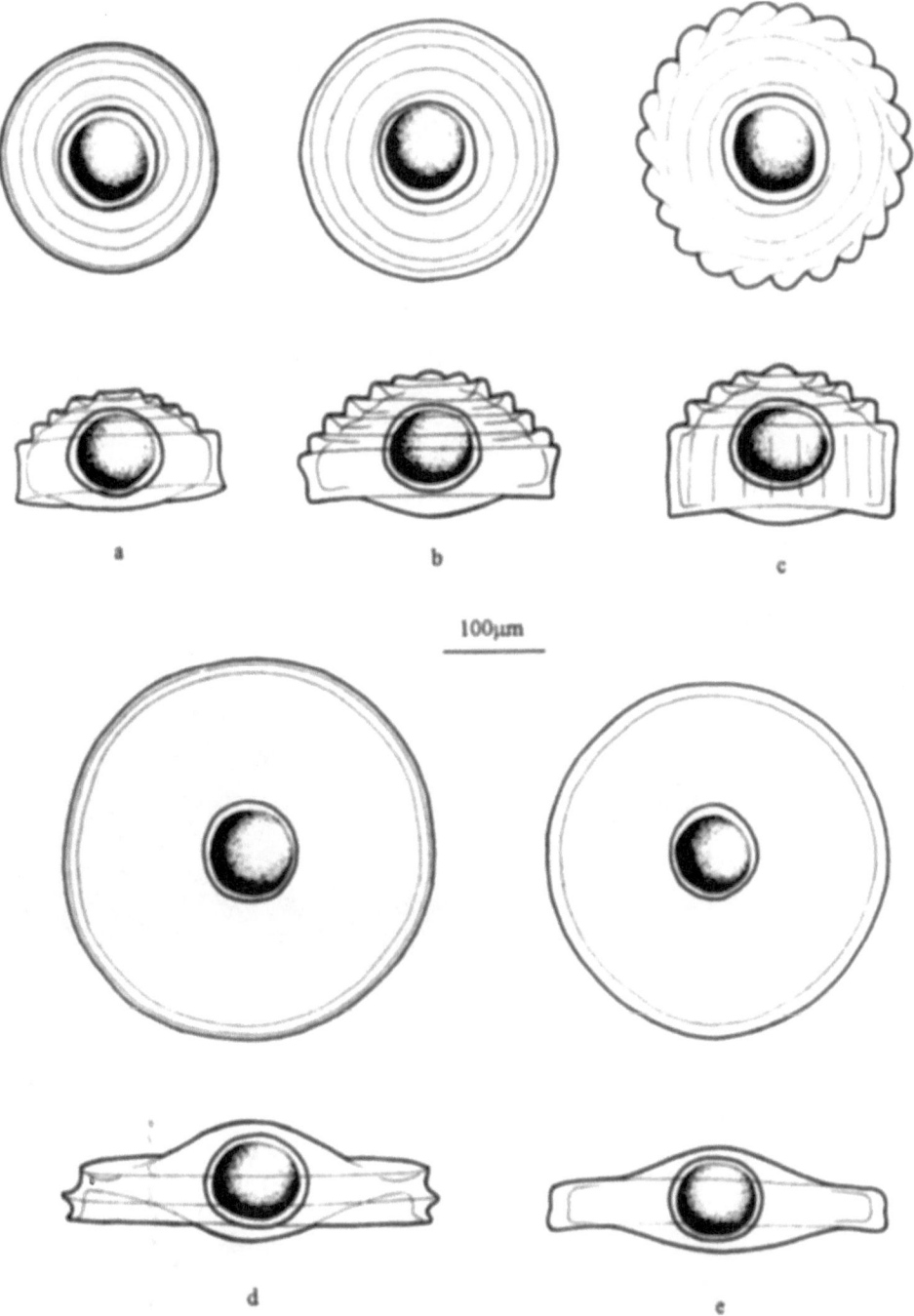

*Fig. 2.* Top and lateral views of the egg capsules of the five rocky shore littorinids (a) *Nodilittorina trochoides*, (b) *N. radiata*, (c) *N. vidua*, (d) *Peasiella roepstorffiana* and (e) *Littoraria articulata*.

Berry (1986). *Nodilittorina trochoides* and *N. radiata* egg capsules can be distinguished only by the size of the capsules, the former having smaller and fewer concentric rings on the convex side than the latter.

*Peasiella roepstorffiana* and *Littorina articulata* egg capsules can be distinguished by the structure of the circumferential rim, with the former having two layers and the latter a thickened rim.The sizes of the ovum

58

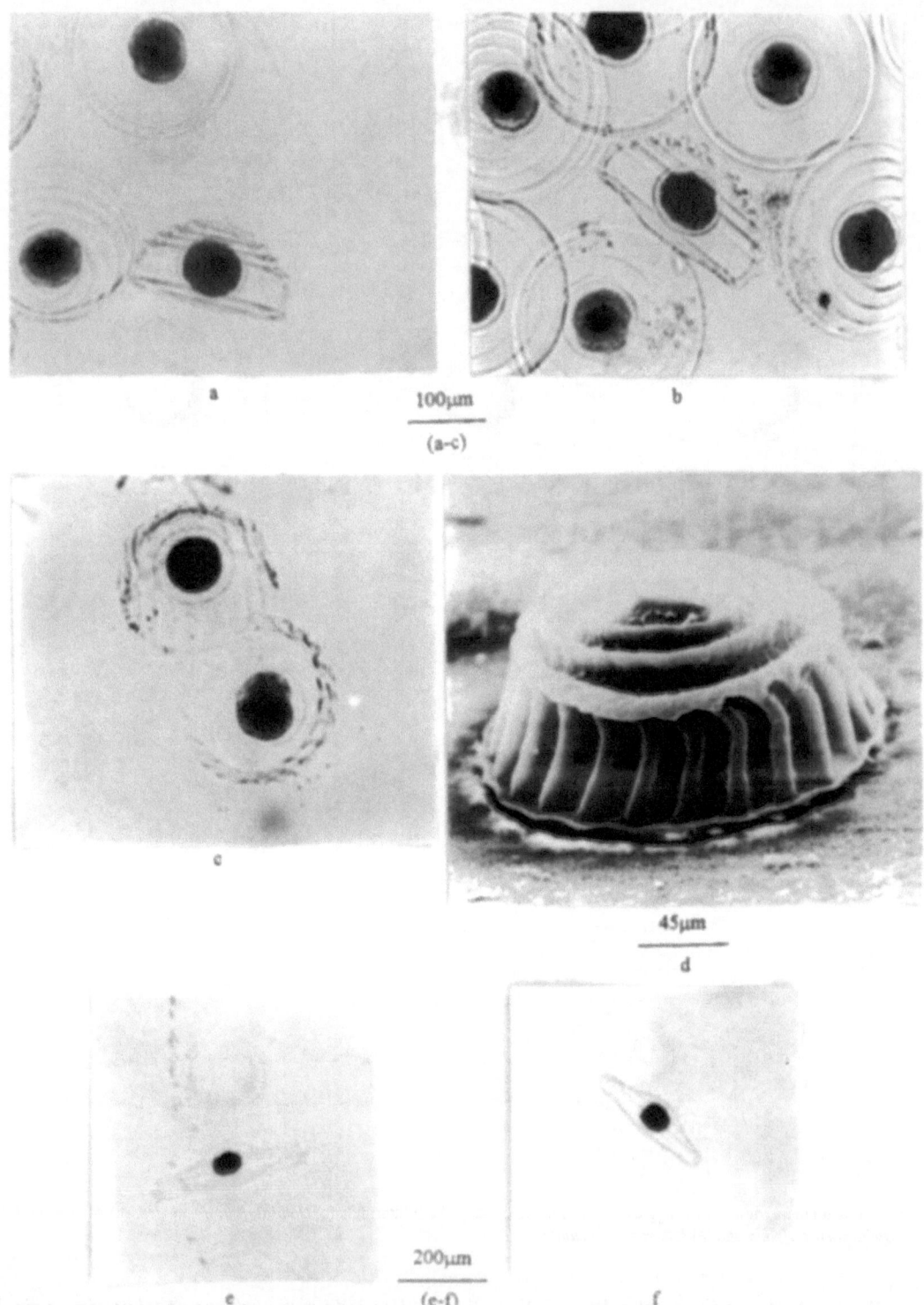

*Fig. 3.* Photographs of the egg capsules of the five rocky shore littorinids (a) *Nodilittorina trochoides*, (b) *N. radiata*, (c) *N. vidua* (d) SEM of *N. vidua*, (e) *Peasiella roepstorffiana* and (f) *Littoraria articulata*.

are similar for the three *Nodilittorina* species, though those of *Peasiella roepstorffiana* and *Littoraria articulata* are slightly larger.

Due to the great differences in the egg capsules (both size and structure) they are ideal as a taxonomic character in identification and as supplementary information to shell morphology. Identification of the five different egg capsules may also be useful in studies of plankton and consequent dispersal patterns of littorinids.

## Acknowledgments

I would like to express my thanks to Dr Gray A. Williams for reading and improving the manuscript. Dr David Reid also helped in the species identification. Attendance at the conference was funded by a conference grant from The University of Hong Kong.

## References

Abbott, R. T., 1954. Review of the Atlantic periwinkles, *Nodilittorina, Echininus* and *Tectarius*. Proc. U.S. natn. Mus. 103: 449–464.

Amio, M., 1963. A comparative embryology of marine gastropods, with ecological consideration. J. Shimonoseki University of Fisheries 12: 229–358.

Bandel, K., 1974. Studies on Littorinidae from the Atlantic. Veliger 17: 92–114.

Bandel, K. & D. Kadolsky, 1982. Western Atlantic species of *Nodilittorina* (Gastropoda: Prosobranchia): comparative morphology and its functional, ecological, phylogenetic and taxonomic implications. Veliger 25: 1–42.

Berry, A. J., 1986. Semi-lunar and lunar spawning periodicity in some tropical littorinid gastropods. J. moll. Stud. 52: 144–149.

Borkowski, T. V., 1971. Reproduction and reproductive periodicities of South Floridian Littorinidae (Gastropoda: Prosobranchia). Bull. mar. Sci. 21: 826–840.

Habe, T., 1955. The breeding of *Nodilittorina granularis* (Gray). Venus 18: 206–207.

Habe, T., 1956. The floating egg capsule of the Japanese periwinkles (Littorinidae). Venus 19: 117–121.

Kojima, Y., 1958a. On the floating egg capsule of periwinkles, *Littorina squalida* Broderip et Sowerby and *Nodilittorina pyramidalis* (Quoy et Gaimard). Venus 19: 233–237.

Kojima, Y., 1958b. On the planktonic egg capsule of *Littorivaga mandschurica* (Schrenck) and *Littoraria strigata* (Lischke). Venus 20: 81–86.

Kojima, Y., 1960. On the reproduction of periwinkles, Littorinidae (Gastropoda). Bull. biol. Atn. Asamushi 10: 117–120.

Murray, T. E., 1979. Evidence for an additional *Littorina* species and a summary of the reproductive biology of *Littorina* from California. Veliger 21: 469–474.

Ohgaki, S., 1985. Distribution of the family Littorinidae (Gastropoda) on Hong Kong rocky shores. In B. Morton & D. Dudgeon (eds), The Malacofauna of Hong Kong and southern China II, Proceedings of the Second International Workshop on the Malacofauna of Hong Kong and Southern China, Hong Kong, 6–24 April 1983. Hong Kong: Hong Kong University Press: 457–464.

Pilkington, M. C., 1971. Eggs, larvae and spawning in *Melarapha cincta* (Quoy and Gaimard) and *M. oliveri* Finlay (Littorinidae: Gastropoda). Aust. J. mar. Freshwat. Res 19: 79–90.

Reid, D. G., 1986. The littorinid molluscs of mangrove forests in the Indo-Pacific region: The genus *Littoraria*. London: British Museum (Natural History).

Reid, D. G., 1989. The comparative morphology, phylogeny and evolution of the gastropod family Littorinidae. Phil. Trans. r. Soc., Lond. B 324: 1–110.

Reid, D. G., 1990. A cladistic phylogeny of the genus *Littorina* (Gastropoda): Implications for evolution of reproductive strategies and for classification. Hydrobiologia 193 (Dev. Hydrobiol. 56): 1–19.

Reid, D. G., 1992. The Gastropod family Littorinidae in Hong Kong. In B. Morton (ed), The marine flora and fauna of Hong Kong and Southern China III Proceedings of the Fourth International Marine Biological Workshop: The Marine Flora and Fauna of Hong Kong and Southern China, Hong Kong, 11–29 April 1989. Hong Kong: Hong Kong University Press: 187–210.

Schmitt, R. J., 1979. Mechanics and timing of egg capsule release by the littoral fridge periwinkle *Littorina planaxis* (Gastropoda: Prosobranchia). Mar Biol. 50: 359–366.

Tokioka, T. & T. Habe, 1953. Droplets from the plankton net XI. A new type of *Littorina-capsula*. Publ. Seto mar. biol. Lab. 3: 55–56.

Winckworth, G., 1922. Nomenclature of British Littorinidae. Proc. malac. Soc. Lond. 15: 95–96.

*Hydrobiologia* **309**: 61–71, 1995.
*P. J. Mill & C. D. McQuaid (eds), Advances in Littorinid Biology.*
©1995 *Kluwer Academic Publishers.*

# Shape variation in the rough periwinkle *Littorina saxatilis* on the west and south coasts of Britain

Peter J. Mill & J. Grahame
*Department of Pure and Applied Biology, University of Leeds, Leeds LS2 9JT, UK*

*Key words: Littorina saxatilis*, shape, morphometrics, clines

## Abstract

Variation in the shape of the shell in *Littorina saxatilis* Olivi has been shown to be due largely to the same variables on both the west and the south coasts of Britain, and it exhibits various clines. Two important aspects are the size of the aperture, which becomes relatively larger from the Isle of Man southwards to Cornwall and eastwards from Devon to the Isle of Wight, and the jugosity of the shell, which increases with distance from Cornwall both northwards as far as the Isle of Man and eastwards as far as Kent. Superimposed on the clines are domains of shape, notably one in Lewis/Harris, where the shells have a relatively large aperture, which is long and narrow, coupled with a rather globose second whorl. The local and geographical aspects of shell shape variation are discussed.

## Introduction

There is a general consensus that within the *Littorina 'saxatilis'* complex the ovoviviparous *Littorina saxatilis* (Olivi) and the oviparous *L. arcana* Hannaford Ellis and *L. nigrolineata* Gray are good species. The status of a fourth taxon, the ovoviviparous *L. neglecta* Bean, is controversial (Johannesson & Johannesson, 1990a, b; Reid, 1993; Caley *et al.*, 1994; Grahame *et al.*, 1994, 1995). In Britain *L. saxatilis* occurs wherever there is a suitable substratum, on both open shores and in estuaries. This includes bedrock, boulders, small stones and even gravel (e.g. in the Fleet at Chesil Beach); also man-made concrete structures and wooden groynes. Thus it is extremely widespread. *L. arcana* has a more limited distribution. It occurs on rocks and boulders on open shores, generally inhabiting the same niche as does *L. saxatilis*, although on exposed shores the latter species may not extend onto the most exposed sites (Grahame & Mill, 1986). It is widespread in western England and Wales but rather more scattered in west Scotland. On the south coast of England it only extends as far eastwards as the centre of Lyme Bay (Mill & Grahame, 1990a). It also occurs on the east coast of Scotland and northern England. *L. nigrolineata* occurs rather lower on the shore than

the previous two species but it may also be found in estuaries. It has an even more limited distribution. It is found in southwest England and both south-west and north-west Wales as well as in a few west Scottish localities; on the east coast it is found only at a couple of sites in south-east Scotland (Mill & Grahame, 1990b, 1992b).

This paper concentrates on the shape characteristics of *L. saxatilis*, which is the most variable species. Much of the observed variation is of a local nature. Thus there is evidence both for a change of shape along a shore from sheltered to exposed conditions, with the more exposed animals having a larger foot area (Grahame & Mill, 1986), and for a clinal change in shape up and down the shore, with animals higher up having relatively taller spires (Grahame, Mill & Brown, 1990). Furthermore, on a given shore, animals from crevices in cliffs have a different shape from those found on and under boulders while, to complicate the picture further, there is sexual dimorphism (Grahame & Mill, 1992). Superimposed on this local variation we have found there to be variation on a geographical scale. Along the south coast of England a cline in shape occurs from west to east (Grahame & Mill, 1992) with a possible character displacement phenomenon associated with the change over from *L. saxatilis* being sympatric

Table 1. Listing of sites and samples used for the 'western' data. Distance, distance north (km) from the origin of the British National Grid.

| No. | Distance | Site | n |
|---|---|---|---|
| 1a | 52 | Trevaunance | 26 |
| 1b | 52 | " | 7 |
| 2 | 86 | Trebarwith | 15 |
| 3 | 111 | Duckpool | 21 |
| 4a | 195 | Mumbles | 14 |
| 4b | 195 | " | 19 |
| 5a | 200 | St Ann's Head | 11 |
| 5b | 200 | " | 8 |
| 5c | 200 | " | 14 |
| 5d | 200 | " | 11 |
| 6a | 206 | Great Castle Head | 29 |
| 6b | 206 | " | 46 |
| 7 | 227 | Porth Mawr | 16 |
| 8 | 306 | Cae du | 27 |
| 9a | 320 | Porth Neigwl | 15 |
| 9b | 320 | " | 19 |
| 9c | 320 | " | 14 |
| 9d | 320 | " | 12 |
| 10 | 327 | Porth Ysgo | 22 |
| 11 | 327 | Porth Llanllawen | 15 |
| 12 | 332 | Porth Iago | 10 |
| 13 | 381 | Penmon Point | 17 |
| 14 | 478 | Narbyl | 24 |
| 15a | 891 | Toe Head | 18 |
| 15b | 891 | " | 8 |
| 16 | 895 | Drinishader | 11 |
| 17 | 937 | Tumpa Head | 10 |
| 18a | 948 | Loch Shawbost | 8 |
| 18b | 948 | " | 13 |
| 19 | 955 | Rubha Bhlanisgaidh | 20 |
| 20a | 965 | Traigh Sands | 16 |
| 20b | 965 | " | 13 |
| 21 | 966 | Port Sto | 23 |
| 22 | 1110 | Grutness Voe | 28 |
| 23 | 1124 | Mousa | 21 |

Table 2. Listing of sites and Samlpes used for the 'southern' data. ♂, samples of males. Distance, distance east (km) from the origin of the British National Grid.

| No. | Distance | Site | n |
|---|---|---|---|
| 1a | 171 | Trevaunance | 26 |
| 1b | 171 | " | 7 |
| 2 | 204 | Trebarwith | 15 |
| 3 | 222 | Duckpool | 21 |
| 24a | 276 | Prawle Point | 29 |
| 24b | 276 | " | 39 |
| 25 | 289 | Stoke Fleming | 20 |
| 26 | 306 | Ladram Bay | 32 |
| 27 | 333 | Pinhay Bay | 26 |
| 28 | 341 | Golden Cap | 18 |
| 29a | 367 | Portland Bill | 11 |
| 29b | 367 | " | 27 |
| 29c | 367 | " | 27 |
| 30a | 396 | St Albans | 10 |
| 30b | 396 | " | 30 |
| 30c | 396 | " | 24 ♂ |
| 31a | 404 | Peveril Point | 21 |
| 31b | 404 | " | 30 |
| 31c | 404 | " | 23 ♂ |
| 31d | 404 | " | 28 ♂ |
| 32 | 430 | Alum Bay | 6 |
| 33 | 448 | St Catherine's Point | 36 |
| 34 | 589 | Southend | 20 |
| 35a | 633 | St Margaret's at Cliffe | 21 |
| 35b | 633 | " | 21 |
| 35c | 633 | " | 13 ♂ |
| 36a | 633 | White Ness | 18 |
| 36b | 633 | " | 13 ♂ |

with *L. arcana* in the west to being allopatric from it in the east (Grahame & Mill, 1989).

## Material and methods

This study is based on 607 animals in 35 samples collected from 23 sites on the west coast of Britain from Cornwall in the south to the Shetlands in the north

(Fig. 1, Table 1) and 612 animals in 27 samples from 16 sites in south England from Cornwall in the west to Essex in the east (Fig. 1, Table 2). The three sites in north Cornwall were used in each set to act as a reference point. The data from 24 sites in south west Wales and southern England were used in a previous study on local and regional variation where they were subjected to Canonical Variate Analysis (Grahame & Mill, 1992).

At most sites only females were used because of the difficulties of classification of the males. However, five samples were of males (10c, 11c, d, 15c and 16b); these were all well into the region of the south coast where *L. saxatilis* is allopatric from *L. arcana* and hence where no mis-identification could occur. Most of the collections were made from exposed parts of

63

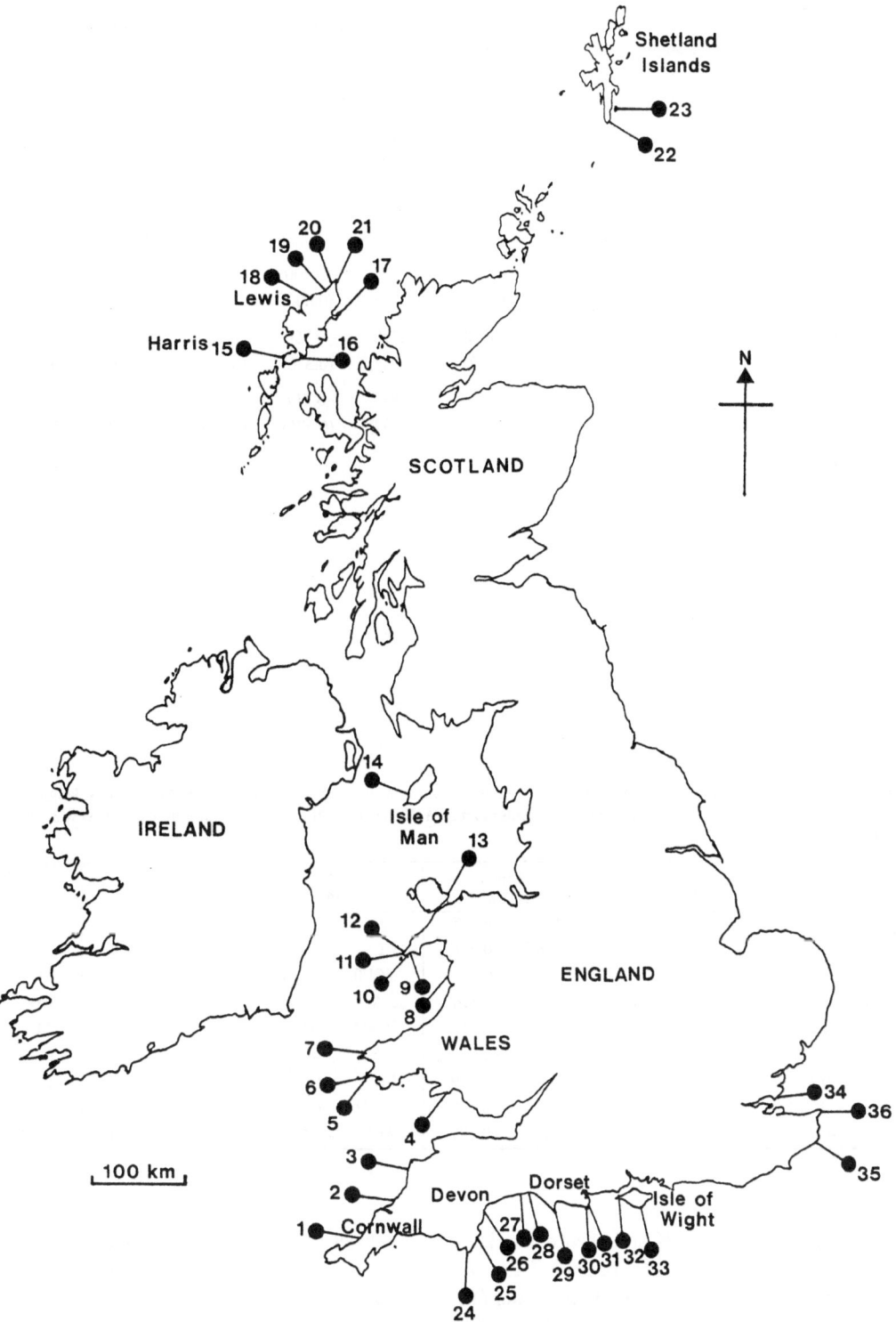

*Fig. 1.* Map to show the location of the sample sites. The numbers are those referred to in Tables 1 and 2. Sites 8, 15–23 and 26–34 are all allopatric for *L. saxatilis*; at the other sites *L. saxatilis* and *L. arcana* both occur.

*Table 3.* Spearman Rank Correlation Coefficients and their probability levels (in parantheses) for corresponding Principal Components derived from the western (PW1, etc.) and southern (PS1, etc.) data sets. PW1, etc, Principal Components for 'western' sites; PS1, etc, Principal Components for 'southern' sites.

|  | PS1 | PS2 | PS3 | PS4 | PS5 | PS6 | PS7 | PS8 | PS9 |
|---|---|---|---|---|---|---|---|---|---|
| PW1 | 0.700 | 0.500 | 0.400 | −0.083 | −0.233 | −0.050 | 0.067 | 0.150 | −0.133 |
|  | (0.036) | (0.125) | (0.286) | (0.831) | (0.546) | (0.898) | (0.865) | (0.700) | (0.732) |
| PW2 |  | 0.767 | 0.017 | −0.250 | 0.167 | −0.067 | 0.117 | 0.233 | 0.250 |
|  |  | (0.016) | (0.967) | (0.517) | (0.765) | (0.865) | (0.765) | (0.546) | (0.517) |
| PW3 |  |  | 0.833 | −0.317 | 0.133 | −0.200 | 0.483 | 0.200 | −0.467 |
|  |  |  | (0.005) | (0.406) | (0.732) | (0.606) | (0.188) | (0.606) | (0.205) |
| PW4 |  |  |  | 0.500 | 0.600 | −0.350 | −0.250 | 0.000 | −0.317 |
|  |  |  |  | (0.171) | (0.088) | (0.356) | (0.517) | (1.000) | (0.406) |
| PW5 |  |  |  |  | −0.817 | −0.217 | −0.100 | 0.250 | −0.233 |
|  |  |  |  |  | (0.007) | (0.576) | (0.798) | (0.517) | (0.546) |
| PW6 |  |  |  |  |  | 0.850 | −0.117 | −0.117 | −0.183 |
|  |  |  |  |  |  | (0.004) | (0.765) | (0.765) | (0.637) |
| PW7 |  |  |  |  |  |  | 0.817 | 0.167 | −0.133 |
|  |  |  |  |  |  |  | (0.007) | (0.668) | (0.732) |
| PW8 |  |  |  |  |  |  |  | 0.963 | −0.283 |
|  |  |  |  |  |  |  |  | (0.0001) | (0.460) |
| PW9 |  |  |  |  |  |  |  |  | 0.950 |
|  |  |  |  |  |  |  |  |  | (0.0001) |

*Table 4.* The first six Principal Components (PCs) for the data from western sites. AA, apical angle, AL, aperture length, AW, aperture width; CL, columella length; LL, lip length; OA, operculum area; WW0, shell width; WW1, width of first whorl; WW2, width of second whorl.

| PC1 38.6% | | PC2 33.7% | | PC3 14.2% | | PC4 4.7% | | PC5 3.4% | | PC6 2.7% | |
|---|---|---|---|---|---|---|---|---|---|---|---|
| OA | 0.493 | WW2 | 0.485 | WW1 | 0.718 | AL | 0.620 | AW | 0.500 | CL | 0.467 |
| AW | 0.486 | CL | 0.481 | WW0 | 0.618 | CL | 0.497 | WW0 | 0.458 | WW1 | 0.422 |
| AL | 0.453 | WW1 | 0.219 | WW2 | 0.137 | WW0 | 0.435 | CL | 0.263 | LL | 0.210 |
| WW1 | 0.154 | OA | 0.110 | AA | 0.114 | AA | 0.083 | OA | 0.154 | AW | 0.172 |
| AA | 0.062 | AW | −0.026 | AW | 0.046 | WW2 | −0.011 | LL | −0.053 | AL | 0.069 |
| CL | −0.164 | AL | −0.114 | OA | −0.085 | OA | −0.063 | AA | −0.254 | AA | −0.012 |
| WW0 | −0.190 | WW0 | −0.272 | CL | −0.109 | LL | −0.181 | WW2 | −0.267 | WW0 | −0.262 |
| WW2 | −0.192 | LL | −0.300 | AL | −0.118 | WW1 | −0.045 | WW1 | −0.360 | OA | −0.349 |
| LL | −0.434 | AA | −0.544 | LL | −0.187 | AW | −0.275 | AL | −0.424 | WW2 | −0.579 |

the coastline, mainly from cliff, solid bedrock or large boulders. However, one of the sites in south Wales was from sheltered bed-rock (4b), while Drinishader (16) in Harris and Grutness Voe (22) in Shetland were very sheltered, small-boulder sites. The two data sets were subjected to separate Principal Component Analysis (PCA) and the Principal Components analysed with respect to those aspects of shape which are of most importance. Since, in a preliminary study, 81% of the variation in the west coast data set was accounted for by size, the analyses reported on in this paper are all based on data after the removal of size as a primary source of variation by use of the geometric mean. Spearman rank correlation coefficients were calculated to deter-

65

*Table 5.* The first six Principal components (PCs) for the data from southern sites. AA, apical angle; AL, aperture length; AW, aperture width; CL, columella length; LL, lip length; OA, operculum area; WW0, shell width; WW1, width of first whorl; WW2, width of second whorl.

| PC1 40.5% | | PC2 28.2% | | PC3 15.3% | | PC4 6.1% | | PC5 4.1% | | PC6 2.9% | |
|---|---|---|---|---|---|---|---|---|---|---|---|
| AL | 0.451 | CL | 0.358 | WW1 | 0.696 | CL | 0.724 | AL | 0.703 | WW1 | 0.621 |
| AW | 0.441 | OA | 0.342 | WW0 | 0.610 | WW0 | 0.496 | WW2 | 0.325 | LL | 0.276 |
| OA | 0.399 | WW2 | 0.223 | AA | 0.097 | AL | 0.209 | AA | 0.293 | CL | 0.222 |
| AA | 0.333 | AW | 0.222 | OA | 0.076 | AW | 0.066 | CL | 0.043 | AA | 0.157 |
| WW0 | −0.036 | WW1 | 0.204 | AW | 0.023 | LL | 0.009 | WW1 | 0.039 | AL | 0.069 |
| LL | −0.164 | AL | 0.105 | WW2 | 0.010 | AA | −0.081 | WW0 | −0.058 | AW | −0.019 |
| WW1 | −0.175 | WW0 | −0.333 | AL | −0.069 | OA | −0.139 | LL | −0.216 | OA | −0.165 |
| CL | −0.286 | AA | −0.442 | CL | −0.211 | WW1 | −0.161 | OA | −0.220 | WW2 | −0.436 |
| WW2 | −0.437 | LL | −0.544 | LL | −0.282 | WW2 | −0.361 | AW | −0.461 | WW0 | −0.492 |

mine the degree of similarity of the ordering of the variables in corresponding PCs from the two data sets. The relationships between some of the shell variables were analysed further by looking at their standardised means for each sample.

The shell variables were measured using a digitiser tablet and microcomputer (Grahame & Mill, 1989). The same variables were measured as in previous studies (Grahame & Mill, 1989, 1992; Grahame, Mill & Brown, 1990) except that lip length and the width of the basal shell whorl were calculated so that the measurements were not subsumed within other measurements, i.e. they were both measured independently (Fig. 2) (Grahame *et al.*, 1994). Measurements were taken from 'landmarks' whose positions were judged by eye. All of the measurements were taken by one investigator to minimise errors. The variables used provided a set of values summarizing the shell attributes of length, width and globosity. Operculum area was also measured; this was used rather than aperture area because the latter is difficult to define reliably.

**Results**

The PC analysis revealed that the calculated Spearman Rank Correlation Coefficients showed significant correlations between the ordering of the variables in the corresponding PCs for the west and south coasts in all cases except for PC4 (Table 3). This is interpreted as indicating that similar variables, with similar degrees of importance, are involved in determining the geographical variation on each coast. No signifi-

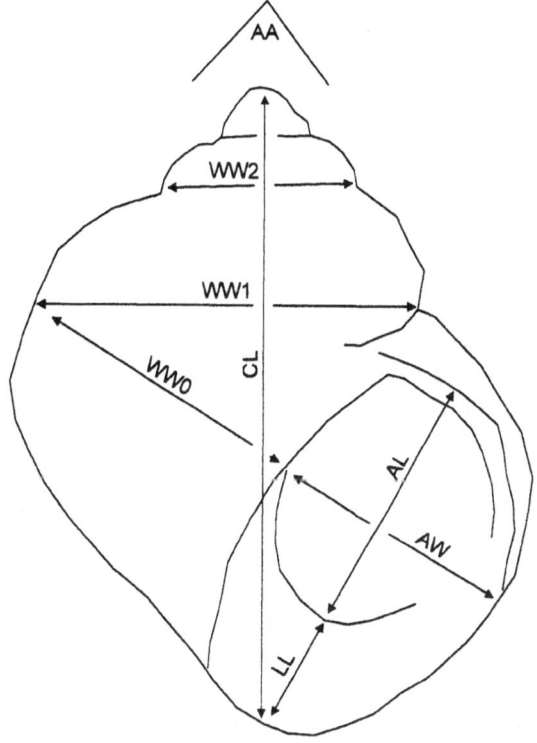

*Fig. 2.* Diagram of shell to show the variables measured. (After Grahame *et al.*, 1994). AA, apical angle, AL, aperture length; AW, aperture width; CL, columella length; LL, lip length; WW0, shell width; WW1, width of first whorl; WW2, width of second whorl.

cant correlation occurred between any other pairing of west coast and south coast PCs.

The first six Principal Components for the west coast data are shown in Table 4. When the scores for each site are plotted against distance north from the origin of the British National Grid a number of trends are revealed. PC1, which accounts for 38.6% of the variation shows a clear trend from Cornwall to the Isle of Man, with the aperture becoming relatively smaller (operculum area and aperture width and length) further north while the lip length increases to give more jugose shells (Fig. 3a). The two points which do not conform to this trend represent a boulder site (1b) and a sheltered site (4b). Shells from Lewis/Harris are similar to those from north Cornwall and south Wales with the exception of those from Drinishader in Harris (site 16 – very sheltered, small boulders) and Rhuba Bhlanisgaidh (site 19 – a boulder site). The two sites in Shetland also have low scores. Grutness Voe (site 22) is a small boulder, sheltered site while the site at Mousa (23) was not markedly exposed and the sample was obtained from medium-sized boulders. Thus there is certainly some indication that small boulder/sheltered populations tend to have relatively small-apertured, jugose shells. There is also some indication of a trend in the same region for PC2 (33.7%) (Fig. 3b). In this case the spire becomes less pointed and WW2 and the shell height (CL) relatively smaller from south to north Wales but the trend appears to reverse from south Wales to Cornwall. Of further interest here is the confirmation of a within-shore cline at St Ann's Head (site 5). Thus 5a and 5b represent sites at the top and middle of a long sloping rock which faces away from the sea, while sites 5c and 5d are at the bottom of the rock face. There is little difference between the top two sites but the shells from them are more pointed than those from lower down.

Although PC5 only accounts for 3.4% of the variation this component does show a trend covering most or all of the west coast, with a tendency for the shell aperture to become relatively narrower and longer from Cornwall to the Outer Hebrides (Fig. 3c). The next component, PC6 (2.7%), shows no obvious trend. Although these components have a low individual importance, a plot of PC6 against PC5 clearly reveals 'fields' of shell shape for some geographical areas. Thus there is an interesting clustering of the animals from Lewis/Harris in the Outer Hebrides (Fig. 4). These sites group together so as to indicate shells with a large aperture (operculum area) which is relatively long and narrow and a rather globose second whorl (WW2). Shells from north Cornwall form a looser grouping in the opposite quadrant. These shells have a relatively small, broad aperture, are relatively tall and have rather globose basal whorls. There is an indication that

shells from Shetland share the aperture characteristics of the Lewis/Harris shells and the shell height/basal whorl characteristics of the north Cornish shells, but only two sites have so far been analysed (Fig. 4).

When the scores for each site on the south coast are treated in a similar way (i.e. plotted against distance east from the origin of the British National Grid) they show similar sorts of trends; the first six Principal Components for these data are shown in Table 5. PC1 (40.5%) shows a possible trend involving aperture size (length, width and operculum area) and globosity of the second whorl (WW2) from south Devon eastwards to the Isle of Wight (Fig. 5a). PC2 (28.2%) and PC3 (15.3%) both reveal trends along the whole south coast (Fig. 5b, c). The former indicates an eastwards decrease in shell height and aperture size, and flatter more jugose shells, while PC3 indicates an eastwards decrease in the globosity of the basal whorls and confirms the increase in jugosity. PC4 shows a slight trend from south Devon eastwards to the Isle of Wight, with shells becoming relatively taller, and with a broader basal whorl and a less globose second whorl, in the east (Fig. 5d). In spite of the previously noted occurrence of sexual dimorphism in the shells of this species (Grahame & Mill, 1992) the within site dimorphism, though evidently real, is relatively small with respect to the overall trends in the above Principal Components at the three sites from which samples of both were used (Fig. 5). PC5 (4.1%) and PC6 (2.9%) show no obvious trends but again a plot of PC6 on PC5 shows some clustering. In this case it is a rather weak clustering of the Kentish shells (Fig. 6), which tend to have relatively long and narrow apertures, as in the Lewis/Harris shells. Also of interest here is that the two samples of male shells from this region both occur lower on the PC6 axis than do the samples of female shells indicating, amongst other things, a reduced globosity of the first whorl; this is in accord with the type of sexual dimorphism noted by Grahame & Mill (1992).

A very clear, negative relationship was observed between operculum area (also aperture length and aperture width) and lip length in both west coast and south coast shells when comparing the standardised means for the variables (Fig. 7).

## Discussion

In earlier studies on the shape of shells from south and south-west Britain, using Canonical Variate Analysis (Grahame & Mill, 1989, 1992), it was demonstrated

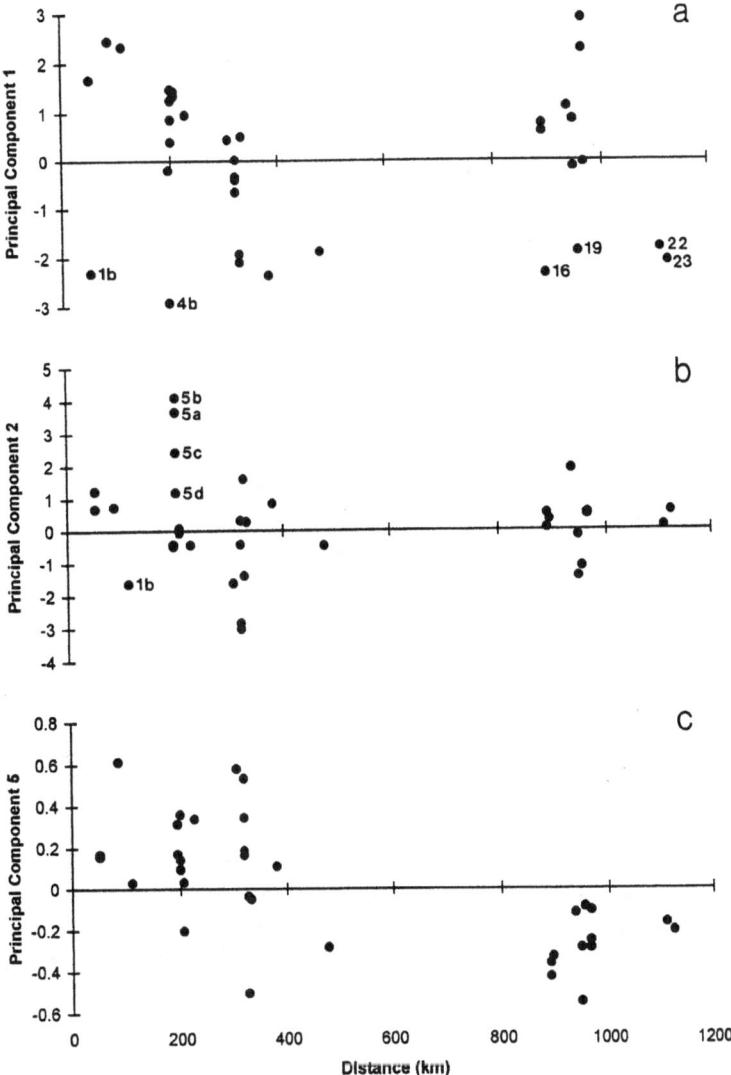

*Fig. 3.* Mean scores for shells from west coast sites plotted against distance north from the origin of the British National Grid. (a) for Principal Component 1, (b) for Principal Component 2, (c) for Principal Component 5. Where points are numbered the numbers refer to Table 1 and Fig. 1.

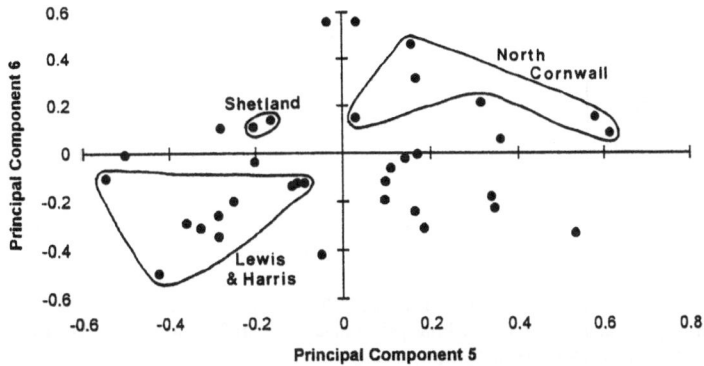

*Fig. 4.* Mean scores for shells from west coast sites. Plot of Principal Component 6 against Principal Component 5.

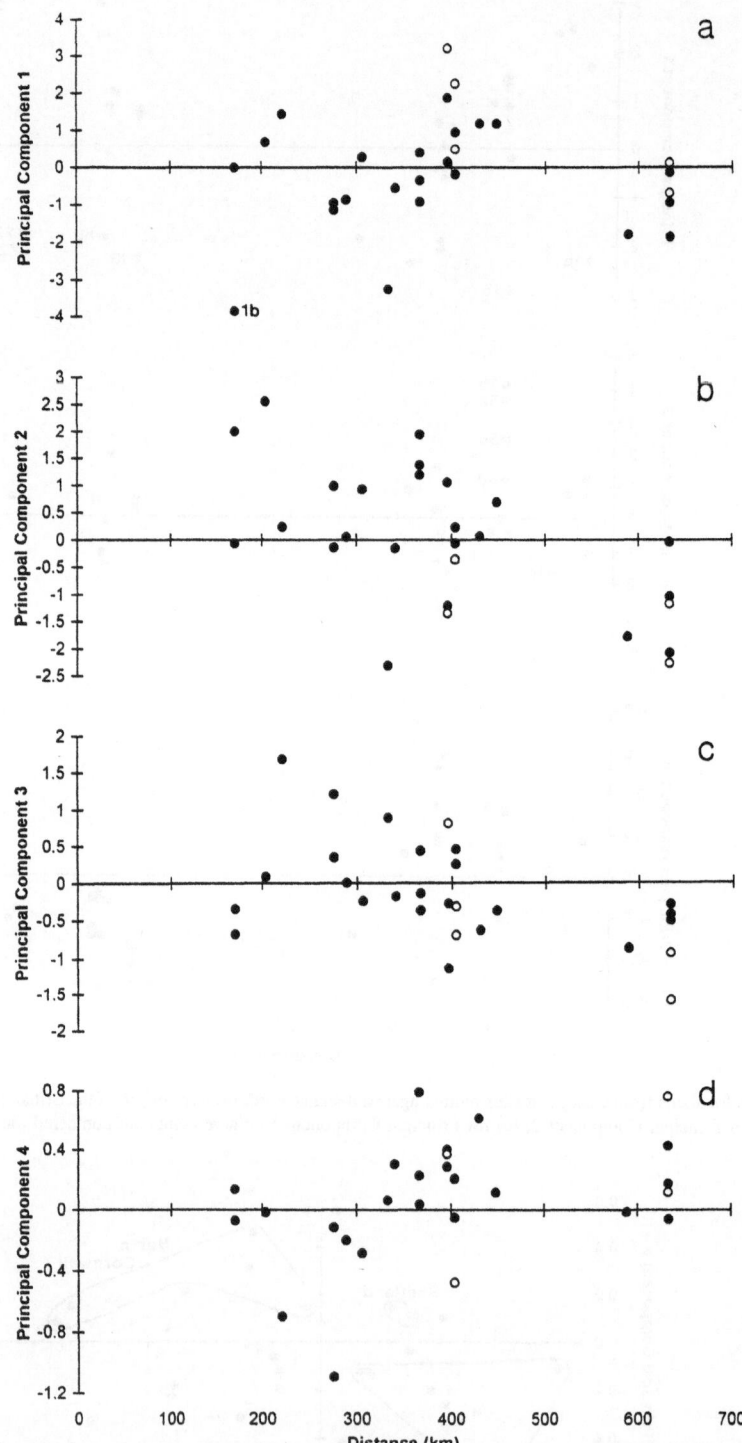

*Fig. 5.* Mean scores for shells from south coast sites plotted against distance east from the origin of the British National Grid. (a) for Principal Component 1, (b) for Principal Component 2, (c) for Principal Component 3, d) for Principal Component 4. Where points are numbered the numbers refer to Table 2 and Fig. 1. o, samples of males.

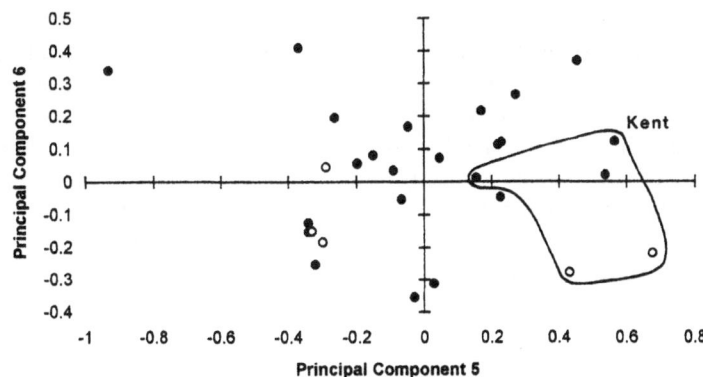

*Fig. 6.* Mean scores for shells from south coast sites. Plot of Principal Component 6 against Principal Component 5. o, samples of males.

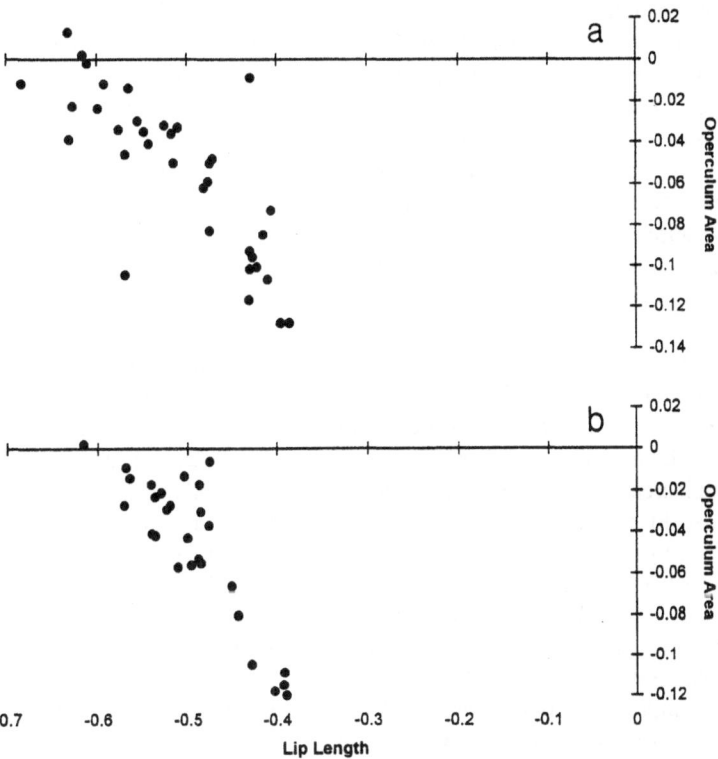

*Fig. 7.* Plots of the standardised means of operculum area against lip length. (a) west coast sites, (b) south coast sites.

that the shells of *Littorina arcana* have a relatively larger aperture than do those of *L. saxatilis*, whereas *L. saxatilis* shells tend to be broader. Furthermore, the shells of *L. saxatilis* found on the same shores as those of *L. arcana* (i.e. sympatric) differ from the shells of *L. saxatilis* allopatric from *L. arcana*, in that the former are more globose and have slightly smaller apertures. The difference in aperture size between sympatric and allopatric *L. saxatilis* led us to consider that character displacement is occurring (Grahame & Mill, 1989),

i.e. that sympatric *L. saxatilis* are more dissimilar from *L. arcana* than are allopatric *L. saxatilis*, as far as this character is concerned. This was further indicated by a discriminant analysis of the shells. However, the reliability of classification of allopatric *L. saxatilis*, using discriminant analysis, appears to increase with distance eastwards from Lyme Regis on the south coast of England, i.e. from the region of transition between sympatry and allopatry. Hence there is a cline involved rather than, or possibly in addition to, character dis-

70

placement. An additional complication occurs when
the shells of male *L. saxatilis* are included, because of
the sexual dimorphism present in this species. Thus,
at two allopatric shores in Dorset some 70 km east of
the sympatric/allopatric divide on the south coast of
England, shells of male *L. saxatilis* more readily misclassified as *L. arcana* than did the shells of females
(Grahame & Mill, 1992). In addition to clines in shell
shape, clines in esterase variability occur along at least
the south coast in both *L. saxatilis* and *L. arcana* (Mill
& Grahame, 1992a).

We have demonstrated that pre-emergence juveniles of *L. saxatilis* show substantially the same pattern
of shell shape variation (Grahame & Mill, 1994), indicating that at least part of this is heritable. Coefficients
of determination as great as 20% provided a crude estimate of this heritability. In *Nodilitorina unifasciata*
there are differences in relative aperture size and shape
and in shell globosity between populations on different
shores (some sheltered and some exposed) (McMahon,
1992). McMahon suggested that these differences are
ecophenotypic because this species has a larval planktonic phase. In *L. saxatilis* Newkirk & Doyle (1975)
showed that, with increasing distance up an estuary
(i.e. increasing shelter), the translation rate of the shell
increased while the rate of whorl expansion and the
circularity of the aperture decreased. The genetic variance of these variables was low except for the rate of
whorl expansion at the most exposed sites.

It is becoming increasingly apparent that there are
clear geographical differences as well as considerable local variation in *L. saxatilis*. The local differences include clines from sheltered to exposed regions
of a shore (Grahame & Mill, 1986), up-shore clinal changes (Grahame, Mill & Brown, 1990), sexual
dimorphism (Grahame & Mill, 1992) and differences
between cliff and boulder populations. Janson & Sundberg (1983) and Sundberg (1988) have shown clines in
the shape of *L. saxatilis* shells between exposed cliff
sites and sheltered boulder sites. However, these presumably have components of both exposed/sheltered
and cliff/boulder. In one of these studies (Sundberg,
1988) most of the variation was shown to be due to
size, shells from exposed sites having smaller, thinner shells than those from sheltered sites. It has also
been demonstrated that shells of *Littorina nigrolineata* from exposed shores are smaller and more globose
than those from sheltered shores (Heller, 1976) and are
also thicker and have a wider, rounder aperture (Naylor & Begon, 1982). Naylor & Begon also showed
that boulder populations had larger, thicker shells than

those inhabiting rock crevices and the latter characteristic was particularly marked in a population on stones.
They thus suggested that increase in thickness may be
a response to increased likelihood of crushing due to
the mobility of the substrate and possibly also as a
protection against increased predation. They also noted that some populations of *L. nigrolineata* showed
sexual dimorphism, with females being squatter and
broader than males (Naylor & Begon, 1982).

The geographical variation shown in this and previous papers (Grahame & Mill, 1989, 1992; Grahame, Mill & Brown, 1990) involves various clines
along both the south and west coasts, the possibility of character displacement in sympatric populations
as opposed to allopatric populations and domains of
shape, as clearly demonstrated, for example, by the
shells from the Outer Hebrides (Lewis and Harris),
where we may be witnessing a founder effect. It is
also clear that the same variables changing in the same
sort of way are of considerable importance in determining changes in shell shape on both the west and the
south coasts. Indeed, certain overall trends are apparent when the results from the two coasts are compared.
Thus, from the PC1s, it is clear that the relative aperture size increases from the Isle of Man southwards to
Cornwall and eastwards from south Devon as far as the
Isle of Wight. More data are needed to ascertain what
happens to this variable around the Cornish peninsula.
The PC2s indicate that shells in the outer Hebrides and
Shetland and in the Isle of Wight are relatively lower,
flatter and more jugose than are those in the south west
of Britain. Is this because there are general clines in
these features away from the south-west or are low, flat,
jugose animals typical of island populations? Further
studies on this are in hand.

*L. saxatilis* is the most variable of the four species
in its sibling group and it is possible that its flexibility
with respect to shape accounts for its success in terms
of the wide variety of habitats in which it occurs and
of its extensive geographical range. There still remains
the possibility, however, that *L. saxatilis* as presently
understood is itself a composite of species; this problem is currently being addressed.

### References

Caley, K. J., J. Grahame & P. J. Mill, 1994. Morphometrics of small
rough periwinkles. Cah. Biol. mar. 35: 240–241.
Grahame, J. & P. J. Mill, 1986. Relative size of the foot of two
species of *Littorina* on a rocky shore in Wales. J. Zool., Lond.
(A) 208: 229–236.

Grahame, J. & P. J. Mill, 1989. Shell shape variation in *Littorina saxatilis* and *L. arcana*: a case of character displacement? J. mar. biol. Ass. U.K. 69: 837–855.

Grahame, J. & P. J. Mill, 1992. Local and regional variation in shell shape of rough periwinkles. In: J. Grahame, P. J. Mill & D. Reid (eds), Proceedings of the 3rd International Symposium on Littorinid Biology. The Malacological Society of London, London: 99–106.

Grahame, J. & P. J. Mill, 1994. Shell shape variation in rough periwinkles: genotypic and phenotypic effects. In J. C. Aldrich (ed.), Quantified Phenotypic Responses in Morphology and Physiology. Proc. 27th Europ. mar. Biol. Symp.: 25–30.

Grahame, J., P. J. Mill & A. C. Brown, 1990. Adaptive and non-adaptive variation in two species of rough periwinkle (*Littorina*) on British shores. In: K. Johanneson, D. G. Rafaelli & C. J. Hannaford Ellis (eds), Progress in Littorinid and Muricid Biology. Developments in Hydrobiology 56. Kluwer Academic Publishers, Dordrecht: 223–231. Reprinted from Hydrobiologia 193.

Grahame, J., P. J. Mill & S. Hull, 1994. Shape, size and enzymes: the problem of *Littorina neglecta*. Cah. Biol. mar. 35: 246–247.

Grahame, J., P. J. Mill, S. Hull & K. J. Caley, 1995. *Littorina neglecta* Bean: ecotype or species? J. nat. Hist., Lond.

Heller, J., 1976. The effects of exposure and predation on the shells of two British winkles. J. Zool, Lond. 179: 201–213.

Janson, K. & P. Sundberg, 1983. Multivariate morphometric analysis of two varieties of *Littorina saxatilis* from the Swedish west coast. Mar. Biol. 74: 49–53.

Johannesson, K. & B. Johannesson, 1990a. *Littorina neglecta* Bean, a morphological form within the variable species *Littorina saxatilis* (Olivi)? Hydrobiologia 193 (Dev. Hydrobiol. 56): 71–87.

Johannesson, K. & B. Johannesson, 1990b. Genetic variation within *Littorina saxatilis* (Olivi) and *Littorina neglecta* Bean: is *L. neglecta* a good species. Hydrobiologia 193 (Dev. Hydrobiol. 56): 89–97.

McMahon, R. F., 1992. Microgeographic variation in the shell morphometrics of *Nodilittorina unifasciata* from southwestern Australia in relation to wave exposure of shore. In: J. Grahame, P. J. Mill & D. Reid (eds), Proceedings of the 3rd International Symposium on Littorinid Biology. The Malacological Society of London, London:107–117.

Mill, P. J. & J. Grahame, 1990a. The distribution of *Littorina saxatilis* (Olivi) and *Littorina arcana* Hannaford Ellis in Lyme Bay, southern England. J. moll. Studs. 56: 133–135.

Mill, P. J. & J. Grahame, 1990b. Distribution of the species of rough periwinkle (*Littorina*) in Great Britain. In: K. Johanneson, D. G. Rafaelli & C. J. Hannaford Ellis (eds), Progress in Littorinid and Muricid Biology. Developments in Hydrobiology 56. Kluwer Academic Publishers, Dordrecht: 21–27. Reprinted from Hydrobiologia 193.

Mill, P. J. & J. Grahame, 1992a. Clinal changes in esterase variability in *Littorina saxatilis* (Olivi) and *L. arcana* Hannaford Ellis in southern Britain. In: J. Grahame, P. J. Mill & D. Reid (eds), Proceedings of the 3rd International Symposium on Littorinid Biology. The Malacological Society of London, London: 31–37.

Mill, P. J. & J. Grahame, 1992b. Distribution of the rough periwinkles (*Littorina* spp.) in Great Britain: an update. In: J. Grahame, P. J. Mill & D. Reid (eds), Proceedings of the 3rd International Symposium on Littorinid Biology. The Malacological Society of London, London: 305–307.

Naylor, R. & M. Begon, 1982. Variation within and between populations of *Littorina nigrolineata* Gray on Holy Island, Anglesey. J. Conch. 31: 17–30.

Newkirk, G. F. & R. W. Doyle, 1975. Genetic analysis of shell-shape variation in *Littorina saxatilis* on an environmental cline. Mar. Biol. 30: 227–237.

Reid, D., 1993. Barnacle-dwelling ecotypes of three British *Littorina* species and the status of *Littorina neglecta* Bean. J. moll. Studs 59: 51–62.

Sundberg, P., 1988. Microgeographic variation in shell characters of *Littorina saxatilis* Olivi – a question mainly of size? Biol. J. linn. Soc. 35: 169–184.

*Hydrobiologia* **309**: 73–87, 1995.
*P. J. Mill & C. D. McQuaid (eds), Advances in Littorinid Biology.*
©1995 *Kluwer Academic Publishers.*

# Microgeographical variation in shell strength in the flat periwinkles *Littorina obtusata* and *Littorina mariae*

C. R. Fletcher
*Department of Pure and Applied Biology, University of Leeds, Leeds LS2 9JT, UK*

*Key words:* sibling species, morphometrics, shell strength, local variation, predator resistance, *Littorina*

## Abstract

The strength of molluscan shells has been shown to vary in adaptive ways in a number of species and one of the main factors thought to be involved is shell-crushing by predators. A recent study found that the sibling species of flat periwinkle *Littorina obtusata* and *Littorina mariae* showed significant differences in the rates at which shell strength increased with shell length in specimens which had been collected from the same location, where the species were sympatric. This paper describes differences between the shells of the two species from a number of localities around Milford Haven in Dyfed, Wales, and local geographical variation in the shells.

*Littorina mariae*, which is normally found at lower tidal levels than *L. obtusata*, matures at a smaller shell length. Both species reinforce the shell as they grow since shell strength, determined as the maximum force applied by a hydraulic tensile testing machine before the shell cracked, is strongly positively allometric; it increases at a rate close to the cube of shell length whilst isometric growth would result in strength increasing in proportion to the square of shell length. Because *L. mariae* matures earlier and reinforces the shell at a smaller size, the mature shell of *L. mariae* is substantially stronger on average than that of a similar sized but immature *L. obtusata*. At maturity the shell strengths of the two species are not very different despite the substantial difference in mean shell length.

Strength varies significantly from shore to shore, and with the level of the shore from which the animals were collected. Strength increases down the shore in both species. Shell strength decreases with exposure to wave action in *L. mariae* but increases with exposure in *L. obtusata*; there is also substantial shore-to-shore variation which is not explained by exposure.

Path analysis was used to explore the relationship between shell strength and other measured shell parameters (mass, length, height, thickness). The best predictor of shell strength in both species is a parameter which is heavily positively loaded on LN (shell mass) and strongly offset by negative loadings on LN (shell length) and LN (shell height). This is logical because for a given shell length a heavier shell will be thicker and stronger, whilst for a given shell mass a bigger shell will be thinner and therefore weaker. Such differential variation of shell mass and shell length explains most of the geographical variation observed in shell strength; shells are stronger in snails collected from one place than from another because, for the same shell length they are heavier or, to put it the other way, because at the same shell mass, they are smaller.

## Introduction

The strength of molluscan shells has been shown to vary in adaptive ways in a number of species and one of the main factors thought to be involved is shell-crushing by predators (Vermeij, 1978; Vermeij & Currey, 1980). Selective thickening of the body whorl or aperture, sculpturing and a reduced apical spire can all help to resist shell-breaking predators or increase handling time (Zipser & Vermeij, 1978; Palmer, 1979; Vermeij & Currey, 1980; Bertness & Cunningham, 1981). However, there is likely to be a cost in growth and fecundity of excessively thickening the shell (Hughes & Elner, 1979; Palmer, 1981). Mechanical strengths of whole shells have been reported for a variety of different gastropods (Currey, 1979; Vermeij & Currey, 1980;

Blundon & Vermeij, 1983; Brandwood, 1985; Lowell, 1985, 1986, 1987; Ash, 1989; Cook & Kenyon, 1993; Lowell *et al.*, 1994).

The morphology of thaiidid snail shells from different shores has been shown to vary according to the degree of exposure to wave action (Cooke, 1895; Kitching & Ebling, 1967; Hughes & Elner, 1979) and the strength also differs (Currey & Hughes, 1982); these changes are believed to be responses to selective pressures of desiccation, hydrodynamic drag caused by wave action and predation, especially by the shore crab *Carcinus maenas* (Crothers, 1985). The edible periwinkle *Littorina littorea* has been reported to display less (Ash, 1989) or no (Currey & Hughes, 1982) such local variation, and this was ascribed to gene flow resulting from a pelagic larval stage (Ash, 1989; Currey & Hughes, 1982). However, gene flow would presumably not prevent an inducible response like that seen in *Nucella lapillus*, the development of which changes in response to the proximity of crabs feeding on conspecific snails (Palmer, 1990). The strengths of different colour morphs of the mangrove snail *Littoraria pallescens* have been found to differ significantly at some localities but not at others (Cook & Kenyon, 1993).

The flat pariwinkles *Littorina obtusata* (L.) and *Littorina mariae* Sacchi & Rastelli are sibling species of littoral prosobranch molluscs and are frequently found on the same British shores. They were originally separated on the basis of size at maturity, detailed shell morphology, penial morphology and correlated physiological and ecological differences (Sacchi & Rastelli, 1966). In *L. mariae* the shell is normally smaller at maturity and the edge of the shell thickens, restricting the throat more markedly, but the clearest difference is in the form of the penis. In *L. obtusata* the penis is elongate and flat and has a short tip and has about 30 glandular papillae on the edge in 2 to 3 rather irregular rows, whilst in *L. mariae* the basal part is more nearly round with about 12 glands in a single row, and it has an elongate tip. The two species have been shown to have distinct but overlapping distributions (Sacchi, 1969; Reimchen, 1982; Watson & Norton, 1987). *L. mariae*, being apparently more tolerant of exposure to wave action but less tolerant of desiccation, is found lower on the shore than is *L. obtusata* and occurs mainly on *Fucus serratus* (Williams, 1990a) but not exclusively so (Nielsen, 1980; Rolan-Alvarez, 1992; Warmoes *et al.*, 1988). *L. obtusata* usually occurs on *F. vesiculosus* and *Ascophyllum nodosum*, although it can be found from the *Pelvetia* zone to the top of the *Laminaria* zone

(Sacchi, 1969; Reimchen, 1982; Watson & Norton, 1987; Williams, 1990b, 1994). The distribution of the two species overlaps more on relatively exposed shores than in extreme shelter (Williams, 1990b). *L. mariae* grazes preferentially on epiphytes and may be beneficial to its host alga, whereas the preferred food of *L. obtusata* is the fucoid macroalgae (Watson & Norton, 1987). *L. mariae* reaches maturity in one year but may live for two or possibly more years (Reimchen, 1979; Williams, 1990b, 1992) whereas *L. obtusata* apparently takes nearly two years to reach maturity and lives three or more years (Goodwin, 1978; Williams, 1990b, 1992). Predation, feeding and other aspects of their biology have been reviewed recently by Lowell *et al.* (1994), Norton *et al.* (1990) and Williams (1990b).

A recent study (Lowell *et al.*, 1994) examined the shell morphology and strength of the sibling species *L. obtusata* and *L. mariae* in specimens that had been collected from the same location, where the species were sympatric, and showed that there were significant differences between the species in the patterns of growth and the development of shell strength. *L. mariae* showed uniformly allometric growth of juveniles into adults. *L. obtusata*, on the other hand, exhibited a distinct change in growth pattern in most parameters on approaching maturity. *L. mariae* showed a more sustained increase in shell thickness both in the body whorl and in the aperture, and this was reflected in increasing overall shell mass and in shell strength, both of which were positively allometric over the whole size range. *L. obtusata*, however, grew to a larger shell length and the shell could accommodate larger bodies at all sizes. The overall effect is that, although *L. obtusata* live longer and grow substantially larger than *L. mariae*, the strength of the shells of the mature animals are less different than would be expected, based on size alone, on that shore.

In that study, both species of animals were deliberately collected from an area of a few square metres where they were sympatric in order to minimise the effects of possible geographical variations in morphometrics and strength. Here I describe a study which shows that there is variation in the strength of shells of these snails from a range of shores all within a few kilometres of each other, and also marked variation in shell strength with tidal level. These variations in strength are shown to be almost completely determined by the mass of the shell in relation to its length.

## Methods

The flat periwinkles *Littorina obtusata* (L.) and *Littorina mariae* Sacchi & Rastelli were used. The area of study was in the region of Milford Haven, Dyfed, Wales. Milford Haven itself is a natural harbour presenting a large number of rocky shores with varying exposures to tidal and wave action; the adjacent Irish sea coasts offer more exposed shores. The shores used for this study are given in Table 1. The degree of exposure to wave action was assessed using Ballatine's Biological Exposure Index (Ballantine, 1961). This scale is based on the presence and abundance of a selection of 'indicator species' of macroalgae and animals; the scale index decreases with exposure. Several of the collection sites used here were Ballantine's own survey stations. On each shore, collections were normally made from three levels; an upper zone dominated by *Fucus vesiculosus*, a middle zone where *F. vesiculosus* and *F. serratus* were of similar abundance, and a lower zone where *F. serratus* reached its peak abundance. On more sheltered shores, collections were also made from *Ascophyllum nodosum*, which occurs at similar tidal levels to *F. vesiculosus* and which in extreme shelter can displace the latter from the lower part of its normal range; collections were also made from *F. spiralis* which occurs as a narrow zone above the *F. vesiculosus* and is here referred to as the top zone. Collecting was haphazard at each site except that an effort was made to include a reasonable number of individuals of both species where both were present. No effort was made to collect the very small specimens often present. Visual identification in the field is not reliable, especially on the more exposed shores, and the species identification was performed after crushing the shells in Leeds as described below. Collections were made between 1st and 5th July 1992 and the snails were held and transported to Leeds at 4°C in polythene bags which were rinsed out daily with fresh sea water. Laboratory work was completed within seven days of collection; there was under 2% mortality with this method of holding.

### Morphological and strength measurements

The crushing force (shell strength) required for breakage was measured with an Instron Dynamic Testing machine in the manner described by Lowell *et al.* (1994). Ash (1989) concluded that shell strength measured in this way was a good indicator of the ability of littoral snails to survive attack by a shell crushing

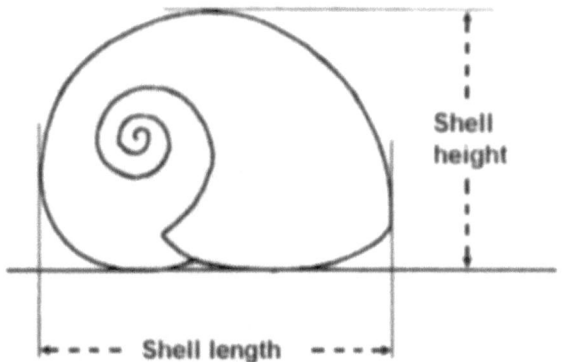

*Fig. 1.* Diagram of a flat periwinkle shell showing the dimensions used.

predatory crab, even though the crab may break the shell of different snails in a different manner. Prior to crushing, vernier callipers were used to measure shell length and shell height to the nearest 0.05 mm. Shell height is here defined as the distance of the highest point of the shell above a flat surface upon which the shell is lying with the aperture facing downwards (Fig. 1), and shell length is used to denote the maximum linear dimension of the shell, which because of the lack of a spire is a nearly horizontal line spanning the aperture and the body whorl (Fig. 1). Shells were classified into colour morphs, including the four used by Sacchi (1969) and Reimchen (1979) for *L. mariae* as well as several other; fusca, (dark brown exterior with a purple interior, common in *L. obtusata*), and less common hues – red, orange, light brown, dark brown, and black. The presence of repaired fracture scars was noted, although such scars have previously been found not to affect the experimentally determined strength of these snails (Lowell *et al.*, 1994). Since crushing stopped as soon as the shell failed, the body of the animal was not unduly affected by the procedures.

Following each test, the winkle was quickly killed in hot water and the soft parts dissected away from the shell. Each animal was examined for sexual maturity and scored from 1 (fully immature) to 3 (fully mature) as described previously (Lowell *et al.*, 1994). They were identified to species by the form of the penis in the males (Sacchi & Rastelli, 1966) and the length of the bursa copulatrix in the females (Reid, 1990a). Very immature animals where identification could not be made with certainty were excluded from the subsequent analysis. Dry weights of the shells were

*Table 1.* Collecting sites, algal zones, and tidal levels used in this study.

| Site | BritishNational Grid Reference | Flucoids[1] | Levels[2] | Exposure index[3] |
|------|-------------------------------|-------------|-----------|-------------------|
| Angle Bay | SM870027$_5$ | a/v/sr | M | 7 |
| Angle Point | SM875$_5$032 | v/sr | M | 6 |
| West Angle 1 | SM851032 | v | U | 5 |
| West Angle 1 | SM851$_5$032$_5$ | a/v/sr | M | 5 |
| West Angle 2 | SM851$_5$033$_5$ | v | U | 5.5 |
| Musselwick 1 | SM819066 | sp,v,a,sr | T,U,M,L | 6.5 |
| Musselwick 2 | SM821062$_5$ | v,sr | U,L | 5 |
| Musselwick 3 | SM820063 | v,a,sr | U,M,L | 5.5 |
| Musselwick 4 | SM819$_5$064 | sp,a,v,sr | T,U,L | 6 |
| Gt. Castle Bay | SM799055 | a/v/sr | M | 5 |

Key [1] a – *Ascophyllum nodosum*; sp – *Fucus spiralis*; sr – *Fucus serratus*; v – *Fucus vesiculosus*
Key [2] T – top zone (sp); U – upper zone (a/v); M – middle zone (v/sr); L – lower zone (sr)
Key [3] lower numbers indicate stronger wave action.

determined following drying in an oven at 80°C for 24 hours (determined to give constant weight).

*Statistical procedures*

Measurements of shell dimensions and breaking forces were transformed to natural logarithms (LN) prior to analysis so that regression coefficients would represent the coefficients in power-law relationships and, therefore, indicate whether allometric changes were occurring during growth (see discussions in Peters, 1983; Calder, 1984). Shell length is used as the basic measurement to which to relate other variables since it is the most commonly used measure of size for gastropods, is a first order variable, and was measured with good accuracy to the nearest 0.05 mm. Also reported are the relationships between shell strength and measurements of several other shell variables which might contribute to crushing resistance.

Single and multiple regression analyses were carried out using the procedures 'Regression' (PROC REG) and 'General Linear Models' (PROC GLM) in the Statistical Analysis System (SAS) package (SAS Institute Inc., 1985).

Regression relationships were tested for simple curvature by fitting a quadratic term to the residuals from the linear regression. Any such second order term significantly different from zero was taken as evidence of non-linearity. Lowell *et al.* (1994) found that where significant non-linearity was detected it appeared to be occurring at a size between stages 1 and 2. Hence separate analyses for the adult animals (stages 2 and 3) and the juveniles (stage 1) were made to see whether the degree of allometry changed as the snails became mature, as indicated by significant differences of slope between juveniles and adults.

Model II regression coefficients were calculated as the square root of the ratio of the model I regression coefficients of each variable on the other (Ricker, 1973). Model II (also termed reduced major axis or geometric mean; Ricker, 1973; Clarke, 1980) regression coefficients are appropriate to data where the variables cannot be properly classified as dependent and independent (Sokal & Rohlf, 1981, section 14.2). The slopes of model II regression lines were tested (Clarke, 1980) against each other or against the value expected for isometric growth (assuming strength to be proportional to cross-sectional area). To examine the slopes and relative elevations of regression lines, I have sometimes used the tests for differences between slopes and between adjusted means in the analysis of covariance (ANCOVA) routines of PROC GLM. Therefore, in these cases I have reverted to using tests for model I regressions, the equivalent statistics for model II ANCOVA not being available.

To determine the significance of variation of shell mass and of shell strength between different shores, between different zones, and with other factors, the species were examined separately. ANOVA of the transformed data was used, using LN (shell length) as a covariate, which was entered into the analysis first; thus the subsequent factorial analysis was effectively performed on the residuals of the regression. The various factors were then entered and, because the data are unbalanced, the corrected sums of squares were used in the analyses (SAS PROC GLM with Model III sums of squares). LN (shell strength) was also examined using LN (shell mass) as a covariate.

Alternative approaches involving transforming the data by rotation until the model II regression line was horizontal and then using ANOVA, or else calculating the actual distances of data points from the model II regression line and analysing these by ANOVA were also tried.

To investigate the dependence of shell strength on the other shell parameters (length, height and mass) the techniques of multiple regression and path analysis were used, again with logarithmically transformed data (Sokal & Rohlf, 1981, section 16.3). Path diagrams represent the influence of predictor variables on criterion variables, and they are particularly useful when the predictor variables are correlated with each other. The goodness of fit of the model is judged by the similarities between predicted and observed correlation coefficients. Discrepancies between the actual strengths and the strength predicted using the models developed with path analysis were also studied by ANOVA.

## Results

Adult snails form the bulk of the collections and provide the modal values for shell length, shell height, shell mass and crushing force for both species (Fig. 2). The distributions and the modal values for the first three variables are clearly different but the distributions of force to crush the shells of the two species differ to a lesser extent; the mean strength of *Littorina obtusata* is higher than that for *L. mariae* (Table 2) but the modal values are similar.

The morphometric coefficients determined are shown in Table 3; where they differ significantly the morphometric coefficients of juveniles and of adults are also shown separately in Table 3. Shell mass as a function of shell length is shown in Fig. 3. In *Littorina obtusata* the shell mass increases isometrically with shell length, as the model II regression slope does not differ significantly from 3. There is no evidence for non-linearity in the log transformed data, but conversely there is an indication difference of slopes for

*Table 2*. Shell dimensions and the force required to crush the shells of the two species, expressed as mean ± S.E. (N).

| Measure | *L. mariae* | *L. obtusata* |
|---|---|---|
| Shell length, mm | 10.51±0.11 (230) | 14.26±0.13 (328) |
| Shell height, mm | 6.58±0.06 (230) | 9.14±0.07 (3327) |
| Shell mass, g | 0.415±0.013 (230) | 0.866±0.019 (328) |
| Shell strength, N | 283.6±7.9 (230) | 351.2±8.6 (328) |
| Body dry mass, mg | 18.5±0.6 (228) | 52.8±1.3 (328) |

immature and for adult animals ($P < 0.02$). *L. mariae* shows marked positive allometry of shell mass as a function of shell size ($P < 0.001$); thus the species differ markedly as the rates of increase of shell mass with size in *L. mariae* is faster than it is in *L. obtusata* ($P < 0.0001$). In *L. mariae* there is also some evidence of non-linearity as inclusion of a second order term in the regression increases the multiple $R^2$ by a small but significant extent from 0.955 to 0.956 ($P = 0.004$). However, there is no significant evidence for change in regression slope with the onset of maturity, and in any case such slight effects are not likely to be of biological significance.

Shell strength increases as a function of shell length much faster than isometrically (Table 3); the isometric slope would be 2 but the observed slopes are close to 3 if calculated from the combined juvenile and adult data. There is considerable variability in shell strength (Fig. 4); it is shown below that much of this variability is geographical in origin. In *L. mariae* there is evidence of non-linearity in the logarithmic strength – length relationship because incorporating a second order term increases $R^2$ significantly from 0.649 to 0.679 ($P < 0.0001$) but, as with shell mass, the slope does not change significantly at the onset of maturity. Although inclusion of a second order term does not significantly improve the fit of the regression for *L. obtusata* when strength is considered as a function of maturity, the morphometric slope increases from 2.88 for juveniles to 4.31 for adult animals ($P < 0.001$). Similar effects are also seen if shell strength is studied as a function of shell mass; the isometric coefficient is 0.667 but the observed slopes are close to 1 and show distinct differences between juveniles and adults in *L. obtusata* but not in *L. mariae*. In view of the contradictory nature of the results in relation to non-linearity and to changes in slope with maturity, possible departures from linearity have been ignored in the subse-

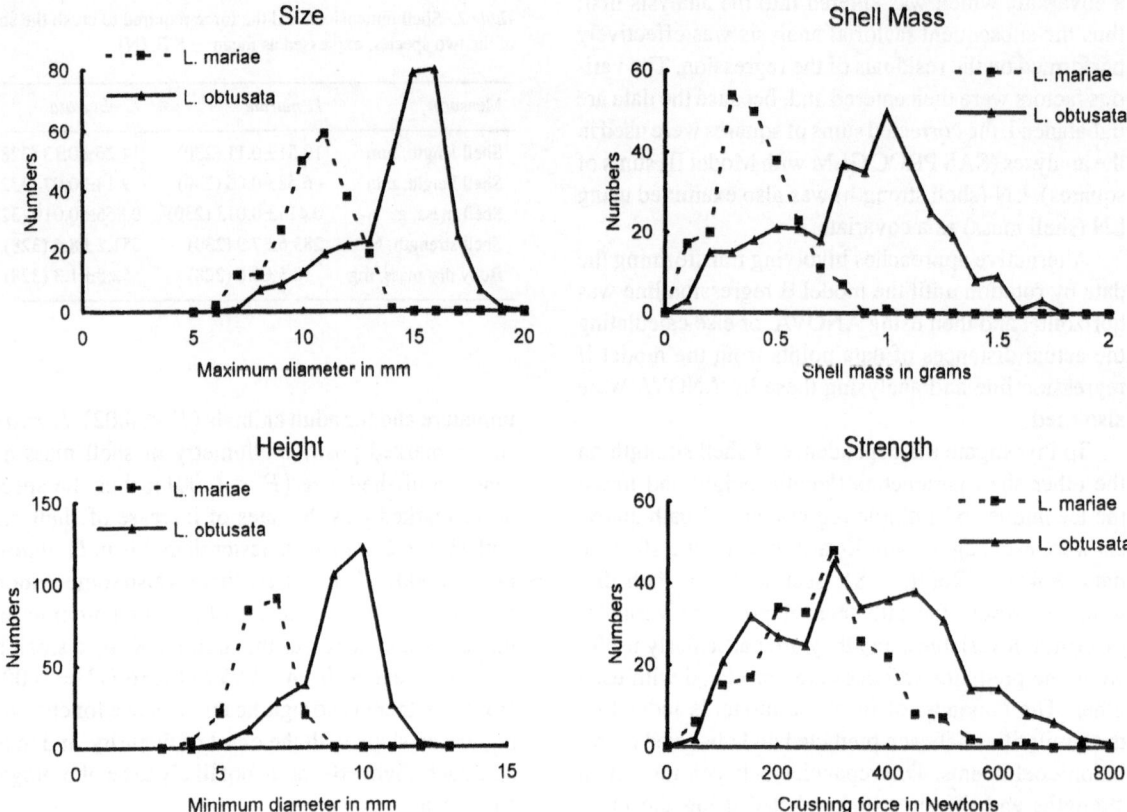

**Fig. 2.** Frequency distributions of size (shell length), shell height, shell mass, and strength (crushing force) in *L. obtusata* and *L. mariae*.

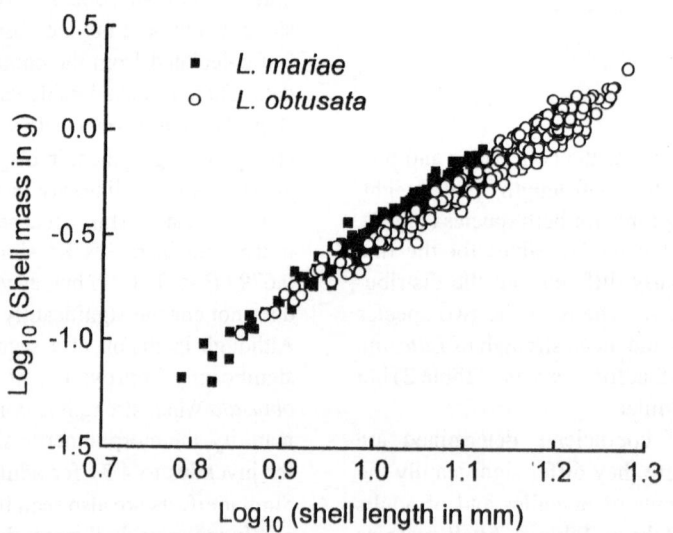

**Fig. 3.** Logarithmic plot of shell mass as a function of size (shell length).

quent analyses and adult and juveniles data have been pooled.

The morphometric coefficient for strength as function of size in the two species does not differ (ANCO-VA, $P = 0.22$) but the strengths of the two species

*Table 3.* Strength and shell mass morphometric coefficients, calculated as the model II regression slope of the variable on the length of the shell and tested for departure from isometry, also shell strength as a function of shell mass. The Pearson product moment correlation coefficient ($r$) is given. Also are the statistic $T$ and its degree of freedom (DF) for difference of observed from isometric slope in model II regression comparisons, calculated after Clarke (1980).

| Variables tested | Isometric slope | Observed slope | $r$ | $N$ | $T$ | DF | $P$ |
|---|---|---|---|---|---|---|---|
| *L. mariae* (maturity categories 1, 2 and 3) | | | | | | | |
| Shell height/shell length | 1.000 | 0.882 | 0.9274 | 219 | 4.96 | 154 | <0.001 |
| Shell mass/shell length | 3.000 | 3.214 | 0.9772 | 219 | 4.81 | 150 | <0.001 |
| Strength/shell length | 2.000 | 3.143 | 0.8056 | 219 | 9.67 | 167 | <0.001 |
| Strength/shell mass | 0.667 | 0.920 | 0.8804 | 219 | 10.07 | 159 | <0.001 |
| *L. obtusata* (maturity categories 1, 2 and 3) | | | | | | | |
| Shell height/shell length | 1.000 | 0.838 | 0.9720 | 335 | 13.68 | 228 | <0.001 |
| Shell mass/shell length | 3.000 | 2.953 | 0.9779 | 335 | 1.37 | 228 | N.S. |
| Strength/shell length | 2.000 | 2.963 | 0.7457 | 335 | 10.78 | 263 | <0.001 |
| Strength/shell mass | 0.667 | 1.003 | 0.8329 | 335 | 13.57 | 250 | <0.001 |
| *L. obtusata* juveniles (maturity category 1) | | | | | | | |
| Shell mass/shell length* | 3.000 | 2.868 | 0.9809 | 63 | 1.82 | 44 | N.S. |
| Strength/shell length** | 2.000 | 2.876 | 0.6526 | 64 | 3.79 | 54 | <0.001 |
| Strength/shell mass*** | 0.667 | 1.003 | 0.7483 | 63 | 4.84 | 50 | <0.001 |
| *L. obtusata* adults (maturity categories 2 and 3) | | | | | | | |
| Shell mass/shell length* | 3.000 | 3.120 | 0.9266 | 272 | 1.72 | 192 | N.S. |
| Strength/shell length** | 2.000 | 4.307 | 0.5492 | 271 | 15.08 | 237 | <0.001 |
| Strength/shell mass*** | 0.667 | 1.380 | 0.7451 | 272 | 18.08 | 214 | <0.001 |

\* $P < 0.002$; \*\* $P < 0.002$; \*\*\* $P < 0.001$ for null hypothesis that adult and juvenile regression slopes are equal within species, using the test of Clarke (1980).

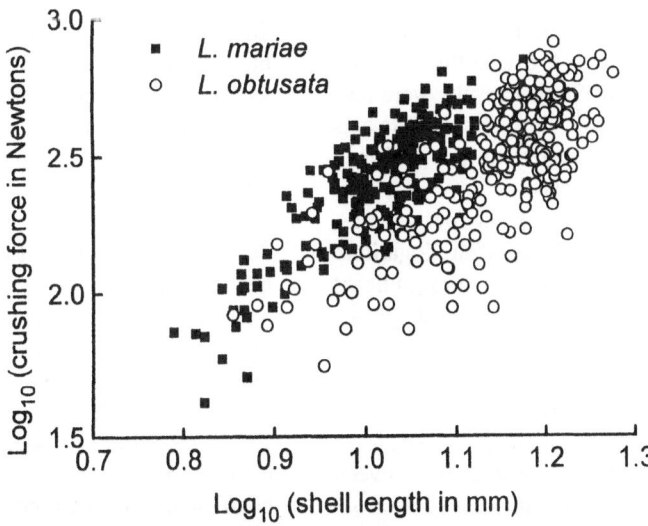

*Fig. 4.* Logarithmic plot of shell strength as a function of size (shell length).

(elevations of the lines) do differ; this is expressed by the mean values of LN (strength) of each species corrected to the grand mean of LN (shell length) of all the shells studied. This is 5.982 for *L. mariae* and 5.480 for *L. obtusata*; these values differ significantly ($P < 0.0001$). Size for size, *Littorina mariae* is stronger than *L. obtusata*.

There is considerable variation of shell mass about the model II regression line as seen in Fig. 3. Analysis of variance of the transformed shell mass using shell length as a covariate shows that a substantial part of the variability is associated with the shore and with the algal zone of the shore from which the winkles were collected (Table 4). Other factors studies are not significant and together account for less than 0.25% of the scatter about the regression line.

The same factors are also significant when shell strength is studied as a function of shell length (Table 5). There is also significant sexual dimorphism in shell strength of *L. mariae* ($P = 0.0002$) but not in *L. obtusata* ($P = 0.09$).

If shell mass is used as the covariate in the above analysis then the covariate accounts for a higher proportion of the total variation (78.0% for *L. mariae* and 69.4% for *L. obtusata* compared to 65.3% and 55.5% respectively for shell length). However, the same factors remain significant determinants of strength; that the average shell mass varies from one location to another does not provide a sufficient explanation for local variations in shell strength.

ANOVA analysis of the residuals of LN (strength) about the model II regression lines gave results in which the significance of the effects studied did not differ materially from those derived from the model I analysis reported in Table 4, except that maturity became a significant factor. Similar results were also obtained after rotating the axes until the ordinate axis coincided with the model II regression line and then using ANOVA on the new abscissa.

The resistance to crushing of the periwinkles, expressed as the least squares mean of the logarithm of strength, for each level of all factors with significant influence is given in Table 6. It is clear that each species is stronger if collected lower on the shore, and the effect is progressive. In *L. mariae* the males are stronger on average than the females at the same size ($P = 0.003$), but sexual dimorphism is not statistically significant in *L. obtusata*. The reasons for the differences between collection sites, which are shown in order of decreasing exposure, is less clear. A Spearman rank correlation between the strength corrected for size and the estimat-

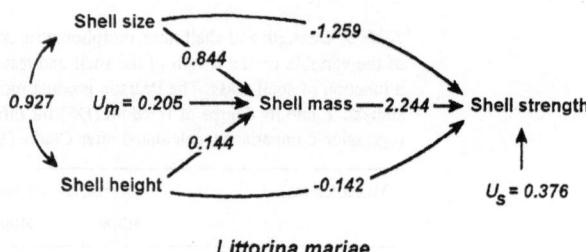

*Littorina mariae*

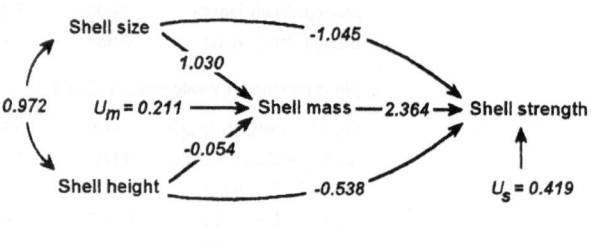

*Littorina obtusata*

*Fig. 5.* Path diagrams showing how shell mass may be predicted from shell length and shell height; and also how shell strength may be predicted from shell length, shell height and shell mass, all of the variables being logarithmically transformed. The values associated with the arrows are path coefficients, the $U$ values are the square root of the coefficient of non-determination of the variable. $U_m$ represents the extent to which shell mass varies independently of shell height and length, and $U_s$ represents variation in shell strength not related to the other three measurements. $U_m$ will include shell thickness whilst $U_s$ will include such factors as mineral structure, defects and shape.

ed exposure index of the collection shore (shown in Table 1) is significant for both species ($P < 0.0001$), but the sign of the correlation differs between the two species. In *L. mariae* $r_s = 0.381$; since the Ballantine index *increases* for sheltered conditions this shows that *L. mariae* is stronger on more sheltered shores. However, for *L. obtusata* $r_s = -0.300$, indicating the reverse. In each species the effect accounts for about half of the overall sum of squares associated with differences between shores.

Path analysis indicated that LN (shell strength) may be represented as a parameter in which a very heavy positive loading on LN (shell mass) is offset by strong negative loadings on LN (shell length) and on LN (shell height) in both species (Fig. 5). The path coefficients, derived from standard partial regression coefficients, may be used to predict the observed correlations between the variables to within 0.005 (the possible range of $r$ being $\pm 1$), showing that the model

*Table 4.* Analysis of variation in shell mass of flat periwinkles using logarithmically transformed data with shell length as a covariate. The covariate was extracted first and then the remaining variation was partitioned, correcting for all other factors. The factor 'notes' includes parasitised, recent damage, or repaired damage scar as well as blank.

ANOVA table of LN (shell mass)

| Source | DF | Sq. SS | Adj. SS | MS | $F$ | $P$ |
|---|---|---|---|---|---|---|
| *Littorina mariae* | | | | | | |
| Covariate | | | | | | |
| LN (length) | 1 | 53.23 | | 53.23 | 7344 | <0.001 |
| Factors | | | | | | |
| shore | 7 | 0.503 | 0.401 | 0.057 | 7.91 | <0.0001 |
| zone | 2 | 0.469 | 0.422 | 0.211 | 29.14 | <0.0001 |
| sex | 1 | 0.007 | 0.021 | 0.021 | 2.95 | N.S. |
| maturity | 2 | 0.048 | 0.019 | 0.009 | 1.30 | N.S. |
| colour | 8 | 0.051 | 0.056 | 0.007 | 0.96 | N.S. |
| notes | 3 | 0.033 | 0.033 | 0.011 | 1.51 | N.S. |
| residual | 192 | 1.391 | 1.391 | 0.007 | | |
| Total | 216 | 55.735 | | | | |
| *Littorina obtusata* | | | | | | |
| Covariate | | | | | | |
| LN (length) | 1 | 88.09 | | 88.09 | 14238 | <0.0001 |
| Factors | | | | | | |
| shore | 8 | 1.039 | 0.561 | 0.070 | 11.33 | <0.0001 |
| zone | 4 | 0.940 | 0.745 | 0.186 | 30.12 | <0.0001 |
| sex | 1 | 0.001 | 0.009 | 0.009 | 1.44 | N.S. |
| maturity | 2 | 0.001 | 0.028 | 0.014 | 2.30 | N.S. |
| colour | 10 | 0.117 | 0.100 | 0.010 | 1.63 | N.S. |
| notes | 3 | 0.070 | 0.071 | 0.024 | 3.82 | N.S. |
| residual | 301 | 1.862 | 1.862 | 0.006 | | |
| Total | 330 | 92.120 | | | | |

is a good one for both species. When the LN (strength) predicted by the path coefficients is compared with the observed values, the residuals may be analysed using ANOVA and the results are given in Table 7.

The use of the path coefficients (effectively a multiple regression) accounts for a much higher proportion of the total variation (expressed as sums of squares) than does either transformed shell length or shell mass alone; 86% for *L. mariae* and 83% for *L. obtusata*. Comparing Tables 5 and 7 it is seen that the variation associated with all of the factors taken together is reduced by 82% for *L. mariae* and by 85% for *L. obtusata*. In this analysis there is not significant variation in shell strength of either species with algal zone; the variation between the different algal zones is thus

explained by the path coefficients. The dependence of strength on the shore from which the winkles were collected is also much reduced, although this does still remain significant. Differences in shell mass for a given size (or equivalently, in shell length and height for a given shell mass) can thus account for the strength differences seen with algal zone on the shore, and also for most of the differences in strength seen from different shores.

The use of the path coefficients also explains some of the apparently random variation; the residual sums of squares are reduced by 44% in *L. mariae* and 36% in *L. obtusata*, and this has the effect of raising some previously insignificant effects to statistical significance. In *L. mariae*, mature specimens are about 10% stronger

82

*Table 5.* Analysis of variation in shell strength of flat periwinkles using logarithmically transformed data with shell length as a covariate. The covariate was extracted first and then the remaining variation was partitioned, correcting for all other factors. The factor 'notes' includes parasitised, recent damage, or repaired damage scar as well as blank.

ANOVA table of LN (shell strength)

| Source | DF | Seq. SS | Adj. SS | MS | F | P |
|---|---|---|---|---|---|---|
| *Littorina mariae* | | | | | | |
| Covariate | | | | | | |
| LN (length) | 1 | 32.42 | | 32.43 | 675 | <0.0001 |
| Factors | | | | | | |
| shore | 7 | 4.51 | 2.57 | 0.37 | 7.45 | <0.0001 |
| zone | 2 | 2.02 | 2.12 | 1.11 | 22.48 | <0.0001 |
| sex | 1 | 0.67 | 0.70 | 0.70 | 14.15 | 0.0002 |
| maturity | 2 | 0.05 | 0.04 | 0.02 | 0.42 | N.S. |
| colour | 8 | 0.34 | 0.34 | 0.04 | 0.86 | N.S. |
| notes | 3 | 0.21 | 0.21 | 0.07 | 1.40 | N.S. |
| residual | 192 | 9.47 | 9.47 | 0.05 | | |
| Total | 216 | 49.67 | | | | |
| *Littorina obtusata* | | | | | | |
| Covariate | | | | | | |
| LN (length) | 1 | 51.76 | | 51.76 | 804 | <0.0001 |
| Factors | | | | | | |
| shore | 8 | 8.18 | 4.51 | 0.56 | 8.78 | <0.0001 |
| zone | 4 | 12.09 | 10.16 | 2.54 | 39.49 | <0.0001 |
| sex | 1 | 0.23 | 0.10 | 0.10 | 1.62 | N.S. |
| maturity | 2 | 0.18 | 0.21 | 0.10 | 1.60 | N.S. |
| colour | 10 | 1.01 | 0.90 | 0.09 | 1.40 | N.S. |
| notes | 3 | 0.40 | 0.40 | 0.13 | 2.05 | N.S. |
| residual | 301 | 19.37 | 19.37 | 0.06 | | |
| Total | 330 | 93.20 | | | | |

than immature specimens of the same shell mass and length; the uncommon colour morphs for this species, brown, black, orange and olivacea seem slightly weaker than the common yellow and reticulate morphs. In *L. obtusata*, shells classified as showing evidence of repaired damage were crushed by forces on average 75% of those not showing evidence of repaired damage.

Path analysis has also been applied to pooled data of the two species and the species then entered as a factor. The resulting path diagram is very similar to those of Fig. 5 and accounts for 83% of the total sum of squares. The loading on shell strength is 2.866 and the loadings on shell length and height are −1.560 and −0.577 respectively. No significant difference remains between the species in this model.

## Discussion

The changes in morphometric coefficients seen with growth and maturity noted in an earlier study (Lowell *et al.*, 1994) are scarcely discernible here; this will be due to the extra variation included in the samples by the diversity of collection sites and this should be no cause for concern. The strong positive allometry of shell strength in both species is confirmed.

*Table 6.* Least Squares means for LN (shell length) of flat periwinkles after correction for variations in shell length about the mean for the species and for other factors. Shores are in order of decreasing exposure, shown in parentheses (lower numbers indicate stronger wave action).

| Shore | Exposure | *L. mariae* | *L. obtusata* |
|---|---|---|---|
| Great Castle Bay | (5) | 5.362 | 5.590 |
| Musselwick 2 | (5) | 5.754 | 6.002 |
| West Angle Bay 1 | (5) | 5.424 | 5.686 |
| Musselwick 3 | (5–6) | 5.710 | 5.746 |
| West Angle Bay 2 | (5–6) | 5.744 | 5.700 |
| Angle Point | (6) | 5.479 | 5.814 |
| Musselwick 4 | (6) | 5.586 | 5.856 |
| Musselwick 1 | (6–7) | 5.787 | 5.771 |
| Angle Bay | (7) | – | 5.326 |

| Algal zone | | *L. mariae* | *L. obtusata* |
|---|---|---|---|
| *F. spiralis* | | – | 5.224 |
| *A. nodosum* | | – | 5.645 |
| *F. vesiculosus* | | 5.376 | 5.718 |
| mixed zone | | 5.702 | 5.938 |
| *F. serratus* | | 5.739 | 6.082 |

| Sex | | *L. mariae* | *L. obtusata* |
|---|---|---|---|
| Males | | 5.675 | 5.740 |
| Females | | 5.536 | 5.702 |

Local variation in various aspects of the morphology, biochemistry and genetics has now been described in most of the north Atlantic littorinids: *Littorina nigrolineata* (Naylor & Begon, 1982; Knight & Ward, 1986); *L. saxatilis* (Janson, 1987; Janson & Ward, 1984; Brandwood, 1985; Knight & Ward, 1986; Johannesson, 1986; Grahame & Mill, 1989, 1992; Grahame *et al.*, 1990; Sundberg *et al.*, 1990; Dytham *et al.*, 1990; Mill & Grahame, 1992); *L. mariae* (Goodwin & Fish, 1977; Reimchen, 1981); *L. obtusata* (Sacchi & Rastelli, 1966; Goodwin & Fish, 1977). Only one member of the genus *Littorina* in the north Atlantic has planktotrophic larvae, *L. littorea* (Reid, 1990b), and this species shows much less local genetic variation (reviewed by Ward, 1990) although phenotypic shell variation is seen in this species (Kemp & Bertness, 1984; Tresierra-Aquilar, 1985; Crothers, 1992) as it is in *Nodilittorina unifasciata* which also has a planktonic larva (McMahon, 1992). In *L. littorea* there is evidence that phenotypic shell plasticity is related to growth rate; it is as though shell deposition is less sen-

sitive to food availability than is body growth (Kemp & Bertness, 1984; Boulding & Hay, 1993).

To this list of environmentally related variation is now added variation in the strength of the shells of both *L. mariae* and *L. obtusata*. There is clear evidence of strength varying from shore to shore. To what extent this is related to exposure to wave action is less certain because the Spearman rank correlations do include rather few shores and therefore a lot of pseudoreplication. In any case the exposure effect appears to operate in opposite directions in the two species, as does change of size with exposure (Goodwin & Fish, 1977). An additional problem is accurate identification of exposure. Angle Point presented particular difficulty; in physical aspect it is very sheltered but it is scoured by a strong tide and this affects the fauna and flora. Interestingly, *L. mariae* also shows marked genetic differentiation between exposed and sheltered shores (A. Tatarenkov, personal communication).

Crabs are more common in the littoral region on sheltered shores (Ash, 1989). Selection by crab predation could be the reason for the changes exhibited by *L. mariae*, which become stronger on sheltered shores, but it cannot explain the opposite changes seen in *L. obtusata*. However, the latter species lives further up the shore where crabs are present only for a short period around high tide, and they grow into a 'size refuge' where they appear to be too big for the common littoral crab on British shores, *Carcinus maenas*, to handle successfully (Ash, 1989; Williams, 1992). As the larger of the two species of flat periwinkle, the risk of *L. obtusata* being dislodged from the weed by wave action will be higher, and a thinner shell which will accommodate a proportionately larger body and foot may be selected. Alternatively these variations could be an effect of growth rate on shell strength rather than of selection. *L. obtusata* is smaller on more exposed shores (Goodwin & Fish, 1977; Reimchen, 1982), which suggests that exposure slows down the growth rate. This would lead to a thicker and stronger shell if the rate of deposition of shell material was less affected than growth of the body, as in *L. littorea* (Kemp & Bertness, 1984; Boulding & Hay, 1993). *L. mariae* is larger on more exposed shores (Goodwin & Fish, 1977; Reimchen, 1982), suggesting that it grows faster there. This would lead to a proportionately thinner shell if mineral deposition was a limiting factor.

Variation with tidal level is less commonly recorded but there are a number of records for littorinids. The planktotrophic *L. littorea* shows marked variation in shell thickness with tidal level on the shore (Tresierra-

*Table 7.* Analysis of variation of shell strength of flat periwinkles after correction for shell length, height and mass using loadings deduced from path analysis. The remaining variation was partitioned, correcting for all other factors. The factor 'notes' includes parasitised, recent damage, or repaired scar as well as blank.

ANOVA table of LN (shell strength)

| Source | DF | Seq. SS | Adj. SS | MS | $F$ | $P$ |
|---|---|---|---|---|---|---|
| *Littorina mariae* | | | | | | |
| Covariates | | | | | | |
| (path) | 3 | 40.44 | | 13.48 | 484 | <0.0001 |
| Factors | | | | | | |
| shore | 7 | 0.731 | 0.506 | 0.072 | 2.65 | 0.012 |
| zone | 2 | 0.021 | 0.040 | 0.020 | 0.74 | N.S. |
| sex | 1 | 0.151 | 0.097 | 0.012 | 0.44 | N.S. |
| maturity | 2 | 0.206 | 0.181 | 0.181 | 6.62 | 0.011* |
| colour | 8 | 0.140 | 0.184 | 0.185 | 6.78 | <0.001* |
| notes | 3 | 0.127 | 0.127 | 0.042 | 1.56 | N.S. |
| residual | 190 | 5.290 | 5.290 | 0.028 | | |
| Total | 215 | 47.11 | | | | |
| *Littorina obtusata* | | | | | | |
| Covariates | | | | | | |
| (path) | 3 | 77.40 | | 25.80 | 623 | <0.0001 |
| Factors | | | | | | |
| shore | 8 | 0.574 | 0.671 | 0.084 | 2.03 | 0.042 |
| zone | 4 | 1.733 | 1.749 | 0.437 | 10.60 | N.S. |
| sex | 1 | 0.463 | 0.406 | 0.041 | 0.98 | N.S. |
| maturity | 2 | 0.106 | 0.103 | 0.103 | 2.50 | N.S. |
| colour | 10 | 0.006 | 0.000 | 0.000 | 0.00 | N.S. |
| notes | 3 | 0.469 | 0.469 | 0.156 | 3.79 | 0.011* |
| residual | 300 | 12.423 | 12.423 | 0.041 | | |
| Total | 330 | 93.17 | | | | |

* Details of these effects are given in the text.

Aquilar, 1985). Rolan-Alvarez *et al.* (1994) record that *L. saxatilis* on the coast of Galicia is markedly polymorphic, with a banded form occupying the upper shore and a smooth and unbanded morph on the lower shore. Gene flow between these two morphs is restricted by zonation, with a narrow zone of hybridization, by assortitive mating and by sexual selection in females. The shells of the same species on a single rock slope at St. Ann's Head, Dyfed vary regularly with position on the slope (Grahame *et al.*, 1990). Biochemical differences with position on the shore are seen in this species and in *L. arcana* both at St. Ann's Head and on the Yorkshire coast (Hull *et al.*, 1994; Smith, 1995). *L. striata*, believed to be rather distantly related to the other north Atlantic members of the genus (Reid, 1989),

also shows genetic differentiation with tidal level (H. De Wolf, pers. com.).

*L. mariae* from the bottom of its range is on average 1.44 times as strong as it is from its upper limit on the mid-shore; *L. obtusata* is 2.31 times as strong from the bottom of its range as it is from the top (Table 6). Whether the variation in shell strength with tidal level in flat periwinkles is due to differential survival at different levels of the shore from a variable juvenile stock, or to growth patterns which depend on the location on the shore, or whether it is genetically determined is not known. They have been shown to move significant distances and to display homing behaviour (Williams, 1995), which would tend to disrupt both ecophenotypic and genetically based clines down the shore unless the level to which they homed varied according to their morphology or genotype, thereby stabilising the zonation. Whatever the mechanism, the adaptive advantage is likely to be the ability to withstand attacks by stronger crabs such as *Liocarcinus puber* and *Cancer pagurus* as well as larger specimens of *Carcinus maenas* which are encountered lower on the shore.

When corrected for shell length *L. mariae* is 65% stronger than *L. obtusata* (Table 4). The earlier study found a smaller difference between the shell strength of the two species than is described here (Lowell *et al.*, 1994). We chose in the earlier study to compare the species from a site where their distributions overlapped. By doing so we inadvertently chose to compare *L. mariae* from the top of their range where they were weakest with *L. obtusata* well down their range where they were comparatively strong, and as a result observed relatively little difference in strength between the species.

In the mangrove snail *Littoraria pallescens*, the strengths of the different colour morphs may differ in a way which could be related to other aspects of niche separation (Cook & Kenyon, 1993). Reimchen (1981) described differentiation in *Littorina mariae* populations which may, on some shores be found in two discrete forms: a smaller and often yellow one found in the sandy run-off channels and a larger form found on the rock platforms, which is usually the dark reticulate colour morph. This distribution was not observed in the collections used in this study. Furthermore there was no significant association of strength with shell colour in *L. obtusata*; only a slight association was present in *L. mariae*, significant only after correction for shell mass and size simultaneously.

The path analysis is interesting in that it reveals a useful way of predicting shell strength. The negative loadings on the linear dimensions were initially a surprise but are actually logical. Comparing two shells of the same mass, the longer or higher one will be thinner and therefore weaker. Such a combination of shell mass, shell length and shell height accounts for 86% of the variation in the strength of *L. mariae* compared with 65% using shell length alone, and the variation associated with the factors studied is reduced by 80%. Similar treatment accounts for 83% of the variation in the strength of *L. obtusata* compared with 56% using shell length alone, and the variation associated with the various factors is reduced by 85%. Thus the path models account for a large part of the variation in shell strength in terms of the three measured shell variables; mass, length and height with appropriate loadings.

The nearly spherical external shape of the flat periwinkles suggests that they might be modelled theoretically as brittle hollow spheroids. Theoretical attempts to predict the path coefficients for the crushing resistance of brittle hollow spheroids have not been completed, but the similarity of the coefficients in the two species does suggest that there should be a good theoretical basis for these. The similarity of the path coefficients also suggests that a common path model could be used to predict the strength of shells of both species jointly, and that is the case. The residuals from such a common description are almost as small as those for separate descriptions and there is no significant unexplained difference between the species. Thus shell mass together with shell size, as measured by shell length and height, combined in the appropriate way, provide a good explanation of shell strength variation in both species of flat periwinkles, at least around the localities studied here.

## References

Ash, V. B., 1989. Resistance to shell breaking in two intertidal snails. J. linn. Soc., Zool. 96: 167–184.

Ballantine, W. J., 1961. A biologically defined exposure scale for the comparative description of rocky shores. Field Studies 1: 1–19.

Bertness, M. D. & C. Cunningham, 1981. Crab shell-crushing predation and gastropod architectural defense. J. exp. mar. Biol. Ecol. 50: 213–230.

Blundon, J. A. & G. J. Vermeij, 1983. Effect of shell repair on shell strength in the gastropod *Littorina irrorata*. Mar. Biol. 76: 41–45.

Boulding, E. G. & T. K. Hay, 1993. Quantitative genetics of shell form of an intertidal snail: constraints on short-term response to selection. Evolution 47: 576–592.

Brandwood, A., 1985. The effects of environment upon shell construction and strength in the rough periwinkle *Littorina rudis* Maton (Mollusca: Gastropoda). J. Zool., Lond. 206: 551–566.

Calder, W. A., 1984. Size, function, and life history. Harvard University Press, Massachusetts.

Clarke, M. R. B., 1980. The reduced major axis of a bivariate sample. Biometrika 67: 441–6.

Cook, L. M. & G. Kenyon, 1993. Shell strength of colour morphs of the mangrove snail *Littoraria pallescens*. J. moll. Stud. 59: 29–34.

Cooke, A. H., 1985. Molluscs. In: Cambridge Natural History, Vol. III, Molluscs and Brachiopods. Macmillan, London.

Crothers, J. H., 1985. Dog-whelks: an introduction to the biology of *Nucella lapillus* (L.). Field Studies 6: 291–360.

Crothers, J. H., 1992. Shell size and shape variation in *Littorina littorea* (L.) from west Somerset. In: J. Grahame, P. J. Mill & D. G. Reid (eds), Proceedings of the 3rd International Symposium on Littorinid Biology. The Malacological Society of London, London: 91–97.

Currey, J. D., 1979. The effect of drying on the strength of mollusc shells. J. Zool., Lond. 188: 301–308.

86

Currey, J. D. & R. N. Hughes, 1982. Strength of the dogwhelk *Nucella lapillus* and the winkle *Littorina littorea* from different habitats. J. anim. Ecol. 51: 47–56.

Dytham, C., J. Grahame & P. J. Mill, 1990. Distribution, abundance and shell morphology of *Littorina saxatilis* (Olivi) and *Littorina arcana* Hannaford Ellis at Robin Hood's Bay, North Yorkshire. Hydrobiologia 193 (Dev. Hydrobiol. 56): 233–240.

Goodwin, B. J., 1978. The growth and breeding cycle of *Littorina obtusata* (Gastropoda: Prosobranchiata) from Cardigan Bay. J. moll. Stud. 44: 231–242.

Goodwin, B. J. & J. D. Fish, 1977. Inter- and intraspecific variation in *Littorina obtusata* and *L. mariae* (Gastropoda: Prosobranchia). J. moll. Stud. 43: 241–254.

Grahame, J. & P. J. Mill, 1989. Shell shape variation in *Littorina saxatilis* and *L. arcana*: a case of character displacement? J. mar. biol. Ass. U.K. 69: 837–855.

Grahame, J. & P. J. Mill, 1992. Local and regional variation in the shape of rough periwinkles in southern Britain. In: J. Grahame, P. J. Mill & D. G. Reid (eds), Proceedings of the 3rd International Symposium on Littorinid Biology. The Malacological Society of London, London: 99–106.

Grahame, J., P. J. Mill & A. C. Brown, 1990. Adaptive and non-adaptive variation in two species of rough periwinkle *(Littorina)* on British shores. Hydrobiologia 193 (Dev. Hydrobiol. 56): 223–231.

Hughes, R. N. & R. W. Elner, 1979. Tactics of a predator *Carcinus meanas* and morphological responses of the prey, *Nucella lapillus*. J. anim. Ecol. 48: 65–78.

Hull, S. L., P. J. Mill & J. Grahame, 1995. Aminotransferases in Littorina: is there a functional story? [Abstract]. Cahiers de Biologie Marine 35: 248–249. Hydrobiologia.

Janson, K., 1987. Allozyme and shell variation in two marine snails (Littorina, Prosobranchia) with different dispersal abilities. Biol. J. linn. Soc. 30: 245–256.

Janson, K. & R. D. Ward, 1984. Microgeographic variation in allozyme and shell characters in *Littorina saxatilis* Olivi (Prosobranchia: Littorinidae). Biol. J. linn. Soc. 22: 289–307.

Johannesson, B., 1986. Shell morphology of *Littorina saxatilis* Olivi: the relative importance of physical factors and predation. J. exp. mar. Biol. Ecol. 102: 183–195.

Kemp, P. & M. D. Bertness, 1984. Snail shape and growth rates: evidence for plastic shell allometry in *Littorina littorea*. Proc. Natn. Acad. Sci. USA 81: 811–813.

Kitching, J. A. & J. F. Ebling, 1967. Ecological studies at Lough Ine. Adv. ecol. Res. 4: 197–311.

Knight, A. J. & R. D. Ward, 1986. Purine nucleoside phosphorylase polymorphism in the genus *Littorina* (Prosobranchia: Mollusca). Biochem. Gen. 24: 405–413.

Lowell, R. B., 1985. Selection for increased safety factors of biological structures as environmental unpredictability increases. Science 228: 1009–1011.

Lowell, R. B., 1986. Crab prediction on limpets: predator behavior and defensive features of the shell morphology of the prey. Biol. Bull. 171: 577–596.

Lowell, R. B., 1987. Safety factors of tropical versus temperate limpet shells: multiple selection pressures on a single structure. Evolution 41: 638–650.

Lowell, R. B., C. R. Fletcher, J. Grahame & P. J. Mill, 1994. Ontogeny of shell morphology and shell strength of the marine snails *Littorina obtusata* and *Littorina mariae*: different defence strategies in a pair of sympatric, sibling species. J. Zool., Lond. 234: 149–164.

McMahon, R. F., 1992. Microgeographic variation in the shell morphometrics of *Nodilittorina unifasciata* from southwestern Australia in relation to wave exposure of shore. In: J. Grahame, P. J. Mill & D. G. Reid (eds), Proceedings of the 3rd International Symposium on Littorinid Biology. The Malacological Society of London, London: 107–117.

Mill, P. J. & J. Grahame, 1992. Clinal changes in esterase variability in *Littorina saxatilis* (Olivi) and *L. arcana* Hannaford Ellis in southern Britain. In: J. Grahame, P. J. Mill & D. G. Reid (eds), Proceedings of the 3rd International Symposium on Littorinid Biology. The Malacological Society of London, London: 31–37.

Naylor, R. & M. Begon, 1982. Variation within and between populations of *Littorina nigrolineata* Gray on Holy Island, Anglesey. J. Conch. 31: 17–30.

Nielsen, C., 1980. On the occurrence of the prosobranchs *Littorina neritoides L. mariae* and *L. obtusata* in Denmark. J. moll. Stud. 46: 312–316.

Norton, T. A., S. J. Hawkins, N. L. Manley, G. A. Williams & D. C. Watson, 1990. Scraping a living: a review of littorinid grazing. Hydrobiologia 193 (Dev. Hydrobiol. 56): 117–138.

Palmer, A. R., 1979. Fish predation and evolution of gastropod shell sculpture: experimental and geographic evidence. Evolution 33: 697–713.

Palmer, A. R., 1981. Do carbonate skeletons limit the rate of body growth? Nature, Lond. 292: 150–152.

Palmer, A. R., 1990. Effect of crab effluent and scent of damaged conspecifics on feeding, growth, and shell morphology of the Atlantic dogwhelk *Nucella lapillus* (L.). Hydrobiologia 193 (Dev. Hydrobiol. 56): 155–182.

Peters, R. H., 1983. The ecological implications of body size. Cambridge University Press, Cambridge, England.

Reid, D. G., 1989. The comparative morphology, phylogeny and evolution of the gastropod family Littorinidae. Phil. Trans. Soc., Lond. B 324: 1–110.

Reid, D. G., 1990a. Note on the discrimination of females of *Littorina mariae* Sacchi & Rastrelli and *L. obtusata* (Linnaeus). J. moll. Stud. 56: 113–114.

Reid, D. G., 1990b. A cladistic phylogeny of the genus *Littorina* (Gastropoda): implications for evolution of reproductive strategies and for classification. Hydrobiologia 193 (Dev. Hydrobiol. 56): 1–19.

Reimchen, T. E., 1979. Substratum heterogeneity, crypsis, and colour polymorphism in an intertidal snail *(Littorina mariae)*. Can. J. Zool. 57: 1070–1085.

Reimchen, T. E., 1981. Microgeographical variation in *Littorina mariae* Sacchi & Rastelli and a taxanomic consideration. J. Conch. 30: 341–350.

Reimchen, T. E., 1982. Shell size divergence in *Littorina mariae* and *L. obtusata* and predation by crabs. Can. J. Zoo. 60: 687–695.

Ricker, W. E., 1973. Linear regressions in fishery research. J. Fish. Res. Bd. Can. 30: 409–434.

Rolán-Alvarez, E., 1992. A method of breeding *Littorina obtusata* (L.) and *L. mariae* Sacchi & Rastelli: preliminary results. In: J. Grahame, P. J. Mill & D. G. Reid (eds), Proceedings of the 3rd International Symposium on Littorinid Biology. The Malacological Society of London, London: 163–180.

Rolán-Alvarez, E., K. Johannesson & A. Ekendahl, 1995. Frequency- and density-dependent sexual selection in natural populations of Galician *Littorina saxatilis* (Olivi). Hydrobiologia 309 (Dev. Hydrobiol. 111): 167–172.

Sacchi, C. F., 1969. Recherches sur L'écologie comparée de *Littorina obtusata* (L.) et de *Littorina mariae* Sacchi et Rastelli (Gastropoda, Prosobranchia) en Galice et en Bretagne. Investigación Pesquera 33: 381–414.

Sacchi, C. F. & M. Rastelli, 1966. *Littorina mariae*, nov. sp.: les différences morphologiques et écologiques entre 'nains' en 'normeaux' chez l''espece' *L. obtusata* (L.) (Gastr. Prosobr.) et leur significance adaptive et évolutive. Atti della Società Italiana de Scienze Naturali e del Museo Civico di Storia Naturale di Milano 105: 351–370.

Smith, D., P. J. Mill & J. Grahame, 1995. Environmentally induced variation in uric acid concentration and xanthine dehydrogenase activity in *Littorina saxatilis* (Olivi) and *L. arcana* Hannaford Ellis. Hydrobiologia 309 (Dev. Hydrobiol. 111): 111–116.

Sokal, R. R. & F. J. Rohlf, 1981. Biometry, 2nd ed. Freeman & Co., New York.

Sundberg, P., A. J. Knight, R. D. Ward & K. Johannesson, 1990. Estimating the phylogeny in mollusc *Littorina saxatilis* (Olivi) from enzyme data: methodological considerations. Hydrobiologia 193 (Dev. Hydrobiol. 56): 29–40.

Tresierra-Aquilar, A., 1985. Between site variation in shell and body measures of *Littorina littorea*. Biol. Bull. 169: 537–538.

Vermeij, G. J., 1978. Biogeography and Adaptation. Harvard University Press, Cambridge.

Vermeij, G. J. & J. D. Currey, 1980. Geographical variation in the strength of thaiid snail shells. Biol. Bull. 158: 383–389.

Ward, R. D., 1990. Biochemical genetic variation in the genus *Littorina* (Prosobranchia: Mollusca). Hydrobiologia 193 (Dev. Hydrobiol. 56): 53–69.

Warmoes, T., T. Backeljau & L. de Bruyn, 1988. The littorinid fauna of the Belgian coast (Mollusca, Gastropoda). Bull. Inst. r. Sci. nat. Belg. 58: 51–70.

Watson, D. C. & T. A. Norton, 1987. The habitat and feeding preferences of *Littorina obtusata* (L.) and *L. mariae* Sacchi et Rastelli. J. exp. mar. Biol. Ecol. 112: 61–72.

Williams, G. A., 1990a. *Littorina mariae* – a factor structuring low shore communities? Hydrobiologia 193 (Dev. Hydrobiol. 56): 139–146.

Williams, G. A., 1990b. The comparative ecology of the flat periwinkles, *Littorina obtusata* (L.) and *L. mariae* Sacchi et Rastelli. Field Studies 7: 469–482.

Williams, G. A., 1992. The effect of predation on the life histories of *Littorina obtusata* and *L. mariae*. J. mar. biol. Ass. U.K. 72: 403–416.

Williams, G. A., 1995. Maintenance of zonation patterns in two species of flat periwinkle, *Littorina obtusata* and *L. mariae*. Hydrobiologia 309 (Dev. Hydrobiol. 111): 143–150.

Zipser, E. & G. J. Vermeij, 1978. Crushing behaviour of tropical and temperate crabs. J. exp. mar. Biol. Ecol. 31: 155–172.

*Hydrobiologia* **309**: 89–100, 1995.
*P. J. Mill & C. D. McQuaid (eds), Advances in Littorinid Biology.*
©1995 *Kluwer Academic Publishers.*

# Lack of metabolic temperature compensation in the intertidal gastropods, *Littorina saxatilis* (Olivi) and *L. obtusata* (L.)

Robert F. McMahon[1,2], W. D. Russell-Hunter[1,3,5] & David W. Aldridge[1,4]

[1]*Marine Biological Laboratory, Woods Hole, Massachusetts 02543, USA*
[2]*Section of Comparative Physiology, Department of Biology, Box 19498, The University of Texas at Arlington, Arlington, Texas 76019-0498, USA (Reprint request address)*
[3]*Department of Biology, Syracuse University, Syracuse, New York 13210, USA*
[4]*Department of Biology, North Carolina Agricultural and Technical State University, Greensboro, North Carolina 27411, USA*
[5]*Present address: 711 Howard Street, Easton, Maryland 21601-3934, USA*

*Key words:* acclimation, *Littorina obtusata*, *Littorina saxatilis*, Littorinidae, oxygen consumption, respiration, temperature effects

## Abstract

Two intertidal snails, *Littorina saxatilis* (Olivi, 1792) (upper eulittoral fringe/maritime zone) and *Littorina obtusata* (Linnaeus, 1758) (lower eulittoral) were collected from a boulder shore on Nobska Point, Cape Cod, Massachusetts, in July and acclimated for 15–20 days at 4 ° or 21 °C. Oxygen consumption rate ($\dot{V}o_2$) was determined for 11–15 subsamples of individuals at 4 °, 11 ° and 21 °C with silver/platinum oxygen electrodes. Multiple factor analysis of variance (MFANOVA) of $log_{10}$ transformed values of whole animal $\dot{V}o_2$ with $log_{10}$ dry tissue weight (DTW) as a covariant revealed that increased test temperature induced a significant increase in $\dot{V}o_2$ in both species ($P<0.00001$). In contrast, MFANOVA revealed that temperature acclimation did not affect $\dot{V}o_2$ in either *L. saxatilis* ($P = 0.35$) or *L. obtusata* ($P = 0.095$). Thus, neither species displayed a capacity for the typical metabolic temperature compensation marked by an increase in $\dot{V}o_2$ at any one test temperature in individuals acclimated to a lower temperature that is characteristic of most ectothermic animals. Lack of capacity for metabolic temperature acclimation has also been reported in other littorinid snail species, and may be characteristic of the group as a whole. Lack of capacity for respiratory temperature acclimation in these two species and other littorinids may reflect the extensive semi-diurnal temperature variation that they are exposed to in their eulittoral and eulittoral fringe/maritime zone habitats. In these habitats, any metabolic benefits derived from longer-term temperature compensation of metabolic rates are negated by extreme daily temperature fluctuations. Instead, littorinid species appear to have evolved mechanisms for immediate metabolic regulation which, in *L. saxatilis* and *L. obtusata* and other littorinids, appear to centre on a unique ability for near instantaneous suppression of metabolic rate and entrance into short-term metabolic diapause at temperatures above 20–35 °C, making typical seasonal respiratory compensation mechanisms characteristic of most ectotherms of little adaptive value to littorinid species.

## Introduction

Rates of metabolic processes in ectothermic animals vary with environmental temperatures but, in many cases, show compensatory adjustment. Metabolic rate functions, such as oxygen uptake rate, can be compensated to variations in temperature regime over three distinct time scales: (a) directly, within minutes or hours, (b) by compensatory acclimation over days or weeks and (c) by natural selection over many generations. For a general account of temperature compensation in ectothermic animals see Prosser & Heath (1991), and specifically for marine intertidal species see Newell (1979).

In this paper, we report the effects of acute ambient temperature variation and temperature acclimation on the rates of oxygen consumption in two common Atlantic intertidal littorinid snails: *Littorina saxatilis* (Olivi) occupying the upper eulittoral fringe/maritime zone of rocky North Atlantic shores at or above the mean high water neap tide mark (MHWNT), and *Littorina obtusata* (L.) occupying the lower eulittoral from the mean low water spring tide mark (MLWST) to just below the mean tide mark (MT).

Earlier studies have indicated that the common rocky shore, mid-eulittoral, North Atlantic littorinid snail, *Littorina littorea* (L.), has a well developed capacity for acute thermoregulation of metabolic rate (McMahon & Russell-Hunter, 1977) but little capacity for temperature acclimation of active metabolic rates (Newell & Roy, 1973). Within the intertidal zone of Northern Atlantic rocky shores, *L. saxatilis* occurs above the mid-eulittoral habitat of *L. littorea* with a slight overlap at MHWNT, while *L. obtusata* occurs below *L. littorea* with a more considerable overlap down to MLWNT. *L. littorea* is generally restricted to rocky substrata while *L. obtusata* inhabits the fronds of sympatric, low-shore, macrophytic, fucoid algae such as *Fucus* and *Ascophyllum* (Newell, 1979; McMahon & Russell-Hunter, 1977). Environmental temperature and tidal emersion conditions differ between the intertidal zones of these three species and appear to be associated with differences in their patterns of respiratory response to acute temperature variation (McMahon & Russell-Hunter, 1977). Also differing are their respiratory responses to progressive hypoxia and their capacities to compensate oxygen consumption rates to prolonged hypoxia (McMahon & Russell-Hunter, 1978).

In order to further elucidate the respiratory adaptations of Northern Atlantic littorinid snails to acute and long-term temperature variation, we studied the oxygen uptake rates of specimens of *L. saxatilis* and *L. obtusata* in response to both temperature acclimation to either 4 ° or 21 °C and to test temperatures of 4 °, 11 ° and 21 °C. The respiratory responses of these two species were compared to available information for *L. littorea* and other littorinid and non-littorinid intertidal molluscs. The results are discussed in terms of both the adaptive significance of the unique capacity of littorinid gastropods to thermoregulate metabolic rates in response to acute temperature variation and how their evolution of short-term metabolic thermoregulation may have lead to a loss of capacity for long-term metabolic acclimation. A preliminary account of this study was published as an abstract by Russell-Hunter *et al.* (1980).

A companion paper by Aldridge *et al.* (1995) published in this volume, describes the effects of temperature acclimation and acute temperature variation on nitrogen metabolism in *L. saxatilis* and *L. obtusata*. It reports differential catabolism of protein and non-protein energy stores under the same temperature and acclimation conditions in which oxygen consumption rates were recorded for this paper.

## Materials and methods

Specimens of *L. obtusata* [shell length (SL) range = 4.0–12.5 mm] and of *L. saxatilis* (SL range = 4.1–9.4 mm) were collected from a steeply sloping boulder beach on 15 July at the tip of Nobska Point on the southern shore of Cape Cod near Woods Hole, Massachusetts (41 ° 39.4′ N: 70 ° 39.3′ W). Specimens of *L. obtusata* were taken from the macrophytic algae, *Ascophyllum* and *Fucus*, growing on boulders in the lower littoral zone, while specimens of *L. saxatilis* were obtained directly from boulder substratum in the eulittoral fringe.

Immediately after collection, the snails were subdivided equally into glass culture dishes (20 cm diameter, seawater to a depth of 4 cm) and held in constant temperature rooms at 4 ° or 21 °C (± 0.5 °C). Specimens of *L. obtusata* were held at these temperatures for 15–16 days and of *L. saxatilis*, for 19–20 days. Seawater was replaced every two days. No mortality occurred in either species under either temperature over the acclimation period.

After temperature acclimation, the aquatic oxygen consumption rates ($\dot{V}_{O_2} = \mu l\ O_2\ animal^{-1}\ h^{-1}$) of 13–15 subsamples of each species acclimation group were determined. Subsamples consisted of one to four individuals depending on size, measured as the greatest distance from the tip of the spire to the anterior margin of the aperture (SL). SL was measured to the nearest 0.1 mm with dial calipers. Individuals in subsamples were selected to fall within ± 2.5 mm of a specific modal sample SL. Modal subsample shell lengths were representative of the SL range of adult field populations.

The $\dot{V}_{O_2}$ of each subsample was monitored with silver-platinum, polarographic oxygen electrodes (Yellow Springs Instrument Company, Model 53). Respiration chambers were held at constant experimental temperatures ($\pm 0.1$ °C) with a refrigerated constant temperature circulator. The glass respiration chambers (6.9 cm high $\times$ 2.04 cm internal diameter) of the YSI Model 53 oxygen electrodes were modified for determination of gastropod $\dot{V}o_2$ by fitting them with an internal plastic snap ring anchored by expansion against the chamber walls approximately 1 cm above the chamber floor. Subsample specimens were placed on the chamber floor and allowed to pedally attach. The plastic snap ring held a magnetic stirring disk above them during $\dot{V}o_2$ determinations. Under these conditions, specimens of both species displayed normal behavior, crawling on the walls and floor of the chamber throughout $\dot{V}o_2$ monitoring periods.

To determine $\dot{V}o_2$, each experimental subsample was initially placed in a chamber with 4 ml of MBL (Marine Biological Laboratory) artificial seawater (salinity = 31‰, Cavanaugh, 1956) at their respective acclimation temperature (4 ° or 21 °C). Natural seawater salinity at Nobska Point was approximately 31‰ during the July collection period. For 4 °C acclimated subsamples, $\dot{V}o_2$ was monitored at 4 °C or chamber temperature was slowly elevated ($\approx 0.5$ °C min$^{-1}$) to either 11 °C or 21 °C and $\dot{V}o_2$ determined. For 21 °C acclimated subsamples, $\dot{V}o_2$ was determined at 21 °C or chamber temperature was slowly lowered ($\approx 0.5$ °C min$^{-1}$) to either 11 °C or 4 °C and $\dot{V}o_2$ determined. Preceding each determination of $\dot{V}o_2$, the electrode was equilibrated in a blank chamber with 4 ml of seawater, but without snails. For any one determination, oxygen consumption was monitored continuously from full air $O_2$ saturation ($Po_2 \approx 159$ torr) until chamber $O_2$ concentration declined to at least 90% of full air $O_2$ saturation or for at least 30 min if a 10% reduction in $O_2$ concentration was not achieved within that period.

After each $\dot{V}o_2$ determination, the shells of all individuals in the subsample were dissolved in 7% nitric acid (<6 h). Repeated tests in the authors' laboratories with many different gastropod species have shown that shell dissolution in 7% nitric acid does not significantly affect dry tissue weight. Remaining tissues were rinsed twice ($\approx 10$ min each) in distilled water and dried to constant weight (DWT) at 95 °C (>24 h).

The mean whole animal $\dot{V}o_2$ of each subsample was computed as $\mu$l $O_2$ (ind$^{-1}$ h$^{-1}$). The log$_{10}$ of mean whole animal $\dot{V}o_2$ values for subsamples of each acclimation group of each species were then fitted to least squares linear regressions versus the log$_{10}$ of subsample mean DTW as the independent variable at each experimental temperature. The weight specific $\dot{V}o_2$ ($\mu$l $O_2$ mg DTW$^{-1}$ h$^{-1}$) of each subsample was computed by dividing its whole animal $\dot{V}o_2$ by the mean subsample DTW. Least squares linear regressions relating weight specific $\dot{V}o_2$ as the dependent variable to individual DTW were computed for subsamples of each species at each acclimation temperature/test temperature combination.

## Results

With the exception of that for 4 °C acclimated specimens of *L. saxatilis* at a test temperature of 4 °C ($r = 0.30$, $P > 0.2$), the slopes of the least squares linear regressions of the log$_{10}$ whole animal $\dot{V}o_2$ versus log$_{10}$ subsample mean DTW for each species acclimation group were significantly different from zero ($P < 0.05$) at all test temperatures (Table 1). The slopes of these regressions ranged from 0.65 to 1.25 and 0.37 to 0.74 in 21 ° and 4 °C acclimated specimens, respectively (Table 1).

Multiple factor analysis of variance (MFANOVA) with subsample mean DTW as a covariate indicated no significant differences ($P > 0.05$) between the whole animal $\dot{V}o_2$ of 4 ° and 21 °C acclimation groups of either species. In contrast, increasing test temperature over 4 °C to 21 °C was associated with a significant increase in whole animal $\dot{V}o_2$ in both species ($P < 0.00001$) (Tables 2 and 3). Lack of difference in the whole animal $\dot{V}o_2$ of 4 ° and 21 °C acclimation groups suggested that temperature acclimation had no significant effect on acute respiratory response over the tested 4–21 °C temperature range in either species. The apparent lack of respiratory compensation after temperature acclimation in both species was further demonstrated by the high degree of overlap in individual respiration rates and continuity of regression lines for each species at each experimental temperature (Table 1, Fig. 1).

Weight specific $\dot{V}o_2$ ($\mu$l $O_2$ mg DTW$^{-1}$ h$^{-1}$) values were fitted as the dependent variable to a least squares linear regression versus mean subsample DTW in specimens of *L. saxatilis* and *L. obtusata* for each acclimation temperature/test temperature combination under which oxygen consumption rates were determined. Of the 12 regressions, the slopes of only three were found to be significantly different from zero ($P < 0.05$) including: specimens of *L. obtusata* acclimated to 4 °C and tested at 11 ° and 21 °C and acclimated to 21 °C

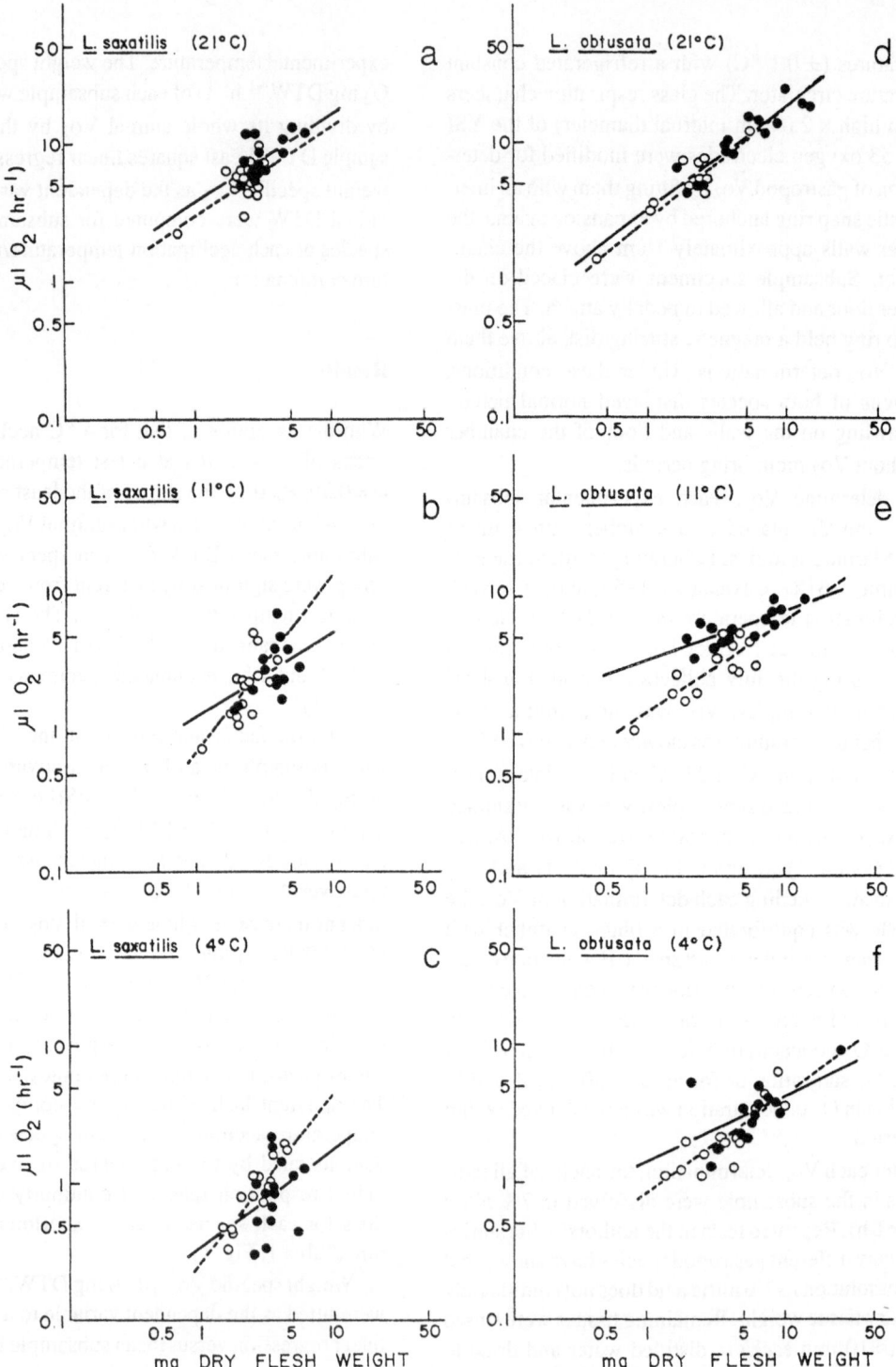

*Fig. 1.* Whole animal oxygen consumption rate in relation to dry tissue weight for *Littorina saxatilis* and *Littorina obtusata* acclimated to 4 °C (solid circles) or 21 °C (open circles). Axes in all figures are $log_{10}$ scale. Horizontal axes are mg dry tissue weight and vertical axes are whole animal oxygen consumption rate or $\dot{V}o_2$ in $\mu l\ O_2\ h^{-1}$. Solid and dashed lines represent the best fit of least squares linear regressions relating the $log_{10}\ \dot{V}o_2$ as the dependent variable to dry tissue weight in 4 °C and 21 °C acclimated individuals, respectively. (a–c) whole animal $\dot{V}o_2$ in *L. saxatilis* at test temperatures of 21 °, 11 ° and 4 °C, respectively. (d–f) whole animal $\dot{V}o_2$ in *L. obtusata* at test temperatures of 21 °, 11 ° and 4 °C, respectively.

*Table 1.* Parameters for least squares linear regressions relating $\log_{10}$ whole animal oxygen consumption rate ($\dot{V}o_2$)(dependent variable, $\mu l\ O_2\ mg^{-1}\ h^{-1}$) to $\log_{10}$ dry tissue weight (independent variable, mg) for specimens of *Littorina saxatilis* and *Littorina obtusata*. P, significance of regression slope.

| Acclimation temperature | Test temperature | $n$ | Intercept | Slope | $r$ | $P$ |
|---|---|---|---|---|---|---|
| | | | *Littorina saxatilis* | | | |
| 4°C | 4°C | 15 | −0.463 | 0.705 | 0.30 | >0.2 |
| 4°C | 11°C | 15 | 0.128 | 0.597 | 0.53 | <0.05* |
| 4°C | 21°C | 15 | 0.599 | 0.619 | 0.62 | <0.05* |
| 21°C | 4°C | 15 | −0.578 | 1.227 | 0.84 | <0.001* |
| 21°C | 11°C | 15 | −0.125 | 1.251 | 0.69 | <0.005* |
| 21°C | 21°C | 15 | 0.493 | 0.649 | 0.66 | <0.01* |
| | | | *Littorina obtusata* | | | |
| 4°C | 4°C | 14 | 0.159 | 0.475 | 0.59 | <0.05* |
| 4°C | 11°C | 15 | 0.543 | 0.368 | 0.78 | <0.001* |
| 4°C | 21°C | 15 | 0.513 | 0.743 | 0.92 | <0.001* |
| 21°C | 4°C | 14 | −0.074 | 0.711 | 0.84 | <0.001* |
| 21°C | 11°C | 13 | 0.168 | 0.668 | 0.82 | <0.001* |
| 21°C | 21°C | 11 | 0.480 | 0.776 | 0.91 | <0.001* |

*Table 2.* Multiple factor analysis of variance table of the effects of acclimation temperature and test temperature on $\log_{10}$ whole animal oxygen consumption rates ($\dot{V}o_2$) with $\log_{10}$ dry tissue weight as a covariate in specimens of *Littorina saxatilis*.

| Source of variation | Sum of squares | Degrees of freedom | Mean square | F-ratio | Significance level |
|---|---|---|---|---|---|
| Dry tissue weight (Covariate) | 3.13 | 1 | 3.13 | 13.07 | 0.0005* |
| Acclimation temperature | 0.23 | 1 | 0.23 | 0.94 | 0.35 |
| Test temperature | 75.88 | 2 | 37.94 | 158.21 | <0.00001* |
| Acclimation by test Temperature interaction | 0.40 | 2 | 0.20 | 0.84 | 0.44 |
| Residual | 19.90 | 83 | 0.24 | | |
| TOTAL | 99.54 | 89 | | | |

and tested at 4 °C. For samples with nonsignificant regressions, mean weight specific $\dot{V}o_2$, standard deviations and standard errors were computed (Table 4 and Fig. 2). For the three acclimation temperature/test temperature combinations in which significant correlations between weight specific $\dot{V}o_2$ and subsample mean DTW were recorded in *L. obtusata*, weight specific $\dot{V}o_2$, standard errors and 95% confidence limits were estimated from the regressions of $\dot{V}o_2$ ver-

sus DTW for a standard individual whose DTW was equivalent to the mean DTW of all test specimens. The mean individual subsample DTW of all specimens of *L. saxatilis* utilized in $\dot{V}o_2$ determinations was 3.02 mg (s.d. = ± 1.07, $n=90$), while that for *L. obtusata* was 4.94 mg (s.d. = ± 3.83, $n=82$). Mean weight specific $\dot{V}o_2$ for specimens of *L. saxatilis* and of *L. obtusata* acclimated to 4 °C and 21 °C appeared to be nearly identical, with standard errors of the means of 4 ° and

*Table 3.* Multiple factor analysis of variance table for the effects of acclimation temperature and test temperature on $\log_{10}$ whole animal oxygen consumption rate ($\dot{V}o_2$) with $\log_{10}$ dry tissue weight as a covariate in specimens of *Littorina obtusata*.

| Source of variation | Sum of squares | Degrees of freedom | Mean square | F-ratio | Significance level |
|---|---|---|---|---|---|
| Dry tissue weight (Covariate) | 10.11 | 1 | 10.11 | 41.24 | <0.00001* |
| Acclimation temperature | 0.70 | 1 | 0.70 | 2.86 | 0.095 |
| Test temperature | 38.04 | 2 | 19.02 | 77.58 | <0.00001 * |
| Acclimation by test Temperature interaction | 1.60 | 2 | 0.80 | 3.26 | 0.44 |
| Residual | 18.39 | 75 | 0.25 | | |
| TOTAL | 68.84 | 81 | | | |

*Table 4.* Mean weight specific oxygen uptake rates ($\mu l\ O_2\ mg^{-1}\ h^{-1}$) and $Q_{10}$ values for individuals of *Littorina saxatilis* and *Littorina obtusata* acclimated to 4°C or 21°C, determined at test temperatures of 4°, 11° and 21°C.

| Species | Accl. Temp. | $\mu l\ O_2\ mg^{-1}\ h^{-1} \pm$ s.d. | | | $Q_{10}$ 4°–11°C | $Q_{10}$ 11°–21°C |
|---|---|---|---|---|---|---|
| | | 4°C | 11°C | 21°C | | |
| *Littorina* | 4°C | 0.28±0.16 | 0.85±0.25 | 2.61±0.88 | 4.89 | 3.07 |
| *saxatilis* | 21°C | 0.34±0.10 | 1.00±0.46 | 2.45±0.64 | 4.66 | 2.45 |
| *Littorina* | 4°C | 0.67±0.57 | *1.52±0.78 | *2.24±1.18 | 3.22 | 1.47 |
| *obtusata* | 21°C | 0.61±0.21 | *0.88±0.48 | 2.71±0.74 | 1.69 | 3.08 |

* Means identified by an asterisk and associated standard deviations were estimated for individuals of standard dry tissue weight (*L. saxatilis* = 3.02 mg; *L. obtusata* = 4.94 mg) from least squares linear regressions of weight specific oxygen uptake rate versus dry tissue weight in which slopes were significantly different from zero (P < 0.05). In all other cases, slope values were insignificantly different from zero, allowing sample mean values to be reported.

21 °C acclimation groups either overlapping or lying close enough to each other to allow 95% confidence limits to overlap at all three test temperatures (Fig. 2). The high degree of similarity between the weight specific $\dot{V}o_2$-temperature curves of 4 ° and 21 °C acclimated individuals of both species was taken as another indication that neither species had the capacity for respiratory temperature compensation.

Mean weight specific $\dot{V}o_2$ values were utilized to calculate $Q_{10}$ values representative of the increase in $\dot{V}o_2$ with increase in experimental temperature [$Q_{10}$ is a standardized measure of physiological response to temperature increase measured as the factor by which $\dot{V}o_2$ would have increased over a 10 °C temperature range (Prosser & Heath, 1991)]. $Q_{10}$ values for 4 °C acclimated individuals of *L. saxatilis* were 4.89

between 4 °and 11 °C and 3.07 between 11 ° and 21 °C. $Q_{10}$ values for 21 °C acclimated individuals were 4.66 (4 °–11 °C) and 2.45 (11 °–21 °C). $Q_{10}$ values for 4 °C acclimated individuals of *L. obtusata* were 3.22 (4 °–11 °C) and 1.47 (11 °–21 °C) and for 21 °C acclimated individuals, were 1.69 (4 °–11 °C) and 3.08 (11 °–21 °C) (Table 4). The relative similarity of the $Q_{10}$ values in 4 °C and 21 °C acclimated groups of both species also suggested that temperature acclimation had little effect on respiratory response to immediate temperature change in either *L. saxatilis* or *L. obtusata*.

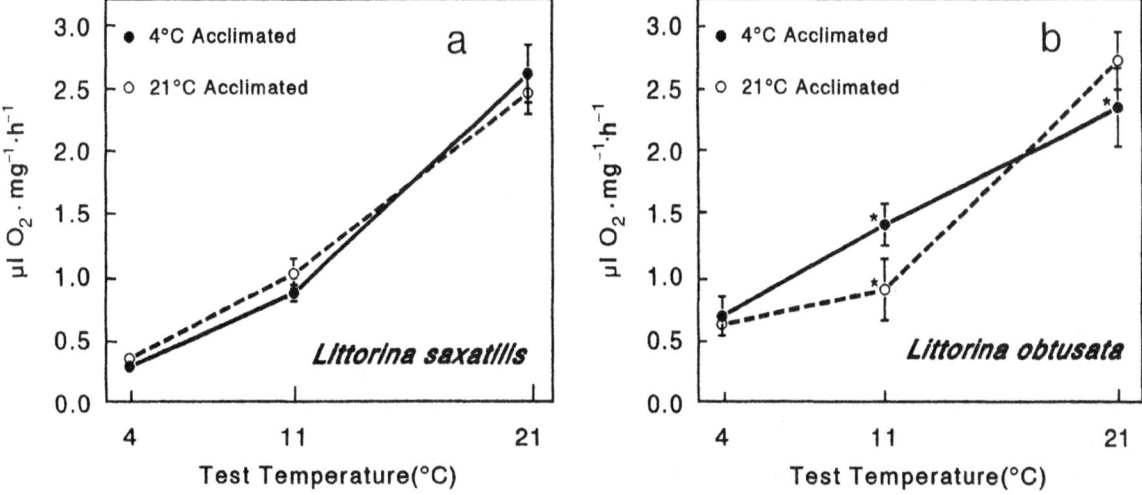

*Fig. 2.* Mean weight specific oxygen consumption rates of 4 °C and 21 °C acclimated individuals of (a) *Littorina saxatilis* and (b) *Littorina obtusata* in relation to test and acclimation temperatures. In both figures, horizontal axes are test temperature (4 °, 11 ° or 21 °C) and vertical axes are mean weight specific, per milligram dry tissue weight, oxygen consumption rates ($\dot{V}_{O_2}$) for subsamples of each species tested at each acclimation temperature/test temperature combination. Solid circles connected by solid lines are mean weight specific $\dot{V}_{O_2}$ values in $\mu l$ $O_2$ mg $DTW^{-1}$ $h^{-1}$) for 4 °C acclimated individuals and open circles connected by dashed lines, for 21 °C acclimated individuals. Vertical bars about the means are standard errors of the means. Means identified by an asterisk (∗) and associated standard errors were estimated for individuals of *L. obtusata* of standard dry tissue weight (= 4.94 mg) from least squares linear regressions of weight specific oxygen uptake rate versus dry tissue weight in which slopes were significantly different from zero (P<0.05). For all other means, slopes were not significantly different from zero.

## Discussion

*L. obtusata* and *L. saxatilis* occupy quite different vertical positions on the shore. *L. obtusata* is restricted to the lower eulittoral, between the mean low water spring tide and mean tide marks. Here, it is submerged with each tidal cycle and exposed to air only 15–45% of the time. In contrast, *L. saxatilis* is restricted to upper shores above the MHWNT mark and extends into the upper eulittoral fringe and maritime zones well above the MHWST mark. In this environment, it may be emersed for prolonged periods, exceeding several days during neap tides, remaining in air for 70%–95% of the time. In North America, a third temperate Atlantic littorinid species, *L. littorea* (L.), occupies the mid-shore, overlapping the ranges of both *L. obtusata* and *L. saxatilis*. It extends between the MLWNT and MHWNT marks in the eulittoral zone where it is submerged with each tidal cycle and emersed for 30–75% of the time (McMahon, 1988; McMahon & Russell-Hunter, 1977; Newell, 1979).

When emersed by receding tides on an exposed, rocky shore, absorption of radiant solar energy by gastropod shells and tissues and conduction of heat from the insolated rock surface through the attached foot can rapidly warm snail body temperatures well above ambient water or air temperatures (Lewis, 1963; Markel, 1971; McMahon, 1988, 1990; Russell-Hunter & Apley, 1965; Vermeij, 1971a). Because increasing height of zonation is associated with increasing duration of emersion and, therefore, increased relative exposure to insolation, it has been assumed that high shore, intertidal species are subject to greater levels of short-term, thermal stress than are low shore species (Newell, 1979; Underwood, 1979; Vernberg & Vernberg, 1972). However, actual measurements of intertidal gastropod tissue and substratum temperatures over the course of emersion indicate that equivalent maximal tissue temperatures may be reached within 0.5 to 1 h of initial tidal emersion regardless of height on the shore (Grainger, 1969; Lewis, 1963; Markel, 1971; McMahon, 1990; Southward, 1958; Vermeij, 1971a, b). Thus, the range of acute temperature variation experienced by intertidal species during tidal emersion may be essentially similar at all levels in the eulittoral and upper eulittoral fringe/maritime zones, with only the duration of exposure to elevated tissue temperatures being correlated with height of habitation (for a review see McMahon, 1990).

Insolation during tidal emersion may elevate the tissue temperatures of intertidal gastropods and their exposed rocky substrata 15 °–25 °C above that of ambient sea water (Grainger, 1969; Lewis, 1963; Markel, 1971; McMahon, 1990; Southward, 1958; Vermeij, 1971a; Wolcott, 1973). Thus, the acute temperature variation experienced by intertidal species with each tidal emersion/submersion cycle can approach or even be greater than the total annual variation in mean daily ambient sea water or air temperature irrespective of a species' vertical range within the intertidal zone (McMahon, 1990).

Typical metabolic temperature compensation (i.e., acclimation) for seasonal ambient temperature variation (i.e., type 1 full compensation or type 2 partial compensation curves, Precht et al., 1973) is marked by cold-acclimated individuals having higher metabolic rates over a wide ambient temperature range relative to those of warm-acclimated individuals. It is viewed as a mechanism whereby ectothermic species can maintain relatively stable, efficient metabolic and activity rates over a wide range of annual, ambient, temperature variation. Laboratory studies indicate that such adjustment of metabolic rates (usually measured as oxygen consumption rates) is usually completely achieved after 7–21 days exposure to a new ambient temperature regime and is generally associated with a parallel shift (i.e., translation) and/or rotation of the $\dot{V}o_2$-temperature curves of cold and warm acclimated individuals relative to each other, such that cold acclimated groups have higher metabolic rates or $\dot{V}o_2$ and/or different $Q_{10}$ values than do warm acclimated individuals.

By inducing a general increase in metabolic rate at any one temperature in cold acclimated individuals and a general decrease in metabolic rate at any one temperature in warm acclimated individuals, metabolic temperature acclimation greatly reduces the annual range of metabolic rate variation that would be experienced by an ectothermic animal. Annual variation in metabolic rate would be much greater in an individual incapable of metabolic temperature compensation and thus able to respond only acutely to temperature change such that metabolic rate would increase or decrease by a factor of 2–3 with each 10 °C increase or decrease in ambient temperature (Precht et al., 1973).

Respiratory temperature acclimation is associated at the cellular level with changes in enzyme and isozyme concentrations (Hochachka & Somero, 1984) and in the fluidity of cell and cellular organelle membranes which directly affects the activities of enzymes anchored in those membranes. This latter process has been termed 'homeoviscous adaptation.' In cold-acclimated individuals, membrane fluidity increases, in turn inducing an increase in the activity of membrane anchored enzymes and vice-versa (Cossins, 1981; Cossins et al., 1981; Cossins & Bowler, 1987).

It has been assumed that the majority of intertidal species would display such a typical pattern of seasonal temperature compensation (Precht et al., 1973, type 1 or 2). However, very few such studies have been carried out in intertidal species, and in particular for rocky shore molluscs (for reviews of temperature acclimation in intertidal species see Newell, 1979; Newell & Bayne, 1973; Vernberg & Vernberg, 1972). In nature, metabolic temperature acclimation requires exposure to relatively stable and predictable temperature regimes with limited short-term temperature variation. Relatively small daily temperature variation (i.e., daily, acute, temperature variation is considerably less than annual variation in mean daily temperature) allows acclimatory adjustment of metabolic rate to relatively efficient levels over a wide seasonal ambient temperature range. In contrast, in habitats such as the intertidal region, subject to extensive and unpredictable daily temperature variation (i.e., daily, acute, temperature variation can approach or be greater than mean annual daily water temperature variation), temperature acclimation would be of little adaptive advantage as acute temperature variation would be too great and/or too unpredictable to allow compensation of metabolic rate to an efficient level at any one ambient temperature. Furthermore, $Q_{10}$ effects could result in massive overshooting or undershooting of efficient metabolic rates if the level of metabolic compensation was not suitable for exposures to the daily temperature extremes experienced in highly temperature variable habitats (Cossins & Bowler, 1987).

Whereas the vast majority of terrestrial, freshwater and subtidal marine habitats have relatively stable temperature regimes, giving metabolic temperature compensation mechanisms adaptive value, intertidal habitats are clearly an exception. The semidiurnal temperature variations to which intertidal animals are exposed by tidal immersion/emersion cycles are often of greater extent than seasonal variation in mean daily ambient water temperatures. Thus, intertidal species are not exposed to sustained temperature regimes long enough to allow metabolic temperature acclimation processes to occur. This is particularly true of the rocky shore habitats of the Littorina species studied because exposure to the warming effects of insolation is almost immediate after emersion without the insulating or thermal buffering effects that result from burrowing in the soft sediments of intertidal beaches, estuaries and salt marshes (McMahon & Wilson, 1981). Indeed, as described above, the unpredictable thermal regimes of rocky intertidal habitats not only reduce the adaptive advantage of acclimatory mechanisms but, due to $Q_{10}$ effects, could actually result in negative selection pressure associated with extreme short-term variation in metabolic rates.

Thus, it is not surprising that a number of studies have revealed a relatively poor capacity for metabolic temperature acclimation among intertidal species compared to animals from freshwater or subtidal habitats. In our study, specimens of both *L. saxatilis* and *L. obtusata* held for long periods at either 4 ° or 21 °C showed no capacity for respiratory temperature acclimation at test temperatures of 4 °, 11 ° or 21 °C. Similarly, the intertidal sea anemones, *Actinia equina* L. and *Anemonia natalensis* Galg. (Griffiths, 1977a, b), the intertidal bivalves, *Mya arenaria* L., *Mulina lateralis* (Say) (Kennedy & Mihursky, 1972), *Macoma balthica* (L.) (McMahon & Wilson, 1981), *Ostrea edulis* L. (Newell *et al.*, 1977), *Crassostrea virginica* (Gmelin) (Shumway & Koehn, 1982) and *Mytilus edulis* L. (Bayne, *et al.*, 1973) all show little capacity for metabolic temperature compensation. Only relatively small and inconsistent differences were noted in the $\dot{V}o_2$ of specimens of the fiddler crab, *Uca pugilator* (Bosc), acclimated to 10 ° or 25 °C (Vernberg, 1969; Vernberg & Vernberg, 1972). Like *L. saxatilis* and *L. obtusata* in this study, 'active' specimens of the eulittoral littorinid, *L. littorea*, have little capacity for metabolic temperature compensation (Newell & Pye, 1970a; Newell & Roy, 1973). However, 'inactive specimens' of *L. littorea* did display a typical pattern of positive, seasonal, respiratory temperature compensation (Newell & Pye, 1970a; Newell & Roy, 1973).

As littorinid snails such as *L. saxatilis*, *L. littorea* and *L. obtusata* graze actively, both when submersed and when emersed on moist substratum (McMahon, 1990), the lack of acclimatory response in the 'active' metabolic rates of *L. littorea* to temperature acclimation could be much more adaptationally significant than are acclimatory differences in the respiratory temperature responses of quiescent, perhaps withdrawn individuals. Indeed, the significance of 'standard' or 'inactive' metabolic rates (Newell & Pye, 1970a, for a review see Newell, 1979) as measured in intertidal molluscs remains an open question (Cossins & Bowler, 1987; McMahon & Russell-Hunter, 1977) with one author (Boyden, 1972) suggesting that 'standard' oxy-

gen consumption rates are equivalent to rates measured when bivalves have closed valves or gastropods are withdrawn into their shells, relying heavily on anaerobic rather than aerobic metabolism.

Gastropods from more temperature stable habitats appear to have relatively good capacities for metabolic temperature compensation. Thus, the subtidal prosobranch, *Crepidula fornicata* (L.), which remains continually immersed and, therefore, not subject to wide tidal temperature variation, displays a typical pattern of metabolic acclimation (type 2, Precht *et al.*, 1973) which allows it to stabilize metabolic rates at efficient levels throughout the year (Newell & Kofoed, 1977). The maritime zone, salt marsh, pulmonate snail, *Melampus bidentatus* (Say), inhabits such high elevations in salt marshes that it is rarely tidally immersed and thus lives in a relatively temperature stable, near aerial environment (Apley, 1970; McMahon & Russell-Hunter, 1981). This species has a 'reverse' or 'inverse' pattern of metabolic temperature compensation (type 5, Precht, *et al.*, 1973) in which cold-acclimated specimens further depress metabolic rates beyond that induced by initial movement to colder conditions. Such a reverse pattern of metabolic temperature compensation is believed to allow this species to conserve organic energy stores during the long, nonfeeding, overwintering, period spent burrowed into hypoxic substrata, by greatly depressing aerobic catabolic demands (McMahon & Russell-Hunter, 1981).

It has been suggested that lack of capacity for metabolic temperature acclimation in intertidal molluscs such as *L. saxatilis* and *L. obtusata* may be compensated for by other adaptations that allow maintenance of relatively stable metabolic rates under acutely variable, short-term, ambient temperature fluctuations (McMahon, 1988, 1990, 1992; McMahon & Russell-Hunter, 1977; McMahon & Wilson, 1981). Many intertidal molluscs can maintain a near constant metabolic rate or $\dot{V}o_2$ ($Q_{10} \approx 1.0$) over a relatively wide range of acutely varying temperature. Such thermoregulation has been described for a number of species of intertidal littorinid snails including *L. littorea*, *L. saxatilis* and *L. obtusata* (McMahon & Russell-Hunter, 1977), *Nodilittorina interrupta* (C. B. Adams) (McMahon, 1992) and *Littoraria irrorata* (Say) (Shirley *et al.*, 1978; McMahon, 1992). It has also been reported in the intertidal pulmonate limpet, *Siphonaria pectinata* (L.) (McMahon, 1992) and the marine intertidal bivalves, *Cerastoderma edule* (L.) (McMahon & Wilson, 1981) and *M. edulis* (Newell & Pye, 1971). Such capacity for acute thermoregulation of metabolic rate does not appear to occur in subtidal gastropods (McMahon, 1992; McMahon & Russell-Hunter, 1977).

Capacity to regulate metabolic rates at near constant levels during acute ambient temperature fluctuation allows maintenance of those rates at efficient levels without the invocation of other, longer-term, acclimatory responses. Indeed, in such metabolically thermoregulating species, only the range of temperatures over which metabolic thermoregulatory control is exercised seems to change with season or temperature acclimation, while metabolic rate within the thermally regulated temperature range remains equivalent regardless of acclimation temperature (Newell & Pye, 1970a, b). However, in the vast majority of tested intertidal snail species, the temperature range in which temperature regulation of $\dot{V}o_2$ operates generally falls only within the upper portions of a species' normal ambient range. Typically, $\dot{V}o_2$ increases with increasing temperature in the lower and middle portions of the normal ambient temperature range until a temperature is reached at which metabolic rate does not greatly further increase with further increase in ambient temperature (McMahon & Russell-Hunter, 1977; McMahon, 1988, 1990, 1992).

Metabolic thermoregulation only in the upper portion of the ambient temperature range suggests that it is an adaptation primarily associated with prevention of energetically wasteful excessive elevation of metabolic rate when tissue temperatures rise well above ambient levels due to insolation during tidal emersion. Thus, the relatively high respiratory $Q_{10}$ values recorded for both *L. saxatilis* and *L. obtusata* (range = 1.47–4.89, see Table 4) over 4 °–11 °C and 11 °–21 °C indicate little, if any, capacity to thermally regulate $\dot{V}o_2$ with acute temperature variation over this tested temperature range. Similar lack of thermoregulation of $\dot{V}o_2$ at temperatures below 20 °C has been reported for *L. littorea* (McMahon & Russell-Hunter, 1977; Newell & Roy, 1973). However, all three species, and a number of other littorinid and nonlittorinid intertidal snails, regulate $\dot{V}o_2$ at temperatures above 20 °–25 °C, with many demonstrating a capacity to suppress metabolic rates and enter short-term aestivation above an upper threshold temperature (McMahon, 1988, 1990, 1992; McMahon & Russell-Hunter, 1977, 1981).

Over test temperatures of 4 °–11 °C and 11 °–21 °C, the $Q_{10}$ values of *L. obtusata* ranged from 1.47–3.22 (Table 4), corresponding to a passive metabolic response to increasing temperatures following simple chemical reaction kinetics (Prosser & Heath, 1991). In contrast, for *L. saxatilis*, $Q_{10}$ values were generally greater than 3.0 (range = 2.45–4.89, Table 4), suggesting that this species actively increases metabolic rate with increasing temperature over at least the tested temperature range. Similar rapid increases in $\dot{V}o_2$ over 5–20 °C, followed by metabolic thermoregulation at higher temperatures, have been recorded in eulittoral fringe/maritime littorinids including *L. saxatilis*, (McMahon & Russell-Hunter, 1977) and *L. irrorata* and *N. interrupta* (McMahon, 1992). The high $Q_{10}$ values for respiratory response to temperature over the lower end of the ambient temperature range suggest that these high shore littorinid species rapidly increase metabolic activity up to a specific, presumably efficient, level as temperatures warm to 20–25 °C and, thereafter, maintain metabolic rates at that level in spite of further increases in ambient temperature. Conversely, it would also allow rapid entry into a relatively inactive state as ambient temperatures cooled below 15 °C. Rapid increase in metabolic rate with increasing temperature would allow high shore littorinids (McMahon, 1990, 1992) to take advantage of the warm tissue temperatures generated by insolation during tidal emersion, such that emersed individuals could graze periphyton at relatively high rates during periods when their tidally emersed substratum surfaces remain moist and aquatic predators such as crabs are absent (for reviews see McMahon, 1988, 1990).

The ability of all tested littorinid species to regulate $\dot{V}o_2$ at temperatures exceeding normal maximal ambient water temperatures and to suppress metabolic rate and enter aestivation on exposure to extremely high temperatures prevents them from experiencing energetically wasteful elevated metabolic rates induced by tissue warming due to insolation during tidal emersion (for a review see McMahon, 1990). It is the ability of both *L. saxatilis* and *L. obtusata*, and indeed of all tested littorinids (McMahon, 1990; 1992; McMahon & Russell-Hunter, 1977), to actively regulate and even suppress metabolic rates at elevated environmental temperatures that has freed them from the requirement for longer-term, temperature acclimatory mechanisms for stabilization of metabolic rates. These species experience relatively small changes in metabolic rate over the lower part (i.e., 0 °–20 ° or 25 °C) of their ambient temperature ranges as it is only in this region that their metabolism is responsive to temperature. Above this range, $\dot{V}o_2$ is highly regulated even when ambient temperatures fluctuate widely and acutely. Thus, metabolic rate appears to be thermally regulated to remain at presumably efficient levels throughout the acute, transient excursions of elevated tissue temperatures that occur when tidally emersed individuals are subjected to insolation. This same mechanism also prevents massive changes in metabolic rate with seasonal, ambient, temperature variation. Indeed, within the regulated temperature range, seasonal variation in metabolic rates should be no greater than that experienced during the diurnal tidal cycle.

## Conclusions

Littorinid snails dominate the upper eulittoral fringe/maritime zones of rocky shores throughout the world's oceans (McMahon, 1990; Reid, 1989). Their success and adaptive radiation is almost certainly due, in part, to their seemly unique capacity to thermally regulate metabolic rates during acute temperature variation, thus avoiding energetically wasteful elevation of metabolic rates during periods of insolatory tissue warming on tidal emersion (McMahon, 1990, 1992). As we have shown for *L. saxatilis* and *L. obtusata* and may also be the case in *L. littorea* (Newell & Pye, 1970a, b), evolution of short-term thermoregulatory ability in littorinid snails has also apparently resulted in loss of capacity for the longer-term metabolic temperature compensation common in most other ectothermic animals, but of little adaptive value in acutely temperature variable intertidal habitats.

Members of the genus *Littorina* appear to have invaded Northern Atlantic shores via dispersal from the northern Pacific Ocean across an open Arctic corridor in the late Cenozoic (Reid, 1990). These littorinids may have invaded down-shore niches in the Northern Atlantic that were occupied by other mesogastropod groups in their endemic Indo-Pacific range. The capacity for metabolic thermoregulation with acute tidal temperature variation, associated with a loss of dependence on less efficient, longer-term, thermal metabolic compensation mechanisms, would clearly be of great adaptive advantage to both the high-shore, Atlantic, littorinid species such as *L. saxatilis* (this paper), *N. interrupta* and *L. irrorata* (McMahon, 1992) and their high-shore, Indo-Pacific ancestors. However, the characteristic high-shore, littorinid capacity for acute metabolic thermoregulation also seems to have been retained in *L. littorea* and *L. obtusata* as they evolved into new mid- and low eulittoral niches. Their retention of metabolic thermoregulatory abilities in

lower shore habitats should not be surprising because the rapid insolatory warming of tissues on tidal emersion can cause lower shore littorinid species such as *L. obtusata* and *L. littorea* to experience nearly the same degree of semidiurnal tissue temperature variation as do high shore species like *L. saxatilis*. Thus, regulation of metabolic rate at elevated temperatures may be equally adaptive in lower shore littorinids as in higher shore littorinids.

## Acknowledgments

We wish to thank Colette O'Byrne-McMahon and Dr Daniel E. Buckley for technical assistance with laboratory experiments. This research was partially supported by Organized Research Funds from the University of Texas at Arlington to R. F. McMahon, NSF Grant DB-7810190 to W. D. Russell-Hunter and NSF Grant RII-8305862 to D. W. Aldridge.

## References

Aldridge, D. W., W. D. Russell-Hunter & R. F. McMahon, 1995. Effects of ambient temperature and of temperature acclimation on nitrogen excretion and differential catabolism of protein and nonprotein resources in the intertidal snails, *Littorina saxatilis* (Olivi) and *L. obtusata* (L.). Hydrobiologia 309 (Dev. Hydrobiol. 111): 101–109.

Apley, M. L., 1970. Field studies on life history, gonadal cycle and reproductive periodicity in *Melampus bidentatus* (Pulmonata: Ellobiidae). Malacologia 10: 381–397.

Bayne, B. L., R. J. Thompson & J. Widdows, 1973. Some effects of temperature and food on the rate of oxygen consumption of *Mytilus edulis* L. In W. Wieser (ed.), Effects of Temperature on Ectothermic Organisms. Springer-Verlag, New York (N.Y.); Berlin: 181–193.

Boyden, C. R., 1972. The behavior, survival and respiration of the cockles, *Cerastoderma edule* and *C. glaucum* in air. J. mar. biol. Ass. U.K. 52: 661–680.

Cavanaugh, G. M., 1956. Formulae and methods V. of the Marine Biological Laboratory Chemical Room. Marine Biological Laboratory, Woods Hole (Massachusetts), 87 pp.

Cossins, A. R., 1981. The adaptation of membrane dynamic structure to temperature. In G. J. Morris & A. Clarke (eds), Effects of Low Temperatures on Biological Membranes. Academic Press, London: 83–106.

Cossins, A. R. & K. Bowler, 1987. Temperature biology of animals. Chapman and Hall, London; New York (N.Y.), 339 pp.

Cossins, A. R., K. Bowler & C. L. Prosser, 1981. Homeoviscous adaptations and its effects on membrane bound proteins. J. therm. Biol. 6: 183–187.

Griffiths, R. J., 1977a. Thermal stress and the biology of *Actinia equina* L. (Anthozoa). J. exp. mar. Biol. Ecol. 27: 141–154.

Griffiths, R. J., 1977b. Temperature acclimation in *Actinia equina* L. (Anthozoa). J. exp. mar. Biol. Ecol. 28: 285–292.

100

Grainger, J. N. R., 1969. Factors affecting the body temperature of *Patella*. Verh. dt. zool. Ges. 1968: 479–487.

Hochachka, P. W. & G. N. Somero, 1984. Biochemical adaptation. Princeton University Press, Princeton (N.J.), 537 pp.

Kennedy, V. S. & J. A. Mihursky, 1972. Effects of temperature on the respiratory metabolism of three Chesapeake Bay bivalves. Chesapeake Sci. 13: 1–22.

Lewis, J. B., 1963. Environmental and tissue temperatures of some tropical intertidal marine animals. Biol. Bull. 124: 277–284.

Markel, R. P., 1971. Temperature relations in two species of tropical west American littorines. Ecology 52: 1126–1130.

McMahon, R. F., 1988. Respiratory response to periodic emergence in intertidal molluscs. Am. Zool. 28: 97–114.

McMahon, R. F., 1990. Thermal tolerance, evaporative water loss, air-water oxygen consumption and zonation of intertidal prosobranchs: a new synthesis. Hydrobiologia 193 (Dev. hydrobiol. 56): 241–260.

McMahon, R. F., 1992. Respiratory response to temperature and hypoxia in intertidal gastropods from the Texas coast of the Gulf of Mexico. In J. Grahame, P. J. Mill & D. G. Reid (eds), Proceedings of the 3rd International Symposium on Littorinid Biology. Malacological Society of London, London: 45–59.

McMahon, R. F. & W. D. Russell-Hunter, 1977. Temperature relations of aerial and aquatic respiration in six littoral snails in relation to their vertical zonation. Biol. Bull. 152: 182–198.

McMahon, R. F. & W. D. Russell-Hunter, 1978. Respiratory responses to low oxygen stress in marine littoral and sublittoral snails. Physiol. Zool. 51: 408–424.

McMahon, R. F. & W. D. Russell-Hunter, 1981. The effects of physical variables and acclimation on survival and oxygen consumption in the high littoral salt-marsh snail, *Melampus bidentatus* Say. Biol. Bull. 161: 246–269.

McMahon, R. F. & J. G. Wilson, 1981. Seasonal respiratory responses to temperature and hypoxia in relation to burrowing depth in three intertidal bivalves. J. therm. Biol. 6: 267–277.

Newell, R. C., 1979. Biology of intertidal animals, third edition. Marine Ecological Surveys Ltd., Faversham, Kent, U.K., 781 pp.

Newell, R. C. & B. L. Bayne, 1973. A review on temperature and metabolic adaptation in intertidal marine invertebrates. Neth. J. Sea Res. 7: 421–433.

Newell, R. C. & L. H. Kofoed, 1977. Adjustment of the components of energy balance in the gastropod *Crepidula fornicata* in response to thermal acclimation. Mar. Biol. 44: 275–286.

Newell, R. C. & V. I. Pye, 1970a. The influence of thermal acclimation on the relation between oxygen consumption and temperature in *Littorina littorea* (L.) and *Mytilus edulis* L. Comp. Biochem. Physiol. 34: 367–383.

Newell, R. C. & V. I. Pye, 1970b. Seasonal changes in the effect of temperature on the oxygen consumption of the winkle, *Littorina littorea* (L.) and the mussel *Mytilus edulis* L. Comp. Biochem. Physiol. 34: 367–383.

Newell, R. C. & V. I. Pye, 1971. Quantitative aspects of the relationship between metabolism and temperature in the winkle *Littorina littorea* (L.). Comp. Biochem. Physiol. 38B: 635–650.

Newell, R. C. & A. Roy, 1973. A statistical model relating the oxygen consumption of a mollusk (*Littorina littorea*) to activity, body size, and environmental conditions. Physiol. Zool. 46: 253–275.

Newell, R. C., L. G. Johnson & L. H. Kofoed, 1977. Adjustment of the components of energy balance in response to temperature change in *Ostrea edulis*. Oecologia 30: 97–110.

Precht, H., J. Christophersen, H. Hensel & W. Larcher, 1973. Temperature and life. Springer-Verlag, New York (N.Y.); Berlin, 779 pp.

Prosser, C. L. & J. E. Heath, 1991. Temperature. In C. L. Prosser (ed.), Environmental and metabolic animal physiology. Wiley-Liss, New York (N.Y.); Chichester, Brisbane; Singapore: 109–165.

Reid, D. G., 1989. The comparative morphology, phylogeny and evolution of the gastropod family, Littorinidae. Phil. Trans. r. Soc., Lond. B 324: 1–110.

Reid, D. G., 1990. A cladistic phylogeny of the genus *Littorina* (Gastropoda): implications for evolution of reproductive strategies and for classification. Hydrobiologia. 193 (Dev. Hydrobiol. 56): 1–19.

Russell-Hunter, W. D. & M. L. Apley, 1965. A condition of temporary hyperthermia in a marine littoral snail. Biol. Bull. 129: 408–409.

Russell-Hunter, W. D., R. F. McMahon & D. W. Aldridge, 1980. Lack of respiratory response to temperature acclimation in two littorinid snails. Biol. Bull. 159: 452.

Shirley, T. C., G. J. Denoux & W. B. Stickle, 1978. Seasonal respiration in the marsh periwinkle, *Littorina irrorata*. Biol. Bull. 154: 322–334.

Shumway, S. E. & R. K. Koehn, 1982. Oxygen consumption in the American oyster *Crassostrea virginica*. Mar. Ecol. Prog. Ser. 9: 59–68.

Southward, A. J., 1958. Note on the temperature tolerances of some intertidal animals in relation to environmental temperatures and geographical distribution. J. mar. biol. Ass. U.K. 37: 49–66.

Underwood, A. J., 1979. The ecology of intertidal gastropods. Adv. mar. Biol. 16: 111–210.

Vermeij, G. J., 1971a. Temperature relationships of some tropical Pacific gastropods. Mar. Biol. 10: 308–314.

Vermeij, G. J., 1971b. Substratum relationships of some tropical Pacific intertidal gastropods. Mar. Biol. 10: 315–320.

Vernberg, F. J., 1969. Acclimation of intertidal crabs. Am. Zool. 9: 333–341.

Vernberg, W. B. & F. J. Vernberg, 1972. Environmental physiology of marine animals. Springer-Verlag, New York (N.Y.); Berlin, 346 pp.

Wolcott, T. G., 1973. Physiological ecology and intertidal zonation in limpets (*Acmaea*): a critical look at 'limiting factors'. Biol. Bull. 145: 389–422.

*Hydrobiologia* **309**: 101–109, 1995.
*P. J. Mill & C. D. McQuaid (eds), Advances in Littorinid Biology.*
©1995 *Kluwer Academic Publishers.*

# Effects of ambient temperature and of temperature acclimation on nitrogen excretion and differential catabolism of protein and nonprotein resources in the intertidal snails, *Littorina saxatilis* (Olivi) and *L. obtusata* (L.)

David W. Aldridge[1,2]*, W. D. Russell-Hunter[1,3,5] & Robert F. McMahon[1,4]
[1] *Marine Biological Laboratory, Woods Hole, Massachusetts, USA*
[2] *Department of Biology, North Carolina Agricultural and Technical State University, Greensboro, North Carolina 27411, USA*
[3] *Department of Biology, Syracuse University, Syracuse, New York, USA*
[4] *Section of Comparative Physiology, Box 19498, Department of Biology, University of Texas at Arlington, Arlington, Texas, USA*
[5] *Present address: 711 Howard Street, Easton, Maryland 21601-3934, USA*
(* *Address for correspondence*)

*Key words:* nitrogen excretion, O:N ratios, ammonia excretion, urea excretion, *Littorina saxatilis*, *Littorina obtusata*.

## Abstract

Nitrogenous excretion in two snails, *Littorina saxatilis* (high intertidal) and *L. obtusata* (low intertidal) was studied in relation to temperature acclimation (at 4° and 21°C), including total N excretion rates, the fraction of urea in N excretion, corresponding O:N ratios and the partitioning of deaminated protein between catabolic and anabolic processes at 4°, 11° and 21°C. Aggregate N excretion rates in both species showed no significant compensatory adjustments following acclimation. Total weight specific N excretion rates at 21°C were higher in standard 3 mg *L. saxatilis* (739 ng N $mg^{-1}$ $h^{-1}$) than standard 5 mg *L. obtusata* (257 ng N $mg^{-1}$ $h^{-1}$) for snails acclimated to 21°C. Comparisons of $Q_{10}$ values of total weight specific N excretion to $Q_{10}$ values for weight specific oxygen consumption ($\dot{V}O_2$) between 4° to 11°C and 11° to 21°C indicated that, while total rates of catabolic metabolism ($\dot{V}O_2$) and protein deamination in *L. obtusata* were essentially parallel, the relationship between N excretion and $\dot{V}O_2$ in *L. saxatilis* revealed the partitioning of a larger share of deaminated protein carbon into anabolism at 4° and 21°C than at 11°C. Urea N accounted for a larger share of aggregate N excreted in *L. saxatilis* than in *L. obtusata*, but in both species urea N is a greater proportion of total N excreted when acclimated at 4°C (urea N: ammonia N ratio range: 1 to 2.15) than in snails acclimated to 21°C (urea N: ammonia N ratio range: 0.46 to 1.39). Molar O:N ratios indicate that the proportion of metabolism supported by protein catabolism is greater in *L. saxatilis* (O:N range: 2.5–8.4) than in *L. obtusata* (O:N range: 7.3–13.0). In both species, regardless of acclimation temperature, the O:N ratios are generally lowest (high protein catabolism) at 4°C and highest at 21°C.

## Introduction

Environmental temperature directly and indirectly affects the rates of metabolic processes in ectothermic animals (Newell, 1979). Many temperature-related shifts in metabolic efficiency include changes in the differential catabolism of protein and nonprotein resources (Roller & Stickle, 1989; Wilbur & Hilbish,

1989). These changes can be detected by experiments involving nearly concurrent measurements of oxygen uptake and of nitrogenous excretion (Aldridge *et al.*, 1980; Aldridge, 1980, 1982; Aldridge *et al.*, 1986; Russell-Hunter *et al.*, 1983; Tashiro, 1980, 1982).

We report in this paper the results of such experiments on two species of intertidal snail, *Littorina saxatilis* (Olivi) from the higher levels of the littoral zone and *L. obtusata* (L.) from the lower littoral (for details of their habitats see McMahon & Russell-Hunter, 1977; McMahon *et al.*, 1995). The two species differ markedly in their daily exposure to desiccation and thus would be expected to have different nitrogen excretion strategies (Needham, 1935, 1938). After assessing the effects of ambient temperature and of temperature acclimation on the concurrent rates of nitrogenous excretion and oxygen uptake, we were able to compute differential catabolism of protein and nonprotein resources. The two species differ in details of their nitrogen metabolism. A companion report (McMahon *et al.*, 1995) details temperature effects on oxygen consumption in the same two species and discusses the evolution of the capacity for acclimation. A preliminary account of this study was published as an abstract by Aldridge *et al.* (1980).

## Materials and methods

Specimens of *L. obtusata* and *L. saxatilis* were collected from a boulder beach on Nobska Point, near Woods Hole, Massachusetss (41°39.4' N; 70°39.3' W) on 15 July 1980. Those of *L. obtusata* were collected from *Ascophyllum* and other fucoid seaweeds on boulders in the area of the mean low water spring tide level (MLW-ST) whereas specimens of *L. saxatilis* were obtained directly from the boulder substratum in the upper littoral region above mean high water spring tide level (MHWST). Collected snails were returned immediately to the laboratory where each species sample was divided into two groups that were acclimated to either 4° or 21°C, as described by McMahon *et al.* (1995). Specimens of *L. obtusata* were acclimated unfed for 15–16 days and *L. saxatilis* for 19–20 days.

On completion of acclimation, the molar oxygen consumption rates and total molar nitrogen excretion rates (as ammonia and urea nitrogen) were determined for groups of each species acclimation group. Oxygen consumption methods and the allocation of groups are detailed in McMahon *et al.* (1995).

The urea-N and ammonia-N excretion rates for each species at 4°, 11° and 21°C were determined for 13–15 groups of snails from each acclimation group. The shell lengths (SL) of all snails were taken (to the nearest 0.1 mm) with dial calipers. This variable was measured as the greatest distance from the tip of the spire to the anterior margin of the aperture (McMahon *et al.*, 1995). Snail numbers in each group ranged from 1 to 3 depending on snail size. The size range of each

group was within 0.5 mm of each arbitrarily chosen average group SL.

To measure urea and nitrogen excretion rates, snail groups were placed in vials containing 25 ml of Marine Biological Laboratory artificial sea water (salinity = 31‰ Na, Cavenaugh, 1956) for 3, 5 or 10 hours for runs at 21°C, 11°C and 4°C respectively. A 1 mm mesh nylon screen was used to keep the snails submerged during determinations. After the appropriate incubation time, snails were removed and two 10 ml aliquots were obtained from each group's vial. To one of these aliquots, 20 microunits of Sigma Type III urease (dissolved in distilled water and cold filtered) were added to hydrolyze urea (Aldridge *et al.*, 1986). The ammonia concentration of each aliquot was then assayed using an Orion 9510 ammonia probe and an Orion 407A meter (after the methods of Russell-Hunter *et al.*, 1983). Subtracting the ammonia N concentration of the untreated aliquot from the urease treated aliquot yields a value for the urea-N excreted by the snails in the groups. Control determinations of ammonia in artificial sea water, with and without urease, allowed correction for background levels of ammonia.

On completion of oxygen consumption and nitrogen excretion rate determinations, the shells of individuals in each subsample were dissolved in 7% nitric acid (<6 hr). Remaining tissues were rinsed twice in distilled water and dried to constant weight (DTW) at 95°C (>24 hr) (McMahon & Russell-Hunter, 1977). For each species acclimation group at each experimental temperature, the above rates were initially computed as $\mu l$ $O_2$ ind$^{-1}$ h$^{-1}$ and ng N ind$^{-1}$ h$^{-1}$ respectively.

Individuals of *L. saxatilis* and *L. obtusata* were also collected from the Nobska Point site on 28 June 1980 and their $\dot{V}O_2$ and urea and ammonia-N excretion rates determined as unacclimated controls at 20°C, using the methods outlined above. The ambient sea water temperature was 20°C on 6 and 8 July, i.e. in the period mid-way between the 28 June and 15 July collection dates.

## Results

For each species acclimation group at each experimental temperature and for the unacclimated controls, least squares linear regressions of the log$_{10}$ of either $\dot{V}O_2$ or N excretion rate versus log$_{10}$ of subsample mean individual DTW were computed. With two exceptions, the slopes for all regressions of $\dot{V}O_2$ and N excretion rate (Tables 1 and 2) versus DTW for each species acclimation group at each experimental temperature proved significantly linear at $P < 0.05$. The two exceptions were those relating $\dot{V}O_2$ to DTW for 4°C-acclimated specimens of *L. saxatilis* at 4°C ($P > 0.20$) and for N excretion for 21°C-acclimated *L. saxatilis* at 4°C.

These regressions were used to predict $\dot{V}O_2$ or N excretion rates of individuals of standard DTW for each species acclimation group at each of the three experimental temperatures. The standard DTW for both species was chosen to approximate the mean DTW of all the individuals used in $\dot{V}O_2$ and N excretion determinations. Thus, the standard snail DTW is 3 mg for *L. saxatilis* and 5 mg for *L. obtusata*. Division of the standard whole animal $\dot{V}O_2$ and N excretion rates by the appropriate standard DTW value provided weight specific estimates of these rate values as $\mu l\ O_2\ mg^{-1}\ h^{-1}$ or $ng\ N\ mg^{-1}\ h^{-1}$. These weight specific rates were then converted to molar equivalents (as O and N) and used to compute molar O:N ratios for standard snails (Bayne & Widdows, 1978; Aldridge *et al.*, 1986). The same data were also computed as catabolic rates for protein carbon and nonprotein carbon (see below).

Figure 1 shows total weight specific nitrogen (urea + ammonia) excretion rates for 4°- and 21°C-acclimated standard specimens of *L. obtusata* (5 mg DTW) and of *L. saxatilis* (3 mg DTW) at 4°, 11° and 21°C, along with rates at sea water ambient temperature (20°C) for the unacclimated control standard snails. There was no evidence for significant compensatory adjustments in total N excretion rates in either species following either acclimation regime used in this study. Weight specific total N excretion rates for the two species are similar at 4° and 11°C. At 21°C, however, total N excretion rates are 187% higher in *L. saxatilis* than in *L. obtusata* ($p < 0.05$; two-tailed *t*-test). Nitrogen excretion rates for the unacclimated *L. saxatilis* are 17% lower than the corresponding 21°C-acclimated snails at 21°C ($p > 0.20$; two-tailed *t*-test) while N excretion rates for unacclimated *L. obtusata* are 39% below those of 21°C-acclimated snails at 21°C ($p < 0.05$; two-tailed *t*-test). Hence, starvation appears to affect N excretion more in *L. obtusata* than in *L. saxatilis*.

$Q_{10}$ values for total N excretion rates of standard individuals were highly variable for both species. For *L. obtusata* (5 mg DTW) acclimated to 4°C $Q_{10}$ values were 1.25 between 4-11°C and 2.31 between 11-21°C. $Q_{10}$ values for *L. obtusata* (5 mg DTW) acclimated to

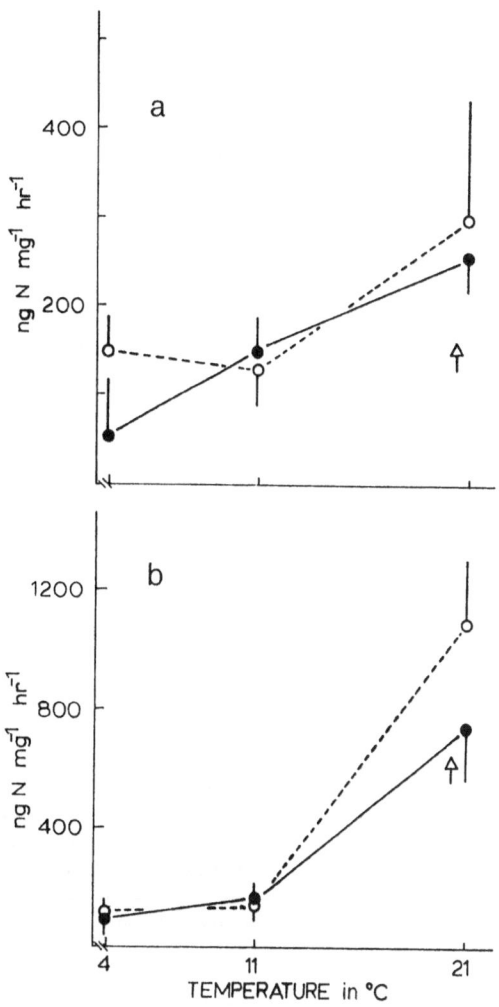

*Fig. 1.* Total nitrogen excretion ($ng\ N\ mg^{-1}\ h^{-1}$) for (a) standard, 5 mg tissue dry weight *Littorina obtusata* and (b) standard, 3 mg tissue dry weight *L. saxatilis* at 4°, 11° and 21°C. ○, 4°C acclimated snails; ●, 21°C acclimated snails; △, unacclimated controls at 20°C. Vertical lines delineate one standard error.

21°C were 4.20 between 4-11°C and 1.71 between 11-21°C. For *L. saxatilis* (3 mg DTW) acclimated to 4°C $Q_{10}$ values were 1.24 between 4-11°C and 7.56 between 11-21°C. $Q_{10}$ values for *L. saxatilis* (3 mg DTW) acclimated to 21°C were 2.19 between 4-11°C and 5.03 between 11-21°C.

The relative contribution of urea to total N excretion is presented as the mean ratio of urea N to ammonia N in Table 3. In unacclimated control *L. obtusata* and *L. saxatilis*, 82% and 88% respectively of the total nitrogen excreted is urea N. In the acclimated snails, *L.*

*Table 1.* Linear least squares regressions of $\log_{10}$ N (ng) excretion rate versus $\log_{10}$ dry tissue weight (mg) (Log N = $a + b$ Log DTW) for acclimated snails.

*Littorina obtusata*

| Accl. temp. | Test temp. | $a$ | $b$ | $n$ | Degrees of freedom | | $f$ |
|---|---|---|---|---|---|---|---|
| | | | | | Regression | Residual | |
| 4°C | 4°C | 2.32 | 0.79 | 15 | 1 | 13 | 12.56* |
| 4°C | 11°C | 2.35 | 0.67 | 13 | 1 | 11 | 8.25* |
| 4°C | 21°C | 2.96 | 0.31 | 14 | 1 | 12 | 5.01* |
| 21°C | 4°C | 2.14 | 0.39 | 13 | 1 | 11 | 4.91* |
| 21°C | 11°C | 1.96 | 1.30 | 15 | 1 | 13 | 33.61* |
| 21°C | 21°C | 2.35 | 1.09 | 14 | 1 | 12 | 58.36* |

*Littorina saxatilis*

| Accl. temp. | Test temp. | $a$ | $b$ | $n$ | Degrees of freedom | | $f$ |
|---|---|---|---|---|---|---|---|
| | | | | | Regression | Residual | |
| 4°C | 4°C | 1.79 | 1.63 | 13 | 1 | 11 | 8.70* |
| 4°C | 11°C | 0.90 | 1.75 | 13 | 1 | 11 | 4.97* |
| 4°C | 21°C | 3.28 | 0.50 | 13 | 1 | 11 | 6.01* |
| 21°C | 4°C | 2.47 | −0.02 | 13 | 1 | 11 | <0.01 |
| 21°C | 11°C | 2.39 | 0.53 | 13 | 1 | 11 | 5.67* |
| 21°C | 21°C | 2.97 | 0.79 | 14 | 1 | 11 | 14.23* |

\* $p < 0.05$

*Table 2.* Linear least squares regressions of $\log_{10}$ $O_2$ uptake ($\mu$l) and $\log_{10}$ N (ng) excretion rate versus $\log_{10}$ dry tissue weight (mg) (Log 'rate' = $a + b$ Log DTW) for unacclimated snails at 20°C.

*Littorina obtusata*

| | $a$ | $b$ | $n$ | Degrees of freedom | | $f$ |
|---|---|---|---|---|---|---|
| | | | | Regression | Residual | |
| ng N excreted/h | 2.53 | 0.52 | 18 | 1 | 15 | 13.85* |
| $\mu$l $O_2$ uptake/h | 0.24 | 0.77 | 18 | 1 | 16 | 43.81* |

*Littorina saxatilis*

| | $a$ | $b$ | $n$ | Degrees of freedom | | $f$ |
|---|---|---|---|---|---|---|
| | | | | Regression | Residual | |
| ng N excreted/h | 3.16 | 0.23 | 18 | 1 | 16 | 5.31* |
| $\mu$l $O_2$ uptake/h | 0.66 | 0.17 | 18 | 1 | 16 | 4.90* |

\* $p < 0.05$

*saxatilis* consistently excreted a greater proportion of its N as urea than did *L. obtusata*. In both species, 4°C-acclimated snails at 4°C were predominantly ureotelic (*L. obtusata* 58% & *L. saxatilis* 65% urea N). At 21°C, *L. saxatilis* acclimated to 21°C still excreted 52% of its total N as urea N while 21°C acclimated *L. obtusata* at 21°C were predominantly ammonotelic (69% ammonia N).

To assess the shifts in metabolism as a result of acclimation regimes in these two snail species, the total N excretion rates for standard individuals were paired with corresponding weight specific $\dot{V}O_2$ val-

*Table 3.* Ratios of urea N: ammonia N excreted ($x \pm$ s.e.) for *Littorina obtusata* and *L. saxatilis* acclimated to 4°C and 21°C, at 4°C, 11°C and 21°C, along with unacclimated controls (20°C).

| Acclimation temperature | Unacclimated control | at 4°C | at 11°C | at 21°C |
|---|---|---|---|---|
| *L. obtusata* | | | | |
| 4°C | | 1.40±0.29 | 1.29±0.21 | 1.01±0.33 |
| | | n = 15 | n = 13 | n = 14 |
| | 4.45±0.29 | | | |
| | n = 15 | | | |
| 21°C | | 1.35±0.25 | 0.74±0.17 | 0.46±0.16 |
| | | n = 13 | n = 15 | n = 14 |
| *L. saxatilis* | | | | |
| 4°C | | 1.82 ±0.15 | 1.86±0.35 | 2.15±0.24 |
| | | n = 13 | n = 13 | n = 13 |
| | 6.99±1.28 | | | |
| | n = 15 | | | |
| 21°C | | 1.39±0.25 | 0.72±0.10 | 1.07±0.21 |
| | | n = 13 | n = 13 | n = 13 |

ues for the same size animals (from McMahon *et al.*, 1995) to produce the mole O to mole N ratios shown in Table 4. These were all relatively low and indicate that both species, are catabolizing large amounts of protein (O:N < 30; Bayne & Widdows, 1978; Aldridge *et al.*, 1986). The O:N ratio for the unacclimated control *L. saxatilis* is 39% of the O:N ratio of the unacclimated *L. obtusata*. Similarly, the O:N ratio of 4°C-acclimated *L. saxatilis* at 4°C is 48% of the ratio for 4°C-acclimated *L. obtusata* at 4°C and the O:N ratio of 21°C-acclimated *L. saxatilis* at 21°C is 34% of the ratio for 21°C-acclimated *L. obtusata* at 21°C. For both species the O:N ratios for 4°C-acclimated snails at 4°C are lower than the ratios for 21°C-acclimated snails at 21°C, indicating a metabolic shift to a greater reliance on protein catabolism in all cold acclimated snails.

To better delineate shifts in the substrates being catabolized, the N excretion and $O_2$ consumption rates were further computed so that all were expressed in carbon terms (i.e. as units of protein carbon or nonprotein carbon). The conversion of these rates into a common carbon currency and the assignment as protein carbon or nonprotein carbon (see Aldridge *et al.*, 1986; Russell-Hunter & Buckley, 1983; Russell-Hunter *et al.*, 1983; Aldridge *et al.*, 1986) are based on several assumptions: in particular that all nitrogen excreted is derived from the breakdown of protein and that all oxygen is consumed in proportion to the breakdown

*Table 4.* Molar O:N ratios of *Littorina obtusata* and *L. saxatilis* acclimated to 4°C and 21°C, at 4°C, 11°C and 21°C, along with unacclimated controls (20°C).

| Acclimation temperature | Unacclimated control | at 4°C | at 11°C | at 21°C |
|---|---|---|---|---|
| *L. obtusata* | | | | |
| 4°C | | 5.20 | 12.08 | 8.96 |
| | 9.50 | | | |
| 21°C | | 12.98 | 7.26 | 10.52 |
| *L. saxatilis* | | | | |
| 4°C | | 2.50 | 7.21 | 3.01 |
| | 3.75 | | | |
| 21°C | | 4.41 | 8.40 | 3.58 |

of each category of organic carbon compound. The conversion factor to estimate the oxygen consumption for protein catabolism from this is 5.92 $\mu$l $O_2$/$\mu$gN (Harper *et al.*, 1977). Thus there is a weight-specific rate of oxygen consumption for the protein fraction of catabolism. Subtraction of this rate from the overall weight-specific oxygen uptake rate gives the rate of oxygen consumption for the nonprotein fraction. Carbon equivalents are obtained from the amount of $CO_2$ given off (0.536 $\mu$gC consumed/$\mu$l $CO_2$) (Harper *et al.*, 1977). The relationship of $CO_2$ evolved to $O_2$ consumed differs for proteins, fats, and carbohydrates. The $CO_2$ given off from protein catabolism can be derived directly from the weight-specific excretion rate (4.75 $\mu$l $CO_2$ evolved/$\mu$g N excreted: Harper et al., 1977). The $CO_2$ given off from the nonprotein catabolism can be estimated for these experiments by multiplying the weight-specific oxygen consumption for the nonprotein fraction by a respiratory quotient of 0.95 (on the basis of 10% fat and 90% carbohydrates being catabolized). Thus, we have calculated for our standard snails estimates of differential metabolism involving three separate terms for carbon catabolism (Table 5): nonprotein carbon catabolism, protein carbon catabolism (carbon content being 3.25 times the nitrogen content of protein; Russell–Hunter & Buckley, 1983), and protein carbon deaminated (being the excess of nitrogen metabolism over total carbon catabolism).

The carbon values in Table 5 show that these snails rely predominantly on protein carbon sources for energy metabolism, and for providing carbon skeletons for nonprotein molecules. For *L. saxatilis*, in all treatments and the unacclimated control, rates of protein deamination exceed the needs of energy catabolism. For example, in *L. saxatilis* acclimated to 4°C at 4°C and acclimated to 21°C at 21°C the rates of protein deamination exceed the needs of energy catabolism by 83% and 69%, respectively. Specimens of *L. obtusata* typically showed less 'excess' deamination of protein compared to those of *L. saxatilis* except in 4°C-acclimated snails at 4°C. For *L. obtusata* acclimated to 21°C at 21°C, protein deamination falls short of supplying the needs of energy catabolism by 15%.

## Discussion

In the well-studied zonation of intertidal prosobranch snails, the upper littoral *Littorina saxatilis* and lower littoral *L. obtusata* occupy markedly different zones, experiencing markedly different temporal regimes of submergence and aerial exposure (McMahon & Russell-Hunter, 1977).

It has become clear that, although these species of littorinid snail show a variety of patterns of thermoregulation of oxygen consumption (McMahon & Russell-Hunter, 1977), they do not show any evidence of capacity for temperature acclimation of respiration to appropriate seasonal temperatures (4°C and 21°C). In discussing this lack of acclimation (McMahon et al., 1995), it was concluded that the large shifts in ambient temperature experienced semidiurnally by these snails effectively denied them the sustained temperature regimes (for say 7–20 days) needed for any acclimation to occur in the field. Indeed, it can be claimed that the capacity for temperature acclimation would be of little selective value under the highly unstable temperature regimes of the intertidal zones (McMahon et al., 1995).

When we investigated direct temperature effects and the results of temperature acclimation on nitrogen excretion in *L. saxatilis* and *L. obtusata*, the patterns which emerged differed from those for oxygen consumption. First, there are significant differences in rates between the two species, but no evidence for acclimation in total N excreted. Secondly, relative nitrogen excretion rates for ammonia and for urea differ between the species with temperature, and with acclimation regime. Thirdly, there are some significant differences in O:N ratios which reflect shifts in differential catabolism of proteinaceous and nonproteinaceous carbon resources.

When $Q_{10}$ values are considered, those for oxygen consumption and nitrogen excretion in *L. obtusata* are all closely correlated, and remain so after acclimation. Thus, in *L. obtusata* the range of $Q_{10}$ variation for both $\dot{V}O_2$ and total N excretion were essentially similar for 4°- and 21°C-acclimated individuals, with $Q_{10}$ values of $\dot{V}O_2$ between 4° and 11°C equal to 2.75 and 2.03 respectively, and between 11° and 21°C, 1.71 and 2.43 respectively (McMahon et al., 1995).

A different situation occurs in *L. saxatilis*, with $Q_{10}$ for total N excretion differing markedly from those for oxygen consumption. Thus, while $Q_{10}$ values for $\dot{V}O_2$ between 4° and 11°C in both 4°- and 21°C-acclimated *L. saxatilis* were elevated (at 5.87 and 4.60, respectively; McMahon et al., 1995), indicating an active increase in metabolic rate between 4° and 11°C, the corresponding $Q_{10}$ values for total N excretion between 4°–11°C were relatively low, which might suggest a passive increase in protein catabolism corresponding to normal chemical kinetics. Thus, for *L. saxatilis*, the major share of this metablic increase is due to the catabolism of a larger percentage of deaminated protein as temperatures rise from 4°C to 11°C. In contrast, analysis of $Q_{10}$ values reveals that total N excretion rates for *L. saxatilis* in both acclimation groups increased more rapidly than $Q_{10}$ values for $\dot{V}O_2$ between 11° and 21°C ($\dot{V}O_2$ $Q_{10}$ values: 4°C-acclimated = 3.02; 21°C-acclimated = 2.95). These elevated $Q_{10}$ values for total N excretion between 11° and 21°C indicate that protein deamination rates are

Table 5. Rates (ng C mg$^{-1}$ h$^{-1}$) of nonprotein and protein carbon catabolized and protein carbon deaminated in excess of total carbon catabolism for standard *Littorina obtusata* (5 mg DTW) and *L. saxatilis* (3 mg DTW) acclimated to 4°C and 21°C, at 4°C, 11°C and 21°C, along with unacclimated controls (20°C).

| | | L. obtusata | | L. saxatilis | |
|---|---|---|---|---|---|
| Unacclimated Controls | Nonprotein C catabolized | 137 | | – – | |
| | Protein C catabolized | 506 | | 1016 | |
| | Protein C deaminated | – – | | 986 | |
| | | Acclimation Temperature | | | |
| | | 4°C | 21°C | 4°C | 21°C |
| at 4°C | Nonprotein C catabolized | – – | 81 | – – | – – |
| | Protein C catabolized | 484 | 166 | 406 | 312 |
| | Protein C deaminated | 238 | – – | 338 | 185 |
| at 11°C | Nonprotein C catabolized | 156 | – – | – – | – – |
| | Protein C catabolized | 423 | 488 | 468 | 478 |
| | Protein C deaminated | – – | 114 | 112 | 44 |
| at 21°C | Nonprotein C catabolized | – – | 150 | – – | – – |
| | Protein C catabolized | 978 | 835 | 3539 | 2399 |
| | Protein C deaminated | 18 | – – | 2709 | 1662 |

being actively increased in excess of the needs of energy catabolism. This provides an increase in the availability of protein carbon skeletons for the synthesis of nonprotein molecules. Very few $Q_{10}$ values for N excretion are available in the literature but Cockcroft (1990) has reported $Q_{10}$ values for N excretion of 2.2 and 2.3, respectively, for *Donax serra* and *D. sordidus* between 15° and 25°C.

There are similar differences between the species in the relative fraction of nitrogenous waste which was excreted as urea. Once again, the more terrestrial *L. saxatilis* excretes proportionately more N as urea than does the more aquatic *L. obtusata*. Increased dependence on urea excretion in *L. saxatilis* is almost certainly an adaptation for preventing the accumulation of ammonia during the prolonged aerial exposure not experienced by *L. obtusata*. Needham (1935, 1938) predicted greater ureotelic excretion in the higher littoral, although, at that time, the pattern hypothesized was distorted by assumptions of extensive uricotelic output based on static measures of excretory organ uric acid content.

Temperature acclimation does have a significant effect on the ratio of urea-N to ammonia-N excreted in both species. During cold acclimation at 4°C, the snails shift the predominant excretory product significantly towards urea. This is somewhat unexpected in relation to the lack of acclimation in oxygen consumption (McMahon *et al.*, 1995) and in aggregate N excretion. The adaptive significance of this acclimatory shift in both species does not relate directly to aerial exposure, nor to presumed water flux rates (Campbell, 1973; Needham, 1938; Prusch & Hall, 1978), and cannot be explained at present. However, the accumulation of urea in haemolymph could, during cold periods, act as a tissue antifreeze during winter tidal emergence. This intriguing possibility warrants further investigation.

When the data are recomputed in terms of O:N ratios or of differential catabolism, segregating nonprotein carbon consumed, protein carbon consumed and protein carbon deaminated, there are interesting shifts with temperature. However, there is also an overall greater relative rate of protein catabolism in the higher littoral *L. saxatilis*. In part, this could reflect dietary differences.

Some problems remain concerning the evolution of the terrestrial structures and functions found in intertidal animals. Most larger littoral invertebrates are alternately aerial and aquatic in their respiration and excretion. Earlier investigations (McMahon & Russell-Hunter, 1977; Newell, 1969, 1979) have stressed the respiratory adaptations of intertidal molluscs to different temperature regimes, each related to degrees of exposure, in turn related to vertical zonation on the seashore. Recent physiological studies, including our companion paper on the absence of acclimatory compensation in respiration in *L. saxatilis* and *L. obtusata* (McMahon *et al.*, 1995) have tended to emphasize the distinctness of the environment provided by the littoral zone and, in particular, the great temperature changes experienced within a few hours during each semidiurnal tidal cycle. Earlier studies on nitrogenous excretion in littoral invertebrates were largely involved with identifying excretory end-products in relation to vertical zonation and aerial exposure (Needham, 1935, 1938; Campbell, 1973; Newell, 1969, 1979), and were not concerned with differential catabolism. The present study has revealed marked differences between one high littoral species of *Littorina* and one low littoral species of *Littorina*, both in terms of the total nitrogen excreted and in their relative excretion rates for ammonia and for urea. Temperature acclimation, while not affecting total N excretion rates, has a significant (demonstrated but not yet elucidated) effect on the relative production of urea. Significant deamination of protein resources is deduced, particularly in the higher littoral species, *L. saxatilis*, and this can be hypothesized as related to diet and reproductive effort. As discussed elsewhere (McMahon & Russell-Hunter, 1977; Russell-Hunter, 1979, 1983) there seems to be a generally anacoluthical evolution of terrestrial structures and different physiological processes in littoral invertebrates as they become adapted to conditions in different vertical zones on the seashore. The co-existence of sub-optimality in one physiological process in an animal species, seemingly optimally adapted to a particular littoral zone in several other aspects of its physiology, should never by surprising.

## Acknowledgements

The authors express their gratitude to Dr Daniel E. Buckley, University of Maine, and Colette O'Byrne-McMahon for assistance in the laboratory, and to Sheere J. Mills, Dr Joseph J. White, Dr Roy J. Coomans and Lonnette G. Edwards for assistance with the manuscript. This research was supported by grants from Organized Research Funds of the University of Texas at Arlington to Robert F. McMahon, by NSF grant DB7810190 to W. D. Russell-Hunter and by NSF grants RII-8305862 and RII-8912685 to David W. Aldridge.

## References

Aldridge, D. W., 1980. Life cycle, reproductive tactics, and bioenergetics in the freshwater prosobranch snail, *Spirodon carinata* (Bruguiere) in upstate New York, Ph.D. Thesis, Syracuse University, Syracuse, N.Y.

Aldridge, D. W., 1982. Reproductive tactics in relation to life-cycle bioenergetics in three natural populations of the freshwater snail, *Leptoxis carinata*. Ecology 63: 196–208.

Aldridge, D. W., R. F. McMahon & W. D. Russell-Hunter, 1980. Effects of temperature acclimation on nitrogen metabolism in two littorinid snails. Biol. Bull. 159: 447.

Aldridge, D. W., W. D. Russell-Hunter & D. E. Buckley, 1986. Age-related differential catabolism in the snail, *Viviparus georgianus*, and its significance in the bioenergetics of sexual dimorphism. Can. J. Zool. 64: 340–346.

Bayne, B. L. & J. Widdows, 1978. The physiological ecology of two populations of *Mytilus edulis* L. Oecologia (Berlin) 37: 137–162.

Campbell, J. W., 1973. Nitrogen excretion. In C. L. Prosser (ed.), Comparative Animal Physiology, 3rd edition, W. B. Saunders, Philadelphia, Pennsylvania: 279–316.

Cavanaugh, G. M., 1956. Formulae and Methods V of the Marine Biological Laboratory Chemical Room. Marine Biological Laboratory, Woods Hole, Mass., 5th edn, 87 pp.

Cockcroft, A. C., 1990. Nitrogen excretion by the surf zone bivalves *Donax serra* and *Donax sordidus*. Mar. Ecol. Prog. Ser. 60: 57–66.

Harper, H. A., V. W. Rodwell & P. Mayes, 1977. Review of Physiological Chemistry. Lange Medical Publications, Los Altos, CA, 867 pp.

McMahon, R. F. & W. D. Russell-Hunter, 1977. Temperature relations of aerial and aquatic respiration in six littoral snails in relation to their vertical zonation. Biol. Bull. 152: 182–198.

McMahon, R. F., W. D. Russell-Hunter & D. W. Aldridge, 1995. Lack of metabolic temperature compensation in the intertidal gastropods, *Littorina saxatilis* (Olivi) and *L. obtusata* (L.). Hydrobiologia 309 (Dev. Hydrobiol. 111): 89–100.

Needham, J., 1935. Problems of nitrogen catabolism in invertebrates. II. Correlation between uricotelic metabolism and habitat in the phylum Mollusca. Biochem. J. 29: 238–251.

Needham, J., 1938. Contributions of chemical physiology to the problem of reversibility in evolution. Biol. Rev. 13: 225–251.

Newell, R. C., 1969. The effect of temperature fluctuation on the metabolism of intertidal invertebrates. Am. Zool. 9: 293–307.

Newell, R. C., 1979. Biology of Intertidal Animals. 3rd edn. Marine Ecological Surveys Ltd., Faversham, Kent.

Prusch, R. D. & C. Hall, 1978. Diffusional water permeability in selected marine bivalves. Biol. Bull. 154: 292–301.

Roller, R. A. & W. B. Stickle, 1989. Temperature and salinity effects on the intracapsular development, metabolic rates, and survival to hatching of *Thais haemastoma canaliculata* (Gray) (Prosobranchia: Muricidae) under laboratory conditions. J. exp. mar. Biol. Ecol. 125: 235–251.

Russell-Hunter, W. D., 1979. *A Life of Invertebrates*. Macmillian, New York, 650 pp.

Russell-Hunter, W. D., 1983. Overview: Planetary distribution of and ecological constraints upon the Mollusca. In W. D. Russell-

Hunter (ed.), The Mollusca, Vol. 6, Ecology, Academic Press, Orlando, Flordia: 1–27.

Russell-Hunter, W. D., D. W. Aldridge, J. S. Tashiro & B. S. Payne, 1983. Oxygen uptake and nitrogenous excretion rates during over winter degrowth conditions in the pulmonate snail, *Helisoma trivolvis*. Comp. Biochem. Physiol. 74A: 491–497.

Russell-Hunter, W. D. & D. E. Buckley, 1983. Actuarial bioenergetics of nonmarine molluscan productivity. In W. D. Russell-Hunter (ed.), The Mollusca, Vol. 6, Ecology Press, Orlando, Florida: 463–503.

Tashiro, J. S., 1980. Bioenergetic background to reproductive partitioning in an iteroparous population of the freshwater prosobranch, *Bithynia tentaculata*. Ph.D. Thesis, Syracuse University, Syracuse, N.Y.

Tashiro, J. S., 1982. Grazing in *Bithynia tentaculata*: Age-specific bioenergetic patterns in reproductive partitioning of ingested carbon and nitrogen. Am. Midl. Nat. 107: 133–150.

Wilbur, A. E. & T. J. Hilbish, 1989. Physiological energetics of the ribbed mussel *Geukensia demissa* (Dillwyn) in response to increased temperature. J. exp. mar. Biol. Ecol. 131: 161–170.

*Hydrobiologia* **309**: 111–116, 1995.
*P. J. Mill & C. D. McQuaid (eds), Advances in Littorinid Biology.*
©1995 *Kluwer Academic Publishers.*

111

# Environmentally induced variation in uric acid concentration and xanthine dehydrogenase activity in *Littorina saxatilis* (Olivi) and *L. arcana* Hannaford Ellis

Delmont C. Smith*, Peter J. Mill & J. Grahame
*Department of Pure and Applied Biology, University of Leeds, Leeds, LS2 9JT, UK*
*Present address: Department of Biological Sciences, State University of New York, Brockport, New York, 14420, USA*

Key words: *Littorina saxatilis, Littorina arcana*; uric acid, xanthine dehydrogenase

## Abstract

*Littorina saxatilis* and *Littorina arcana* collected from a boulder field low in the intertidal zone had a uric acid concentration significantly higher than snails collected from nearby cliff crevices that were at the upper limit of their vertical range on the shore. The absolute concentrations varied with different collections, suggesting a possible seasonal fluctuation in uric acid. *L. arcana* had a greater concentration of uric acid than did *L. saxatilis* when both were from the boulder field, but the two did not differ significantly when taken from crevices. Samples from the two sites were marked and transplanted; by four weeks the transplanted animals showed a tendency to develop a uric acid concentration similar to that of natives of the site, suggesting a physiological rather than a genetic adaptation. Xanthine dehydrogenase activity differed between animals from the two sites in the same manner as the uric acid concentration.

## Introduction

The nature of the end-products of nitrogen metabolism has long been recognised to be a function of the toxicity of ammonia and the water available for its excretion, aquatic species being ammoniotelic and terrestrial species ureotelic or uricotelic (Delaunay, 1931; Needham, 1935). However, apparently all gastropods accumulate uric acid to some degree (Duerr, 1967). Uric acid may be excreted but, because of a short life span, the kidney of accumulation may be a reality in some species (Wilbur, 1983). In *Littorina littorea*, Heil & Eichelberg (1983) noted that, although uric acid, urea and ammonia are all produced, the only metabolite excreted is ammonia; they suggested that uric acid serves as a nitrogen store and has a function in osmoregulation.

The position of littoral molluscs with regard to mode of nitrogen excretion may be thought of as being intermediate between land and aquatic snails, with prosobranch species that live high in the intertidal zone containing more uric acid than species from low-

er down (Needham, 1938; Potts, 1967). We are not aware of previous studies within species of uric acid variation related to position on the shore.

The rough periwinkles *Littorina saxatilis* and *Littorina arcana* occupy a broad range of the upper littoral zone. *L. arcana* is less widely distributed than *L. saxatilis* (Mill & Grahame, 1990) and has a clear preference for exposed sites (Grahame & Mill, 1986). In the present study, samples of both species were collected from two intertidal sites; horizontal fissures in cliffs high on the shore and a boulder field lower on the shore. Uric acid content was measured before and after transplantation of the animals in order to test whether the specific location on the shore is a factor in determining the degree of uricotelism in these two species of periwinkles. Xanthine dehydrogenase (XDH) activity was also measured in samples of *L. saxatilis* collected on one occasion.

## Material and methods

### Field Locations

The primary collection and transplant site was Filey Brigg, North Yorkshire, on the northeast coast of England located at $0°15'W$ $54°13'$ N. Filey Brigg is a rocky peninsula extending about 1 km into the North Sea in a generally eastward direction (Fig. 1). Winkles were collected from two locations: 1) Long, horizontal crevices in the face of a vertical north-east facing cliff (British National Grid Reference TA 132815). This location is high in the zone inhabited by littorinids and, during an average tide and a calm sea, is only splashed with waves. The animals were generally deep within the crevice, and most had to be extracted with forceps or a wire hook. This location is referred to in this paper as 'north Filey'. 2) An extensive boulder field sloping very gradually towards the south (Grid Reference TA 133814). Collections were taken from a single area of about 25 m$^2$. During one tide cycle on 12 May 1992 this area was submerged for about 5 hours (10.00 to 15.00 GMT). This location is referred to as 'south Filey'. The two collection sites are about 200 m apart. All collections were made between 11 February and 12 May 1992.

One collection was made at another site, Porth Neigwl, Gwynedd in north Wales (Grid Reference SH 292254), on 12 April 1992. This site consists of a rock face with horizontal crevices representing the upper limits of the winkles' range and, lower down, a gradually sloping boulder field. Both locations face westward.

### Transplants

On 18 March, 500 animals were collected from north Filey and 225 from south Filey to be transplanted to the opposite site. More were collected from the north since it would be more difficult to find them again after transplanting them to the south boulder field. Each animal was marked on the shell using a small emery wheel on a hobbyist's drill as well as with a small (2–3 mm diameter) spot of red nail varnish. The marked snails were then released at the opposite collection site. A sample of about 75 animals from each site was given a similar mark (but on a different part of the shell) and then returned to their original site to determine whether the handling and marking had an effect on subsequent measurements. Samples of the marked winkles were collected for uric acid determination after 2 weeks (2 April) and after 4 weeks (16 April).

### Species identification and tissue preparation

Field separation of *L. saxatilis* and *L. arcana* is unreliable, so collected animals were returned to the laboratory for identification. The shells were cracked and the body removed and identified under a dissecting microscope by the presence of a brood pouch (*L. saxatilis*) or a jelly gland (*L. arcana*) (Hannaford Ellis, 1979). Only females were used in this study since the separation of the males of these two species is unreliable.

The body was weighed and then homogenised in 100 $\mu$l of phosphate buffer (pH 7.8) at room temperature. After centrifugation the supernatant was removed and frozen for later analysis for uric acid or xanthine dehydrogenase activity.

### Uric acid determination

The supernatant was assayed for uric acid concentration using a modification of the Sigma Diagnostics procedure no. 685. This procedure essentially follows the method of Duerr (1967) and utilises uricase to promote the oxidation of uric acid to allantoin, $CO_2$ and $H_2O_2$. The $H_2O_2$ then reacts, in the presence of peroxidase, with 4-aminoantipyrine and 3,5-dichloro-2-hydroxybenzenesulfonate to form a quinoneimine dye. The intensity of the colour formed is proportional to the concentration of uric acid and is read at 520 nm. Concentration is reported as $\mu$g of uric acid per gram of tissue (wet weight). Since Duerr (1967) expressed his results with reference to dry weight, a large sample of snails ($n = 31$) was weighed and then dried to constant weight to determine the weight reduction.

### Xanthine dehydrogenase (XDH) activity

Tissue supernatant was incubated for 30 min with xanthine substrate in phosphate buffer containing NAD. Absorbance was read at 293 nm at the beginning and end of the incubation period. Since uric acid absorbs light of this wavelength, an increase in absorbance during this time is due to the uric acid generated from xanthine by the action of XDH. The amount of uric acid so formed was determined by comparison with a curve obtained by measuring absorbance at this wavelength of known amounts of uric acid. Activity is reported as $\mu$g of uric acid formed per minute per gram of tissue.

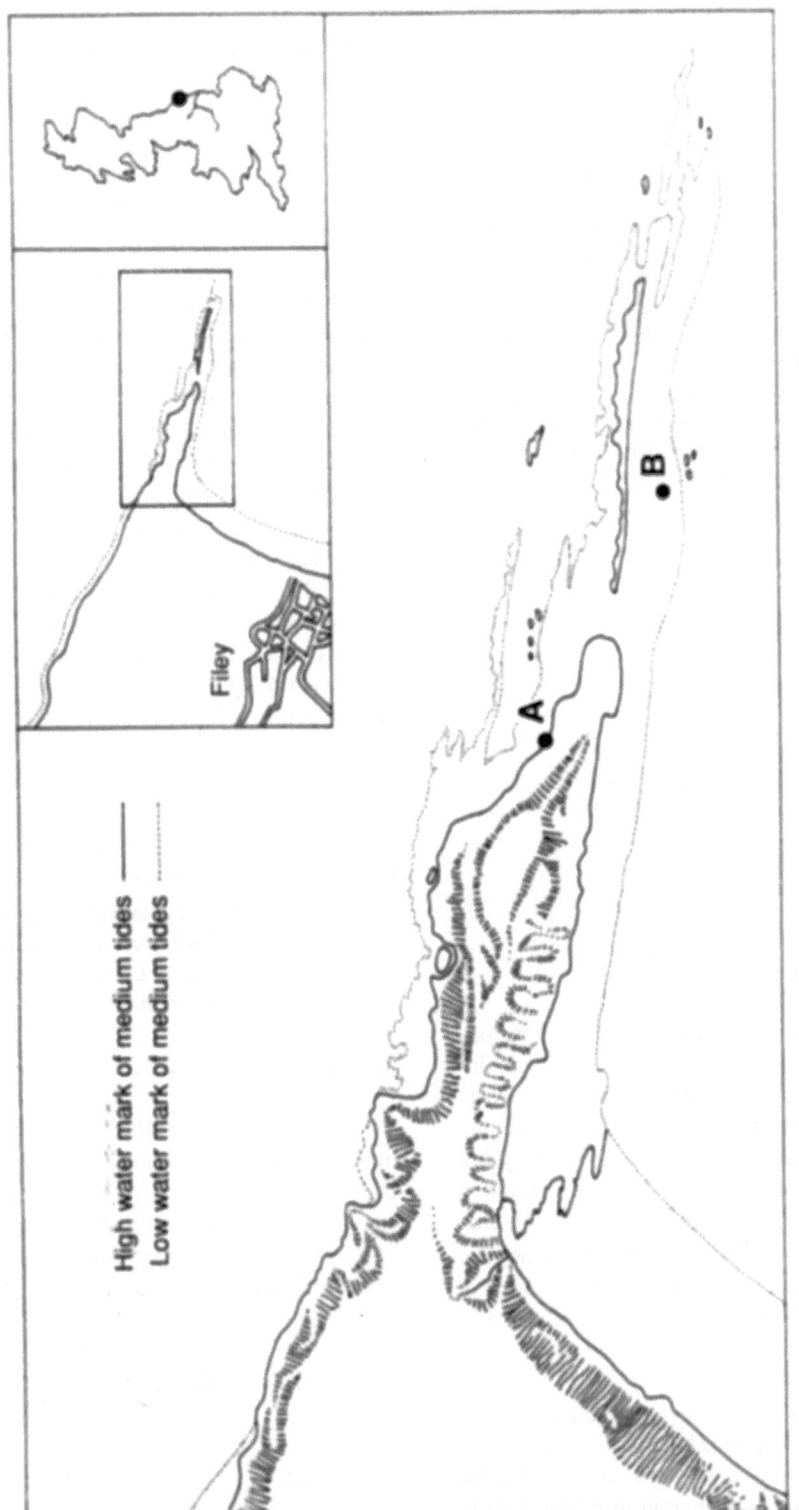

*Fig. 1.* Map of Filey Brigg showing collection sites. A, cliff crevices, referred to in the text as 'north Filey'; B, boulder field, referred to as 'south Filey'. Insets show the Brigg in relation to the town of Filey, and its location on the east coast of England.

114

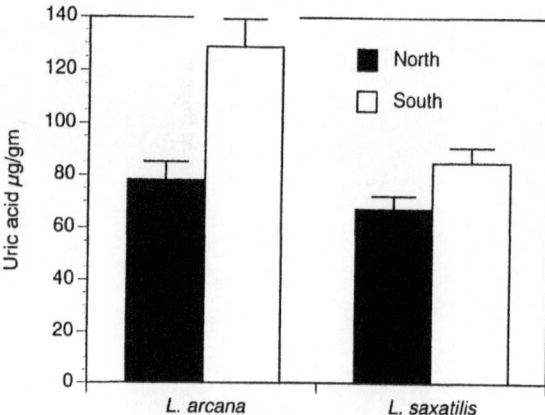

*Fig. 2.* Whole body concentration (mean ± s.e.) of uric acid (µg g⁻¹ wet weight) in winkles collected between 11 February and 12 May, 1992 (composite of all collections). For *L. arcana* north N = 39, south N = 41; for *L. saxatilis* north N = 57, south N = 76.

This procedure is a modification of that of Dykens & Shick (1988).

### Statistics

Differences between means were tested for significance using Student's *t*-test.

## Results

### Uric acid concentration

Figure 2 shows the mean whole-body concentrations of uric acid in the two *Littorina* species after combining the results from all collections at Filey Brigg. Uric acid concentrations in the two species taken from the north Filey (crevice) site do not differ significantly, but both species have a significantly higher concentration in animals collected from south Filey (boulders) than in animals of the same species from north Filey (*L. arcana* P<0.001, *L. saxatilis* P<0.05). Also, *L. arcana* from the south boulder field has a significantly greater (P<0.001) concentration of uric acid than does *L. saxatilis* from the same site.

When the results from individual collections are examined (Fig. 3, left panels) these general relationships are seen to be maintained. However, there appears to be a seasonal change in uric acid content over the course of this study. By 4 March significant differences disappear, to reappear later.

### Transplantation experiment

After four weeks (16 April) there was an apparent shift in uric acid concentration of the transplanted animals in the direction of conformity with the concentration in winkles native to the new location. Thus, by four weeks, *L. saxatilis* that were transferred from south (boulders) to north (cliff crevices) differed significantly (P<0.01) from *L. saxatilis* that were not moved from the southern boulder site, and were no longer significantly different from members of the same species with which they now shared the north Filey site. Transplanted *L. arcana* did not differ significantly from conspecifics from either their original or their new site. Inability to detect a difference in this species could well be due to the much smaller samples recovered than was the case with *L. saxatilis*.

Animals which had been marked, but then returned to their original site rather than transplanted, were also collected at four weeks. Their uric acid concentration was not significantly different from that of unmarked animals collected from that site at the same time, indicating that the process of collecting and marking did not affect uric acid concentration.

### North Wales collection

The uric acid concentrations of winkles collected at Porth Neigwl, north Wales were not significantly different from the mean concentrations for all Filey collections. A notable difference at this location was that very few specimens of *L. arcana* were found in the boulder field low in the intertidal zone.

### Xanthine dehydrogenase (XDH) activity

Two samples of *L. saxatilis* from the 4 March collection were used to determine XDH activity. The samples consisted of 16 winkles from north Filey and 16 from south Filey. The north Filey sample had a mean activity of 0.395 + 0.069 (s.e.) µg uric acid min⁻¹ g⁻¹ of wet body weight (i.e. formed min⁻¹ g⁻¹) and the activity of the south Filey sample was 0.593 + 0.044 µg min⁻¹ g⁻¹. The difference between these two means is statistically significant (P<0.05).

## Discussion

The method used to extract uric acid did not involve heating the homogenate and so would not have extract-

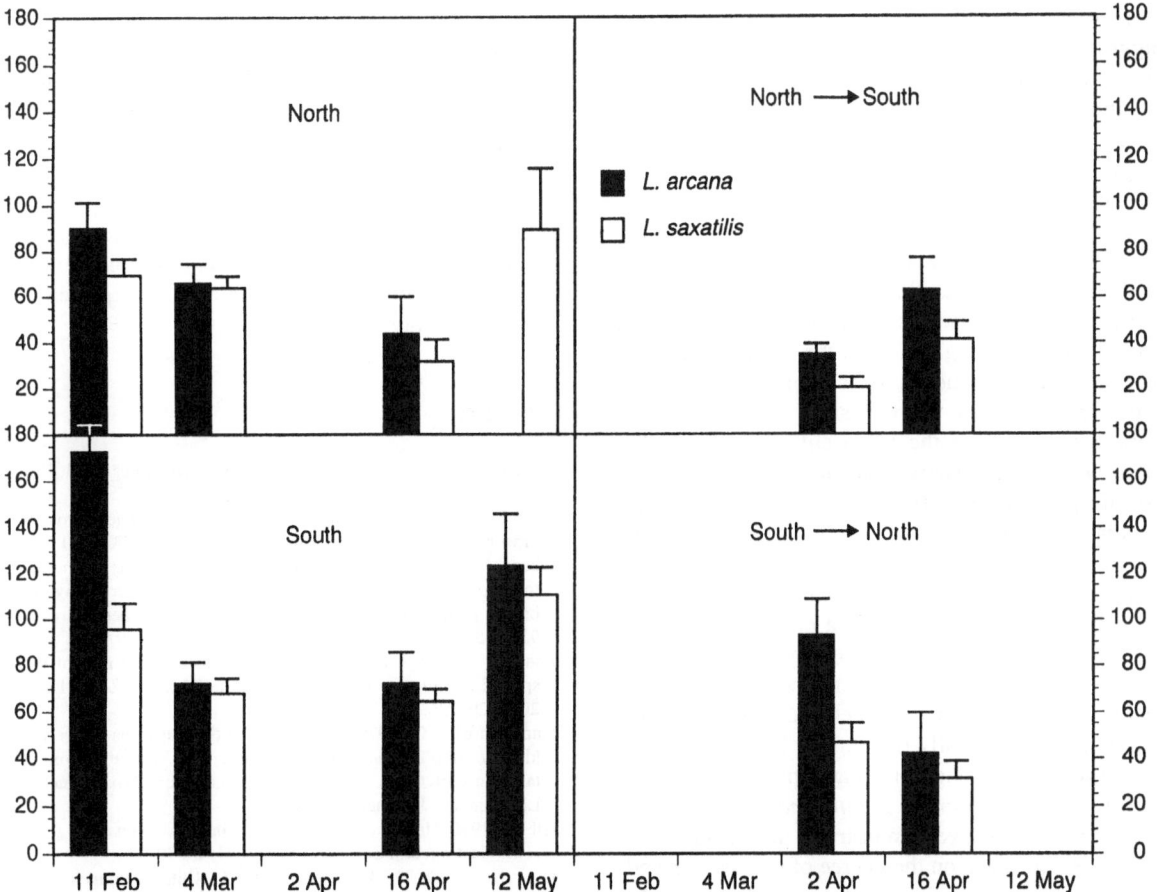

*Fig. 3.* Whole body concentration (mean ± s.e.) of uric acid ($\mu$g g$^{-1}$ wet weight) in winkles collected on five different dates in 1992. Upper left: collections from cliff crevices, north Filey Brigg; upper right: winkles transplanted from north to south Filey Brigg on 18 March; lower left: collections from boulder field, south Filey Brigg; lower right: winkles transplanted from south to north Filey Brigg on 18 March.

ed insoluble, crystalline uric acid. Despite this, the uric acid concentrations of *L. saxatilis* and *L. arcana* obtained in the present study are comparable with those obtained by Duerr (1967), who used lithium carbonate to solubilize any crystalline uric acid. His range for 15 species of snail (i.e. excluding three that were not collected fresh and one, *T. lamellosa*, which had a far higher value than all the others) was 0.035 - 1.33 (mean 0.685) mg g$^{-1}$ dry weight. Converting the results of the current study to dry weight gives a range of 0.246 - 0.740 mg g$^{-1}$.

The uric acid concentrations of *L. saxatilis* and *L. arcana* taken from cliff crevices located above the average high tide level at Filey Brigg did not differ significantly. However, animals of both species from the intertidal boulder field had significantly higher levels than their conspecifics from the crevices. Moreover, the boulder field *L. arcana* had a significantly greater

concentration than *L. saxatilis* from this site, probably indicating that *L. arcana* is capable of responding more vigorously to the conditions in the boulder field that call for greater uric acid production. These differences between sites and between species were maintained over all five collections from February to May, though the absolute concentrations varied between collections. These latter changes suggest a seasonal influence on uric acid concentration which requires further investigation. A seasonal change in uric acid content has been demonstrated in a tropical snail, *Bellamya bengalensis*, which has a higher concentration in the dry than in the rainy season (Kulkarni & Baramativala, 1986).

The difference in uric acid concentration between the two sites is supported by measurements of XDH activity, which show significantly higher activity of this enzyme in animals from the boulder field than from the cliff crevices. Although XDH is not involved in the *de novo* synthesis of uric acid from ammonia, this does suggest that the accumulation of uric acid in the tissues of animals from the boulder field is not entirely due to retention, but at least in part to an increase in synthesis of uric acid. The degree of activity suggests that this enzyme is not a limiting factor in the production of uric acid, a conclusion in agreement with that of Heil (1990).

The difference between the cliff crevice and boulder field populations could be genetic or it could be a physiological response to some factor(s) present in the boulder field which, given time, the animals from the crevices could respond to. To test this, animals were transplanted between the two sites. By four weeks, animals moved from the north cliff crevices to the south boulder field showed an increase in uric acid from the previous collection, suggesting that they were responding to the boulder field stimulus and were accumulating uric acid, as do the snails native to that site. Those animals moved from the south boulder field to the north cliff crevices showed a continuing decline at four weeks and became indistinguishable in uric acid concentrations from native crevice animals. These results indicate that the difference in uric acid concentration is not genetically determined but is a physiological response, though one that apparently required some weeks to accomplish. In *Littorina littorea* Heil & Eichelberg (1983) have demonstrated that the uric acid content is dependent on the degree of submersion and the level of salinity, with animals exposed to less submersion and higher salinity having a higher uric acid content than those exposed to greater submersion and lower salinity. Animals transferred within the intertidal adapted to the changed level of submersion.

The environmental factor(s) resulting in greater uric acid accumulation in snails from the boulder field in comparison to those from the crevices remains unidentified. One would assume that animals in cliff crevices which are rarely (if ever) submerged would be more like terrestrial snails in their nitrogen excretion physiology, and that snails in boulder fields submerged twice daily would be less uricotelic, although we have no independent evidence of greater desiccation for crevice dwelling animals. However, in this study, the reverse was the case. As we did not measure ammonia excretion concomitantly with uric acid accumulation, we cannot conclude that the increase in uric acid reflects its *de novo* synthesis from ammonia as an adaptation to desiccation and a corresponding need to detoxify ammonia. A high uric acid content could also reflect differences in diet or ration, both of which might, like uric acid content, be expected to show seasonality.

## References

Delaunay, H., 1931. L'excretion azotée des invertebres. Biol. Rev. 6: 265–301.

Duerr, F. G., 1967. The uric acid content of several species of marine prosobranch snails. Comp. Biochem. Physiol. 22: 333–340.

Dykens, J. A. & J. M. Shick, 1988. Relevance of purine catabolism to hypoxia and recovery in euroxic and stenoxic marine invertebrates, particularly bivalve molluscs. Comp. Biochem. Physiol. 91C: 35–41.

Grahame, J. & P. J. Mill, 1986. Relative size of the foot of two species of *Littorina* on a rocky shore in Wales. J. Zool., Lond. 208: 229–236.

Hannaford Ellis, C., 1979. Morphology of the oviparous rough winkle, *Littorina arcana*, Hannaford Ellis, 1978, with notes on the taxonomy of the *L. saxatilis* species complex (Prosobranchia: Littorinidae). J. Conch. 30: 43–56.

Heil, K., 1990. Untersuchungen zur xanthindehydrogenase in *Littorina* (Gastropoda). Zool. Jb. Physiol. 94: 19–30.

Heil, H. P. & D. Eichelberg, 1983. Untersuchungen zum Harnsäuremetabolismus von *Littorina littorea* (Gastropoda). Helgoländer Meeresunters. 35: 465–472.

Kulkarni, K. M. & D. Q. Baramativala, 1986. Uric acid storage during season and temperature acclimation stress in snails *Bellamaya bengalensis*. Environ. Ecol. 4: 151–152.

Mill, P. J. & J. Grahame, 1990. Distribution of the species of rough periwinkle (*Littorina*) in Great Britain. Hydrobiologia 193 (Dev. Hydrobiol. 56): 21–27.

Needham, J., 1935. Problems of nitrogen catabolism in invertebrates. II. Correlation between uricotelic metabolism and habitat in the phylum Mollusca. Biochem. J. 29: 238–251.

Needham, J., 1938. Contributions of chemical physiology to the problem of reversibility in evolution. Biol. Revs Cambridge Phil. Soc. 13: 225–251.

Potts, W. T., 1967. Excretion in the molluscs. Biol Rev. 42: 1–41.

Wilbur, K. M., 1983. The Mollusca. Academic Press, New York.

*Hydrobiologia* **309**: 117–121, 1995.
*P. J. Mill & C. D. McQuaid (eds), Advances in Littorinid Biology.*
© 1995 *Kluwer Academic Publishers.*

# Crystalline calcium in littorinid mucus trails

Mark S. Davies[1,2] & Susan J. Hutchinson[2]
[1]*Ecology Centre, University of Sunderland, Sunderland SR1 3SD, UK*
[2]*Department of Environmental Biology, University of Manchester, Manchester M13 9PL, UK*

*Key words:* calcium, granule, heavy metal, *Littorina*, locomotion, mucus

## Abstract

Previous work has shown that the feet of terrestrial and freshwater snails are important in calcium regulation, often secreting granules of $CaCO_3$. This phenomenon has not, until now, been observed in marine snails. Here we report the presence of $CaCO_3$ granules in the trail mucus of *Littorina littorea* (L.), *L. saxatilis* (Olivi) and *L. obtusata* (L.) Fixed mucus trails on plastic coverslips were examined by X-ray microanalysis under the SEM. Of the single-metal granules observed in the mucus trails the most abundant were of calcium (means: *L. littorea*, 440 mm$^{-2}$; *L. saxatilis*, 401 mm$^{-2}$; *L. obtusata*, 348 mm$^{-2}$) followed for each species by silicon (maximum mean density: *L. saxatilis*, 120 mm$^{-2}$) and iron (maximum mean density: *L. saxatilis*, 65 mm$^{-2}$) granules. Single-metal granules of Al, Ti, Mg and P were also found but only in the mucus trails of *L. obtusata*, perhaps reflecting its different collection site from the other two species. The mean size of the calcium granules showed significant interspecific variation (*L. littorea*, 1.32 $\mu$m diameter$\pm 0.08$ $\mu$m, $n = 143$; *L. saxatilis*, 1.80 $\mu$m$\pm 0.12$, $n = 113$; *L. obtusata*, 2.14 $\mu$m$\pm 0.09$, $n = 167$). Most calcium granules (*L. littorea*, 80%, $n = 35$; *L. saxatilis*, 57%, $n = 113$; *L. obtusata*, 69%, $n = 167$) were attached to, or embedded within, microthreads of mucus which tended to run parallel to the direction of locomotion. The significance of this is unknown although it may imply that the $CaCO_3$ granules are secreted with the mucus. It is concluded that calcium losses *via* this route are too small for pedal mucus to function significantly in ionoregulation of calcium. The calcium in the trail may therefore perform other functions, for example indicating trail polarity.

## Introduction

The mucus that is secreted from the foot of littorinids as they crawl and is deposited in a thin layer as a mucus trail is energetically costly (Davies *et al.*, 1992). The main function of this mucus is in locomotion (Denny, 1980), although additional functions have been proposed (see e.g., Denny, 1989) and these may offset the high cost of this form of locomotion. Such functions have included the provision of trail-following cues and ionoregulatory pathways, *i.e.* physical structures in the mucus trail may be used as markers for trail identification by conspecifics or the mucus may be used as a sink for excess calcium and other elements (see Simkiss & Wilbur, 1977).

In terrestrial and freshwater gastropods the foot has been shown to be an important region for calcium metabolism. Subepithelial granules of calcium

have been identified in *Helix aspersa* (Campion, 1961) and in *Pomatias elegans* (Bensalem & Chetail, 1982), often to such an extent that the foot appears white (Campion, 1961). A calcium-staining secretion from mucocytes with a granular content has also been identified in the foot of *Littorina littorea* (Shirbhate & Cook, 1987). However, all these granules were identified only in histological section and were not observed in the mucus trail. A granular component of the trail mucus of *H. aspersa* has been observed (Simkiss & Wilbur, 1977), but its composition was not determined. Despite these findings, the role played by the foot and its mucus in regulating calcium metabolism is unknown.

Here we report positive identification of calcium granules within the mucus trails of three littorine snails (*Littorina littorea* (L.), *L. saxatilis* (Olivi) and *L. obtusata* (L.)) and hypothesise as to their function.

118

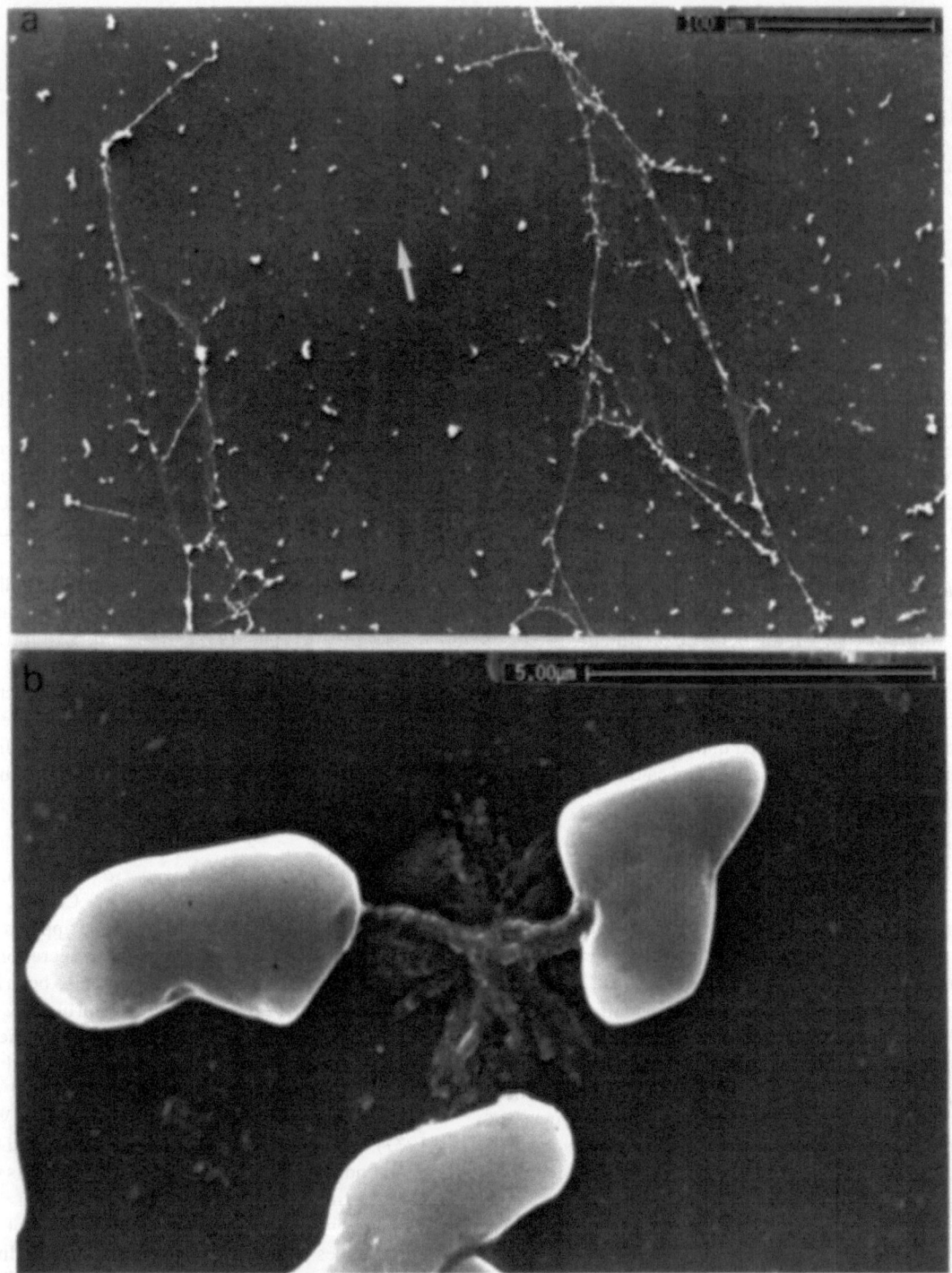

*Fig. 1.* Scanning electron micrographs of littorinid trail mucus. (a) *L. littorea.* Low-power view of mucus microthreads (running parallel to trail direction; arrow indicates direction of trail) and discrete mucus particles. Granules can be seen attached to microthreads. (b) *L. obtusata.* High-power view of three calcium granules clustering around microthreads of mucus.

## Materials and methods

Littorinids were collected from the rocky intertidal at Rhosneigr, Anglesey, Wales (British National Grid reference SH 314729) in March 1990 (*Littorina littorea*) and May 1990 (*L. saxatilis*) and at Derbyhaven, Isle of Man (SC 294685) in May 1990 (*L. obtusata*). Animals were transported to the laboratory at Manchester and used within 10 days.

Snails were placed in small (about $200 \times 100 \times 100$ mm) clear plastic boxes and allowed to crawl over 'thermonox' (Bio-Rad) plastic coverslips. Snails moved only when the humidity was high and to achieve this a small beaker of distilled water was placed in each box. Once a mucus trail had been produced across a coverslip, the trail was immediately washed by a single dip in distilled water and fixed by then plunging the coverslip into liquid nitrogen 'slush' at $-203$ °C. Coverslips were lyophylised overnight, cut into small pieces and mounted for the SEM. Carbon-coated specimens were examined in a Cambridge 360 SEM. Granules present within the mucus trails were analysed for elemental composition using a Link energy dispersive X-ray microanalysis tool and the number of each granule type per unit trail area was calculated. The size of each calcium granule present (determined using an X-ray backscatter detector) was measured using an on-screen facility. Size is expressed as the mean of the longest and shortest visible axes.

At least fifteen snails of each species were used. All data reported here were obtained from the central two-thirds of each trail's width.

## Results

Only data regarding single-element granules are presented here. Granules consisting of more than one identified element (Al + Si, Al + Si + Mg, Al + Si + P, Al + Si + Fe, Na + Cl) were observed, but at a much lower frequency.

Under the SEM, mucus from all three species was visible as microthreads, all $<1$ $\mu$m in width (Fig. 1a). In most cases the microthreads ran parallel to the direction of snail locomotion. Mucus was also present as discrete particles between the microthreads. Microscopical examination revealed no difference in the appearance of mucus between the three species.

Granules of varying composition were found in the trails and they were all $<5$ $\mu$m in diameter. Some granules were completely embedded within the mucus,

their presence only detected using backscattered X-rays. Other granules were attached to the microthreads (Fig. 1b) and others were not associated with mucus. However, most of the calcium granules present within the trail (*L. littorea*, 80%, $n = 35$; *L. saxatilis*, 57%, $n = 113$; *L. obtusata*, 69%, $n = 167$) were attached to, or embedded within, a microthread of mucus. This represents very high localisation in mucus, since the microthreads occupy such little trail area (see Fig. 1a). Figure 1b shows three calcium granules connected to a central thread-like structure of mucus.

Of the single-element granules observed in the mucus trails, the most abundant were of calcium (means: *L. littorea*, 440 mm$^{-2}$; *L. saxatilis*, 401 mm$^{-2}$; *L. obtusata*, 348 mm$^{-2}$) followed for each species by silicon (maximum mean density: *L. saxatilis*, 120 mm$^{-2}$) and iron (maximum mean density: *L. saxatilis*, 65 mm$^{-2}$) granules (Fig. 2). Single-element granules of Al, Ti, Mg, P and Cl were also found but only in the mucus trails of *L. obtusata*. No granules were found on control coverslips which had not been crawled over by snails.

The mean size of the most abundant granule type, calcium, showed significant interspecific variation (Kruskal-Wallis test: $H = 66.3$, $p = 0.000$): *L. littorea*, 1.32 $\mu$m diameter$\pm 0.08$ $\mu$m, $n = 143$; *L. saxatilis*, 1.80 $\mu$m$\pm 0.12$, $n = 113$; *L. obtusata*, 2.14 $\mu$m$\pm 0.09$, $n = 167$ (Fig. 3).

The 'single-element' granules are most likely to be oxides, nitrates or carbonates. This is because the X-ray microanalysis system cannot detect elements with atomic numbers $<9$. Hence the calcium in a granule which appears on analysis to be composed solely of calcium may be combined with any elements with atomic number $<9$. In view of the appearance of the calcium granules and the abundance of carbonate in seawater, these granules are most likely to be of $CaCO_3$.

To address the question of whether loss of calcium in mucus trails may be part of an ionoregulatory process its magnitude has been calculated. Assuming a mucus trail width of 10 mm, that the calcium granules are of calcite ($CaCO_3$) with a density of 2.71 g cm$^{-3}$ and that snails move 2 m d$^{-1}$ (see Davies *et al.*, 1992), then taking the mean size and trail density of calcium granules, the loss of calcium in mucus trails is 11.51 $\mu$g d$^{-1}$ for *L. littorea*, 26.33 $\mu$g d$^{-1}$ for *L. saxatilis* and 39.18 $\mu$g d$^{-1}$ for *L. obtusata*.

120

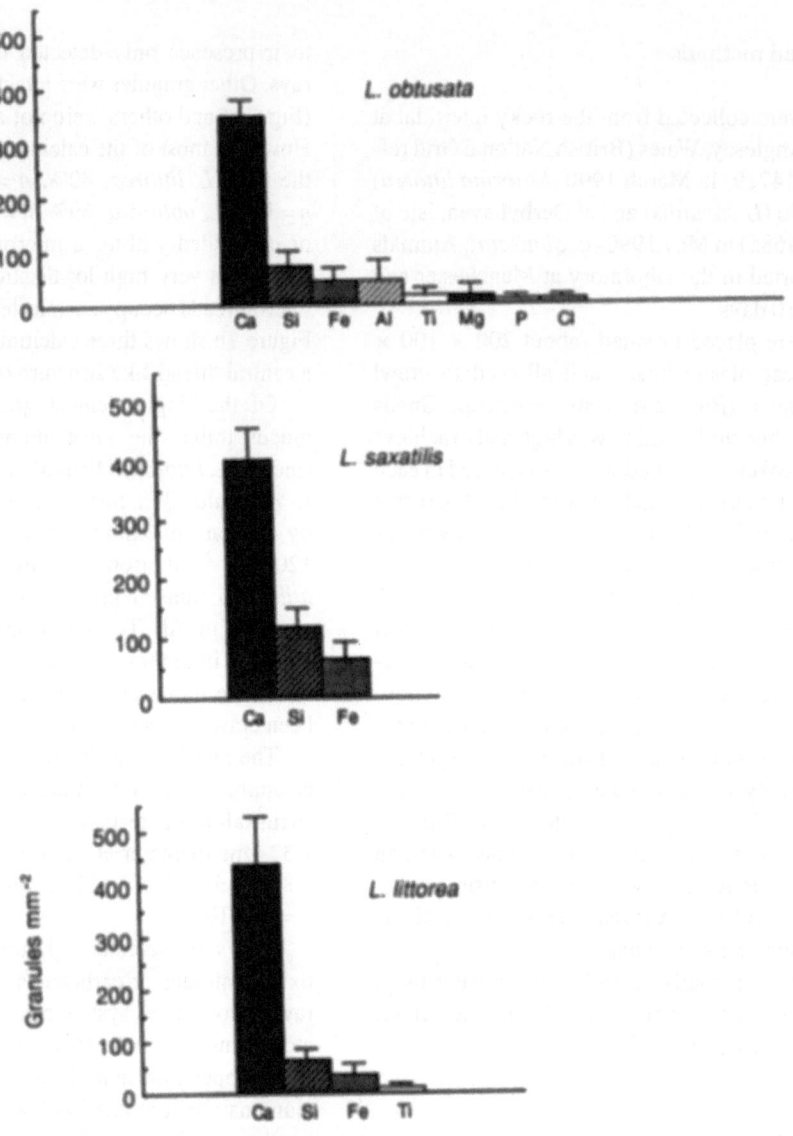

*Fig. 2.* Mean density (±SE) of the various single-element granules recorded by X-ray microanalysis in the mucus trails of three littorinid species.

## Discussion

The close association of calcium granules with mucus microthreads suggests that the mucus and the granules have a common origin within the foot sole, particularly since some granules were embedded within the mucus. However, microthreads of mucus may be artifactual, and if so the process of aggregation of mucus to form microthreads may alter a uniform distribution of granules. Nevertheless, Simkiss & Wilbur (1977) observed similar mucus 'filaments' in the trails of *Helix aspersa* (with associated unidentified granules) under Nomars-

ki optics as did Bretz & Dimock (1983) in the trails of *Ilyanassa obsoleta* using an immunofluorescent technique. Both these groups also noted, as here, that the filaments ran parallel to snail movement.

The second commonest granule in all three species' trails was silicon-based. These particles (sand) were probably attached to the winkles' feet. The iron granules which were also present may be contamination from the metal scissors used to cut the coverslips, although no granules were observed on control coverslips. The greater diversity of granule type in the trails of *L. obtusata* probably reflects the different sam-

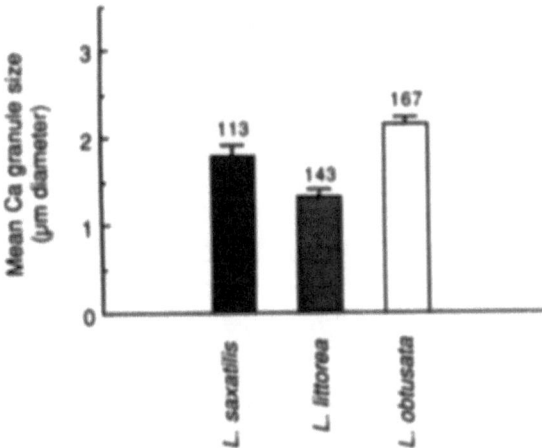

*Fig. 3.* Mean (±SE) granule size (taken as the mean of the longest and shortest visible axes) of the calcium granules recorded by X-ray microanalysis in the mucus trails of three littorinid species. Numbers represent sample sizes. Calcium granule diameter shows significant interspecific variation (see text).

pling site for this species (close to an industrial unit), rather than any interspecific difference in metabolism. This observation warrants further work and may imply that winkles can eliminate a wide variety of metals from their bodies and that the mucus trail can form the end-point of an excretion pathway capable of dealing with numerous elements. The interspecific difference in calcium-based granule size is difficult to explain but could possibly function as a marker of the trail-layer's species or could simply be due to differences in secretory apparatus. Granule size is comparable to those reported as intracellular in *L. littorea* (Mason *et al.*, 1984) and *L. saxatilis* (Brough & White, 1990). It is, however, unlikely that these intracellular granules emerge with the mucus since Walker (1972) showed that $^{45}$Ca in intracellular granules was not transferred to the mucus of the slug *Agriolimax reticulatus*.

The calculations of calcium loss per day in trail mucus indicate losses which appear too small for pedal mucus to be a significant ionoregulatory route for calcium. These losses are of the same order of magnitude as those reported as $\mu mol\,h^{-1}\,cm^{-2}$ of *H. aspersa* foot by Simkiss & Wilbur (1977), if it is assumed that snails move for 2 h d$^{-1}$.

Histological examination of the foot sole of *L. littorea* revealed unidentified granular components in 'cell type L4' (Shirbhate & Cook, 1987). The secretions from this cell type alone also stained positively for calcium. It is therefore likely that this cell type, which is located in the foot sole and is more common anteriorly, is the source of the calcium granules, although granule function remains unclear.

## Acknowledgements

We are grateful to The Royal Society for a travel grant to attend the 4th International Symposium on Littorinid Biology and to Malcolm Haswell for preparing the plate.

## References

Bensalem, M. & M. Chetail, 1982. Hydrocalcic metabolism and pedal glands in *Pomatias elegans* (Müller) (Mollusca, Proso-branchia). Malacologia 22: 293–303.

Bretz, D. D. & R. V. Dimock, 1983. Behaviorally important characteristics of the mucous trail of the marine gastropod *Ilyanassa obsoleta* (Say). J. exp. mar. Biol. Ecol. 71: 181–191.

Brough, C. N. & K. N. White, 1990. Localization of metals in the gastropod *Littorina saxatilis* (Prosobranchia: Littorinoidea) from a polluted site. Acta zool. 71: 77–88.

Campion, M., 1961. The structure and function of the cutaneous glands in *Helix aspersa*. Quart. J. microsc. Sci. 102: 195–216.

Davies, M. S., H. D. Jones & S. J. Hawkins, 1992. Pedal mucus production in *Littorina littorea*. In J. Grahame, P. J. Mill & D. G. Reid (eds), Proceedings of the 3rd International Symposium on Littorinid Biology. The Malacological Society of London, London: 227–233.

Denny, M. W., 1980. The role of gastropod pedal mucus in locomotion. Nature 285: 160–161.

Denny, M. W., 1989. Invertebrate mucous secretions: functional alternatives to vertebrate paradigms. In E. Chantler & N. A. Ratcliffe (eds), Symposia of the Society for Experimental Biology number XLIII. Mucus and related topics. The Company of Biologists Limited, Cambridge: 337–366.

Mason, A. Z., K. Simkiss & K. P. Ryan, 1984. The ultrastructural localization of metals in specimens of *Littorina littorea* collected from clean and polluted sites. J. mar. biol. Ass. UK 64: 699–720.

Shirbhate, R. & A. Cook, 1987. Pedal and opercular secretory glands of *Pomatias*, *Bithynia* and *Littorina*. J. moll. Stud. 53: 79–96.

Simkiss, K. & K. M. Wilbur, 1977. The molluscan epidermis and its secretions. Symp. zool. Soc. Lond. 39: 35–76.

Walker, G., 1972. The digestive system of the slug *Agriolimax reticulatus* (Müller): experiments on phagocytosis and nutrient absorption. Proc. malacol. Soc. Lond. 40: 33–43.

*Hydrobiologia* **309**: 123–128, 1995.
*P. J. Mill & C. D. McQuaid (eds), Advances in Littorinid Biology.*
©1995 *Kluwer Academic Publishers.*

# Allozyme comparison of four littorinid species morphologically similar to *Littorina sitkana*

N. I. Zaslavskaya
*Institute of Marine Biology, Vladivostok 690041, Russia*

*Key words: Littorina sitkana, Littorina kasatka, Littorina subrotundata*, genetic relationships, genetic distance, sibling species

## Abstract

Eight species of the genus *Littorina* were hitherto recognised in the north-western region of the Pacific Ocean: *L. sitkana, L. brevicula, L. mandshurica, L. squalida, L. aleutica, L. naticoides, L. kasatka* and *L. subrotundata*. Using allozyme electrophoresis it has been demonstrated that, in the Kurile Islands, three of these species (*L. sitkana, L. subrotundata* and *L. kasatka*) co-occur, together with a fourth, still undescribed species (*L.* sp.). These four species were compared at 16 loci coding for 13 enzymes. All species were easily distinguished by diagnostic enzyme markers. The mean genetic distances and ranges between species pairs are: *L. sitkana* and *L.* sp. D = 0.622 (0.561–0.741), *L. sitkana* and *L. subrotundata* D = 0.981 (0.821–1.110), *L. subrotundata* and *L.* sp. D = 0.975 (0.955–0.995). The genetic distance between *L. kasatka* and each of the other three species was greater than 1 (range 1.123–2.087). These data suggest that *L. sitkana, L. subrotundata* and *L.* sp could be members of a species complex; according to current classifications these three belong to the subgenus *Neritrema*. However, the genetic distance between *L. kasatka* and *L. sitkana* is much greater than between *L. sitkana* and other *Neritrema* species, and thus supports the classification of *L. kasatka* in the subgenus *Littorina*.

## Introduction

The genus *Littorina* is a taxonomically difficult group. Until recently, shell morphology was used as the main basis for the systematic arrangement of littorinids. Relying on such characters, Golikov & Kusakin (1978) recognised six species of *Littorina* in the north-western region of the Pacific Ocean: *L. squalida* Broderip & Sowerby, 1829; *L. brevicula* (Philippi, 1844); *L. mandshurica* Schrenck, 1861; *L. kurila* Middendorff, 1848; *L. sitkana* Philippi, 1846 and *L. aleutica* Dall, 1872. However, more recently it has been demonstrated that shell morphology may be variable within a species and may be subject to natural selection (e.g. Janson, 1982, 1983; Raffaelli, 1982). Such studies have led to the taxonomic revision of some species (e.g. Reid, 1990).

The use of allozyme markers and anatomical features has resulted in the recognition of sibling species within the Atlantic species of *Littorina* (Johannes-

son & Johannesson, 1990; Reid, 1990; Ward, 1990). Until recently these species were the most extensively studied. However, since 1991, there has been some progress in the study of the Pacific species. Anatomical comparisons of the reproductive system have shown that three of the six species noted by Golikov & Kusakin (1978) (i.e. *L. squalida, L. brevicula* and *L. mandshurica*) are good species (Reid, 1990; Reid & Golikov, 1991; Reid *et al.*, 1991). However, the other three 'species' (*L. kurila, L. sitkana* and *L. aleutica*) have been shown to comprise a mixture of at least five species (*L. sitkana, L. subrotudata* (Carpenter, 1864), *L. naticoides* (Reid & Golikov, 1991), *L. aleutica* and *L. kasatka* (Reid *et al.*, 1991) with *L. kurila* being recognised as a junior synonym of *L. sitkana* (Reid, 1990).

The current investigation was carried out simultaneously with, and independently of, Reid's study (Reid, 1990), with the idea of using allozymes to investigate the population structure of the taxon previously

124

*Table 1.* A designation of samples of the four species collected at different years

| Locality | L. subrotundata | L. sitkana | L. kasatka | L. sp. |
|----------|-----------------|------------|------------|--------|
| **1990** | | | | |
| Jankich Island | sub1 | sit1 | kas1 | |
| | sub2 | sit2 | | |
| Vostok Bay | | sit3 | | |
| | | | | |
| **1991** | | | | |
| Kunashir Island | | sit4 | | |
| | | sit5 | | |
| Iturup Island | | sit6 | | sp |
| | | sit8 | kas2 | |
| | | sit9 | kas3 | |
| | | sit10 | | |
| Komandor Island | | sit7 | | |
| Vostok Bay | | sit11 | | |

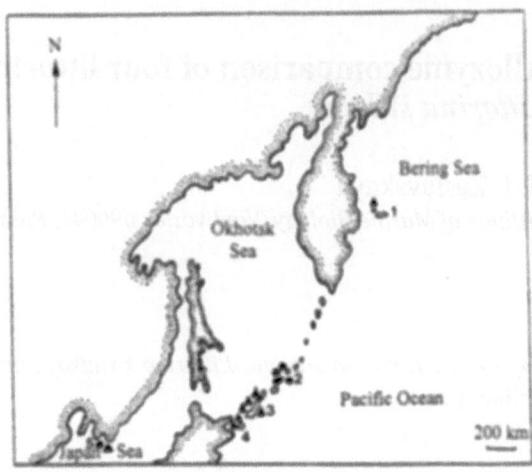

*Fig. 1.* Map to show the sample sites. ●, collections of 1990; ▲, collections of 1991. 1, Komandor Islands; 2, Jankich and Ruponkich Islands; 3, Iturup Island; 4, Kunashir Island; 5, Vostok Bay.

described as *L. kurila.* In 1989, while sorting out samples of supposedly *L. kurila* from Jankich and Ruponkich islands (Kurile Islands) it was suspected that an additional, undescribed species was present in the samples. This new species was later described by Reid *et al.* (1991) as *L. kasatka.* Additional samples from Jankich Island, collected in 1990, were analysed and it was discovered that two of the four samples (from the northern part of the island and from its Pacific coast) consisted of individuals of *L. subrotundata* (D. Reid, personal communication). In 1991 several samples from Iturup, Kunashir and Komandor Islands were studied. The sample from Kasatka Bay (on the Pacific coast of Iturup Island) contained individuals of *L. sitkana*, *L. kasatka* and an additional species which appeared distinct from the former two and from *L. subrotundata.* Specimens of it were collected from the same boulder as *L. sitkana.* A comparison of the allele composition of these animals with that of specimens of *L. sitkana*, *L. kasatka* and *L. subrotundata* points to its status as a new species (Zaslavskaya, in press). However, this taxon cannot yet be described because the pallial oviduct of the females was indistinguishable from that of *L. sitkana* and mature males were absent from the sample (D. Reid, personal communication). Other samples from Iturup Island consisted of a mixture of *L. sitkana* and *L. kasatka.* Samples from Kunashir and Komandor Islands contained only *L. sitkana.*

Hence there were four different species in the samples which were formerly believed to contain only *L. kurila* (referred to as *L. sitkana* in accordance with Reid (1990) throughout the remainder of this paper). The purpose of the current study was to determine the genetic relationships between these species.

## Material and methods

Figure 1 shows the locations of the sampling areas. Samples for analysis were obtained in 1990 and 1991, as follows. 1990: two samples of *L. subrotundata* (sub1, sub2) from Jankich Island, two containing *L. sitkana* (sit1, sit2) and *L. kasatka* (kas1) from Jankich Island and one of *L. sitkana* from Vostok Bay, Japan Sea (sit3). 1991: eight samples of *L. sitkana* from Kunashir, Iturup and Komandor Islands (sit4–sit10) and Vostok Bay (sit11), two of *L. kasatka* (kas2, kas3) and one of *L.* sp. (sp) from Iturup Island (Table 1).

Horizontal starch gel electrophoresis was carried out as described by Zaslavskaya (1989). Data were analysed with the BIOSYS package (Swofford & Selander, 1981). Phenograms reflecting genetic relationships between the species were produced from estimates of Nei's genetic distances (Nei, 1972) by the unweighted pairgroup method (Sokal & Sneath, 1963).

Thirteen enzymes coding for 16 loci were studied in *L. sitkana*, *L. kasatka* and *L.* sp.: alanopine dehydrogenase (Aldh), aldolase (Ald), arginine phosphate kinase

*Table 2.* Allele frequencies for pooled data for *Littorina subrotundata* (sub) *L. sitkana* (sit), *L. kasatka* (kas) and *L.* sp (sp)

| Locus | Species | | | |
|---|---|---|---|---|
| | sub | sit | kas | sp |
| Aldh | | | | |
| (N) | 71 | 91 | 37 | 29 |
| A | 0.000 | 0.302 | 0.000 | 0.000 |
| B | 1.000 | 0.689 | 0.000 | 0.017 |
| C | 0.000 | 0.000 | 1.000 | 0.983 |
| Arpk | | | | |
| (N) | 48 | 129 | 25 | 13 |
| A | 0.000 | 0.190 | 0.000 | 0.962 |
| B | 0.000 | 0.511 | 1.000 | 0.038 |
| C | 0.000 | 0.295 | 0.000 | 0.000 |
| D | 0.948 | 0.004 | 0.000 | 0.000 |
| E | 0.052 | 0.000 | 0.000 | 0.000 |
| Gpt-1 | | | | |
| (N) | 48 | 106 | 29 | 2 |
| A | 1.000 | 1.000 | 0.000 | 1.000 |
| B | 0.000 | 0.000 | 0.448 | 0.000 |
| C | 0.000 | 0.000 | 0.552 | 0.000 |
| Gpt-2 | | | | |
| (N) | 14 | 24 | 8 | 5 |
| A | 0.000 | 0.000 | 1.000 | 0.000 |
| B | 0.000 | 1.000 | 0.000 | 1.000 |
| C | 1.000 | 0.000 | 0.000 | 0.000 |
| Ipp | | | | |
| (N) | 48 | 134 | 44 | 16 |
| A | 0.000 | 0.369 | 0.000 | 0.562 |
| B | 0.000 | 0.552 | 0.011 | 0.000 |
| C | 1.000 | 0.072 | 0.716 | 0.438 |
| D | 0.000 | 0.007 | 0.273 | 0.000 |
| Mdh-1 | | | | |
| (N) | 14 | 24 | 8 | 5 |
| A | 0.000 | 0.000 | 1.000 | 1.000 |
| B | 0.000 | 1.000 | 0.000 | 0.000 |
| C | 1.000 | 0.000 | 0.000 | 0.000 |
| Pep-1 | | | | |
| (N) | 52 | 85 | 33 | 20 |
| A | 0.173 | 0.212 | 0.318 | 0.000 |
| B | 0.135 | 0.112 | 0.682 | 0.000 |
| C | 0.692 | 0.676 | 0.000 | 1.000 |

*Table 2. (cont.).*

| Locus | Species | | | |
|---|---|---|---|---|
| | sub | sit | kas | sp |
| Pep-2 | | | | |
| (N) | 90 | 185 | 63 | 45 |
| A | 0.000 | 0.011 | 0.000 | 0.000 |
| B | 0.011 | 0.759 | 0.000 | 0.033 |
| C | 0.828 | 0.019 | 0.000 | 0.000 |
| D | 0.011 | 0.011 | 0.000 | 0.967 |
| E | 0.150 | 0.200 | 0.000 | 0.000 |
| F | 0.000 | 0.000 | 0.016 | 0.000 |
| G | 0.000 | 0.000 | 0.984 | 0.000 |
| 6-Pgd | | | | |
| (N) | 36 | 67 | 24 | 11 |
| A | 0.403 | 0.940 | 1.000 | 1.000 |
| B | 0.597 | 0.060 | 0.000 | 0.000 |
| Pgi | | | | |
| (N) | 93 | 187 | 49 | 40 |
| A | 0.000 | 0.000 | 0.041 | 0.000 |
| B | 0.000 | 0.112 | 0.959 | 0.000 |
| C | 0.000 | 0.877 | 0.000 | 1.000 |
| D | 0.672 | 0.011 | 0.000 | 0.000 |
| E | 0.323 | 0.000 | 0.000 | 0.000 |
| F | 0.005 | 0.000 | 0.000 | 0.000 |
| Phgk | | | | |
| (N) | 36 | 100 | 43 | 38 |
| A | 1.000 | 0.585 | 0.000 | 0.829 |
| B | 0.000 | 0.405 | 0.000 | 0.171 |
| C | 0.000 | 0.010 | 1.000 | 0.000 |
| Pgm-1 | | | | |
| (N) | 76 | 187 | 48 | 40 |
| A | 0.099 | 0.008 | 0.000 | 0.000 |
| B | 0.000 | 0.000 | 0.000 | 1.000 |
| C | 0.000 | 0.976 | 0.000 | 0.000 |
| D | 0.776 | 0.005 | 0.000 | 0.000 |
| E | 0.125 | 0.011 | 0.042 | 0.000 |
| F | 0.000 | 0.000 | 0.958 | 0.000 |
| Pgm-2 | | | | |
| (N) | 73 | 170 | 19 | 28 |
| A | 0.842 | 0.409 | 0.395 | 1.000 |
| B | 0.158 | 0.588 | 0.605 | 0.000 |
| C | 0.0000 | 0.003 | 0.000 | 0.000 |

(Arpk), glutamate oxaloacetate transaminase (Got), glutamate pyruvate transaminase (Gpt-1, Gpt-2), inorganic pyrophosphatase (Ipp), leucine aminopeptidase (Lap), malate dehydrogenase (Mdh-1), peptidase (the substrate used was leucyl-glycine) (Pep-1, Pep-2), 6-phosphogluconate dehydrogenase (6-Pgd), phosphoglycerate kinase (Phgk), phosphoglucose isomerase (Pgi) and phosphoglucomutase (Pgm-1, Pgm-2). Thirteen loci were assayed in *L. subrotundata* (as above except for Ald, Got and Lap).

*Table 2. (cont.).*

| Locus | Species | | | |
|---|---|---|---|---|
| | sub | sit | kas | sp |
| Ald | | | | |
| (N) | * | 86 | 33 | 29 |
| A | | 0.000 | 1.000 | 0.000 |
| B | | 1.000 | 0.000 | 0.000 |
| C | | 0.000 | 0.000 | 0.948 |
| D | | 0.000 | 0.000 | 0.052 |
| Got | | | | |
| (N) | * | 55 | 12 | 12 |
| A | | 0.048 | 0.000 | 0.000 |
| B | | 0.952 | 0.000 | 0.000 |
| C | | 0.000 | 0.000 | 0.250 |
| D | | 0.000 | 1.000 | 0.750 |
| Lap | | | | |
| (N) | * | 30 | 24 | 15 |
| A | | 0.033 | 0.000 | 0.000 |
| B | | 0.017 | 1.000 | 0.400 |
| C | | 0.783 | 0.000 | 0.533 |
| D | | 0.167 | 0.000 | 0.067 |

Ald, aldolase; Aldh, alanopine dehydrogenase; Arpk, arginine phosphate kinase; Got, glutamate oxaloacetate transaminase; Gpt-1, Gpt-2, glutamate pyruvate transaminase; Ipp, inorganic pyrophosphatase; Lap, leucine aminopetidase; Mdh-1, malate dehydrogenase; Pep-1, Pep-2, peptidase; 6-Pgd, 6-phosphogluconate dehydrogenase; Pgi, phosphoglucose isomerase; Phgk, phosphoglycerate kinase; Pgm-1, Pgm-2, phosphoglucomutase. *, these three loci were not assayed in *L. subrotundata*; N, number of individuals assayed.

## Results

Allele frequencies obtained on the basis of pooled data for comparison of the four species are given in Table 2. Only three species were studied in each year (*L. subrotundata*, *L. sitkana* and *L. kasatka* in 1990 and *L. sitkana*, *L. kasatka* and *L.* sp. in 1991); genetic distances between species are thus shown separately in Fig. 2 for the 1990 and 1991 samples. In each case the within species relationship between sites was close. Hence, to compare all four species and to check that, for a given species, sites between years gave comparable results to sites within years, data from 1991 on *L.* sp., *L. sitkana* (Vostok Bay) and *L. kasatka* (Iturup Island) were pooled with the data from 1990 (Fig. 3A). The genetic distances between the species after pooling

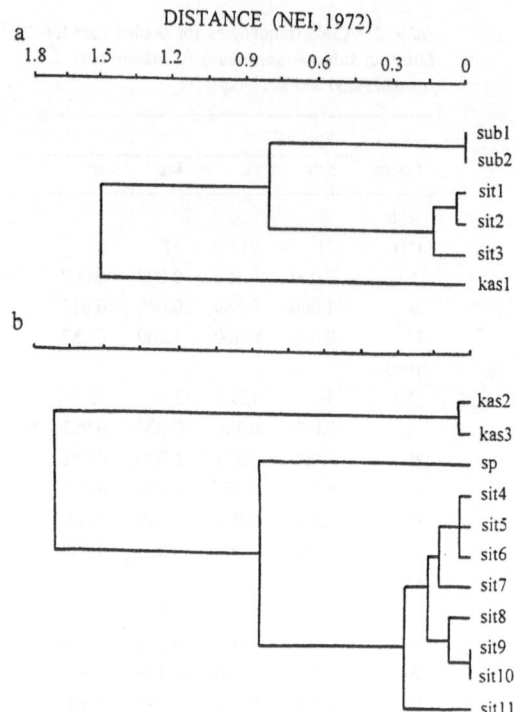

*Fig. 2.* UPGMA trees of genetic relationships between *L. kasatka* (kas), *L. sitkana* (sit), *L. subrotundata* (sub) and *L.* sp. (sp). collected in 1990 (a) and 1991 (b).

these sites for each species are shown in Fig. 3B. Mean estimates of genetic distances (Nei, 1978) between all pairs of species are given in Table 3.

*Littorina sitkana* is a highly polymorphic species, showing both morphological and genetic differentiation. Samples are significantly different over a distance of only several hundred meters (Zaslavskaya *et al.*, in press). These differences may be explained by the lack of a planktonic stage in development and by the limited mobility of adults. Nevertheless, the sample from Vostok Bay differs far less from samples from the Kurile and Komandor Islands than the sample of the new species, *L.* sp., differs from the sample of *L. sitkana* collected from the same stone, the genetic distances being 0.161 and 0.786 respectively.

*Littorina* sp. may be considered as a sibling species of *L. sitkana*. Nevertheless it is quite distinct electrophoretically from *L. sitkana* (Zaslavskaya, in press). At four of the 16 loci studied (Ald, Got, Mdh-1 and Pgm-1) the species share no common alleles. Furthermore, the most common allele present at five other loci (Aldh, Arpk Ipp Pep-2, and Pgm-2) is different in *L.* sp. (Table 2).

*Table 3.* Matrix of distance coefficients (Nei, 1978) averaged by species for *Littorina subrotundata* (sub), *L. sitkana* (sit), *L. kasatka* (kas) and *L.* sp. (sp)

| Species | No. of pops | Species sub | sit | kas | sp. |
|---|---|---|---|---|---|
| sub | 2 | 0.012 (0.012–0.012) | | | |
| sit | 4 | 0.981 (0.821–1.110) | 0.161 (0.017–0.299) | | |
| kas | 2 | 1.911 (1.796–2.087) | 1.660 (1.425–1.878) | 0.120 (0.120–0.120) | |
| sp | 1 | 0.975 (0.955–0.955) | 0.622 (0.561–0.741) | 1.183 (1.123–1.242) | * |

\* - no comparisons

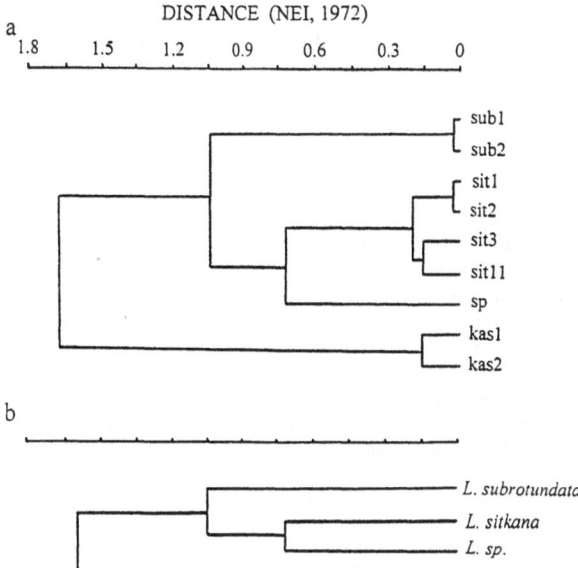

*Fig. 3.* UPGMA trees of genetic relationships between *L. kasatka* (kas), *L. sitkana* (sit), *L. subrotundata* (sub) and *L.* sp. (sp). (a) pooled data for 1990 and 1991; (b) data averaged by species.

*L. subrotundata* clusters next to *L. sitkana* and *L.* sp. (Fig. 3B). Females of this species are different in their reproductive system but males are similar to those of *L. sitkana* (Reid & Golikov, 1991). *L. subrotundata* can be separated easily from *L. sitkana* and *L.* sp. by allozyme characters. At two of the 13 loci studied (Gpt-2 and Mdh-1) *L. subrotundata* and *L. sitkana* share no common alleles and at five (Arpk, Gpt-2, Mdh-1, Pgi and Pgm-1) *L. subrotundata* and *L.* sp. share no common alleles. The genetic distance between *L. subro-*

*tundata* and the other two species (*L. sitkana* and *L.* sp.) is 0.981 and 0.975 respectively.

*Littorina kasatka* is the most different from all of the other species both in morphological characters and in genetic data. The structure of the reproductive system is unique both in males and in females (Reid *et al.*, 1991). It also differs substantially in allozyme characters from other species. Common alleles with *L. sitkana* are absent at seven of 16 assayed loci, with *L. subrotundata* at eight of 13 loci and with *L.* sp. at eight of 16 loci. At three loci (Gpt-1, Gpt-2 and Pep-2) *L. kasatka* had no alleles in common with any of the other three species and at one (Ald) it had none in common with either *L. sitkana* or *L.* sp (*L. subrotundata* was not investigated at this locus); also Pgi allele A and Pgm-1 allele F were unique alleles to *L. sitkana*. The genetic distance between *L. kasatka* and each of the above three species is greater than 1 (Table 3).

## Discussion

On the basis of allozyme comparisons it may be suggested that the three species, *L. sitkana*, *L. subrotundata* and *L.* sp., are closely related and form a species complex. This is confirmed by their shell morphology, and by the similarity of the male reproductive system in *L. subrotundata* and *L. sitkana* and of the female reproductive system in *L.* sp. and *L. sitkana*. Using allozymes, all three species can be separated easily using several diagnostic loci.

According to Reid's classification (Reid, 1990) the above three species are grouped together within the subgenus *Neritrema*, whereas *L. kasatka* is in the sub-

128

genus *Littorina* (Reid *et al.*, 1991). The data from the present study indicate that *L. kasatka* is indeed quite distinct from the other three species studied. Moreover, the genetic distance between *L. kasatka* and *L. sitkana* is much greater than it is between *L. sitkana* and the Atlantic species *L. saxatilis*, *L. mariae* and *L. obtusata* (Zaslavskaya *et al.*, 1991), which are also members of the subgenus *Neritrema*. Furthermore, the values of the genetic distances between *L. kasatka* and the other three species are similar to those reported by Backeljau & Warmoes (1992) for the inter-subgeneric distances between *Neritrema* and *Littorina* (and to the inter-generic distances reported by Janson (1985) for *Littoraria angulifera* and *Nodilittorina ziczac*) and are somewhat greater than inter-subgeneric values provided by Ward (1990). This can be viewed as indirect confirmation of the classification proposed by Reid (1990).

It should be remembered that *L. kasatka* occurs universally on the Kurile Islands together with *L. sitkana*. It seems that *L. kasatka* is a very peculiar species because it has the most common alleles at most loci different from those of all previously studied Atlantic and Pacific species. The peculiarity of this species is accentuated by the unique structure of its male and female reproductive systems (Reid *et al.*, 1991).

*Littorina saxatilis* forms a species complex in the Atlantic Ocean; *L. sitkana* in the Pacific Ocean. Recent studies have clarified the status of the sibling species complexes *L. 'saxatilis'* and *L. 'obtusata'* (Johannesson & Johannesson, 1990; Raffaelli, 1990; Ward, belonging to the *L. 'sitkana'* complex. Recent research on *Littorina* species from the north-eastern Pacific Ocean have also been directed at this question (Boulding *et al.*, 1993). However, the small differences in allele mobility in the different species, a large number of alleles at the same loci and the different loci studied by us (this paper and Zaslavskaya *et al.*, in press) and by Boulding *et al.* (1993) and the different conditions of electrophoresis do not allow comparison of the existing data from the two regions.

## Acknowledgments

Many thanks to Dr D. Reid for his interest in my research and his encouragement. I am also grateful to my colleagues from the Institute of Marine Biology, Drs S. M. Nikiforov, Y. M. Jakovlev and A. V. Martinov for providing me with samples from Kurile and Komandor Islands. I thank Dr A. I. Pudovkin for comments on the manuscript.

## References

Backeljau, T. & T. Warmoes, 1992. The phylogenetic relationships of ten atlantic littorinids assessed by allozyme electrophoresis. In J. Grahame, P. J. Mill & D. G. Reid (eds), Proceedings of the 3rd International Symposium on Littorinid Biology. The Malacological Society of London, London: 9–24.

Boulding, E. G., J. Buckland-Nicks & K. L. V. Alstyne, 1993. Morphological and allozyme variation In *Littorina sitkana* and related *Littorina* species from the Northeastern Pacific. Veliger 36: 43–68.

Golikov, A. N. & O. G. Kusakin, 1978. Recent molluscs of the littoral zone of the USSR. Nauka, Leningrad, 256 pp.

Janson, K., 1982. Phenotypic differentiation in *Littorina saxatilis* Olivi (*Mollusca, Prosobranchia*) in a small area on the Swedish west coast. J. moll. Stud. 48: 167–173.

Janson, K., 1983. Selection and migration in two distinct phenotypes of *Littorina saxatilis* in Sweden. Oecologia, Berlin 59: 58–61.

Janson, K., 1985. A morphologic and genetic analysis of *Littorina saxatilis* (Prosobranchia) from Venice, and on the problem of *saxatilis-rudis* nomenclature. Biol. J. linn. Soc. 24: 51–59.

Johannesson, K. & B. Johannesson, 1990. *Littorina neglecta* Bean, morphological form within the variable species *L. saxatilis* (Olivi)? Hydrobiologia 193 (Dev. Hydrobiol. 56): 71–87.

Nei, M., 1972. Genetic distance between populations. Am. Nat. 106: 283–292.

Nei, M., 1978. Estimation of average heterozygosity and genetic distance from a small number of individuals. Genetics 89: 583–590.

Raffaelli, D. G., 1982. Recent ecological research on some European species of *Littorina*. J. moll. Stud. 48: 342–354.

Raffaelli, D., 1990. Epilogue. Hydrobiologia 193 (Dev. Hydrobiol. 56): 271–273.

Reid, D., 1990. A cladistic phylogeny of the genus *Littorina* (*Gastropoda*): implications for the evolution of reproductive strategies and classification. Hydrobiologia 193 (Dev. Hydrobiol. 56): 1–19.

Reid, D. & A. N. Golikov, 1991. *Littorina naticoides*, new species with notes on the other smooth-shelled species from the northwestern Pacific. Nautilus 105: 7–15.

Reid, D., N. I. Zaslavskaya & S. O. 5Sergievsky, 1991. *Littorina kasatka*, a new species from the Kurile Islands and Okhotsk Sea Nautilus 105: 1–6.

Sokal, R. R. & P. N. A. Sneath, 1963. Principles of numerical taxonomy. Freeman, San Francisco, 359 pp.

Swofford, D. L. & R. B. Selander, 1981. BIOSYS-1: a Fortran program for the comprehensive analysis of electrophoretic data in population genetics and systematics. J. Heredity 72: 281–283.

Ward, R. D., 1990. Biochemical variation in the genus *Littorina* (*Prosobranchia: Mollusca*). Hydrobiologia 193 (Dev. Hydrobiol. 56): 53–69.

Zaslavskaya, N. I., 1989. Genetic variability in four Pacific species of periwinkles (*Mollusca: Gastropoda*). Genetika 25: 1636–1644.

Zaslavskaya; N. I., S. O. Sergievsky & A. N. Tatarenkov, 1991. Allozyme similarity of Atlantic and Pacific species of *Littorina* (*Gastropoda: Littorinidae*). J. moll. Stud. 58: 377–384.

Zaslavskaya, N. I., (in press). Genetic variability in 3 littorinid species morphologically similar to *L. sitkana*. Biologiya Morya.

Zaslavskaya, N. I., B. A. Kalabushkin & A. I. Pudovkin, (in press). Intrademic and interdemic genetic differentiation of the gastropod mollusk *Littorina sitkana*. Genetika.

*Hydrobiologia* **309**: 129–142, 1995.
*P. J. Mill & C. D. McQuaid (eds), Advances in Littorinid Biology.*
©1995 *Kluwer Academic Publishers.*

# The relationship between position on shore and shell ornamentation in two size-dependent morphotypes of *Littorina striata*, with an estimate of evaporative water loss in these morphotypes and in *Melarhaphe neritoides*

Joseph C. Britton

*Department of Biology, Texas Christian University, Fort Worth, TX 76129, USA*

*Key words:* intertidal ecology, size-partitioning, spatial distribution, evaporative water loss, *Littorina striata*, *Melarhaphe neritoides*

## Abstract

Shells of small (< 7 mm) *Littorina striata* are frequently nodulose, but shells of larger individuals are striate. Nodulose *L. striata* dominated the littoral fringe of a black basalt Azorean shore where daytime rock temperatures rose significantly higher than nearby shores of different rock composition or colour. There was no evidence of intraspecific size-partitioning on the latter shores, where the numbers of striate and nodulose *L. striata* were approximately equal between high eulittoral (low-shore) and high littoral fringe (high-shore) localities. The prevalence of small *L. striata* in the littoral fringe is opposite to that usually characteristic of the Littorinidae, where the largest individuals usually occupy the higher positions on the shore. It is hypothesized that small *L. striata* attain a resting posture better able to minimize heat absorption from the substratum than attained by larger individuals. Smaller individuals also take advantage of both posture and a nodulose shell surface to re-radiate absorbed incident radiant thermal energy more effectively to the atmosphere by convection. Thus, small, nodulose *L. striata* are especially well adapted to occupy geologically young basaltic rocks commonly found fringing islands of the mid-Atlantic.

The rate of evaporative water loss was determined for *Melarhaphe neritoides* and striate and nodulose *L. striata* for approximately 11 days emersion. All three groups are exceptionally capable of controlling evaporative water loss. Total percent evaporative water loss by nodulose *L. striata* (17.9%) was significantly greater than that lost by either striate *L. striata* (14.1%) or *M. neritoides* (13.5%) but, among 15 species for which evaporative water loss has been determined by similar methodology, *M. neritoides* and striate *L. striata* are the most capable of conserving body water during 11 days of emersion.

## Introduction

Two species of Littorinidae, *Littorina striata* (King & Broderip, 1832) and *Melarhaphe neritoides* (Linnaeus, 1758), inhabit rocky shores on the island of São Miguel, Azores. Both range from the upper eulittoral to the upper littoral fringe, with the latter species usually occurring slightly higher than the former (Hawkins *et al.*, 1990). Most *L. striata* are sequentially dimorphic with respect to shell ornamentation. Small individuals (< 5 mm) usually produce a nodulose shell but, as they grow, the nodulosity is replaced on subsequent body whorls by spiral striae. Rosewater (1981) cited this change in shell ornamentation as one of several diagnostic features of his subgenus *Liralittorina*,

type-species, *L. striata*, which he placed in the genus *Nodilittorina*. However, the species has been restored subsequently to *Littorina* (Reid, 1989). In contrast, the shell of *Melarhaphe neritoides*, regardless of size or age, is nearly smooth, enclosed in a thickened periostracum, with sculpture restricted to faint spiral and axial growth lines.

Numerous factors influence the spatial distribution of organisms on rocky intertidal shores and they have received considerable study (see Newell, 1979; Underwood, 1979 and McMahon, 1990 for extensive reviews). This paper will focus on only two: desiccation tolerance and morphological adaptations that possibly mediate thermal stress imposed by a substratum heated by radiant solar energy.

Since 1985, the author and R. F. McMahon, either together or alone, have compiled data on the desiccation tolerance of 26 species of intertidal prosobranch gastropods from several parts of the world (McMahon & Britton, 1985, 1991; Britton & McMahon, 1990; McMahon, 1990; Britton, 1992, 1993). These studies have demonstrated fundamental differences between eulittoral and littoral fringe species. This paper adds to the body of literature on desiccation tolerance of littoral fringe littorines and, for the first time, considers possible differences in desiccation tolerance between two size classes of a single species, *Littorina striata*, which also differ with respect to shell ornamentation.

Since the observations of Porter & Gates (1969) on thermal relationships between animals and their external environment, there have been many studies of thermal stress in intertidal environments. Vermeij (1971a; 1973) was among the first to catalogue the several mechanisms by which intertidal gastropods can mediate thermal stress. Eulittoral species, assured of tidal resupply of extrapalial cooling fluids, rely upon evaporative cooling to minimize thermal stress when emersed in air (McMahon, 1990). Higher-zoned intertidal species apparently adjust the size and/or shape of the shell to minimize either the amount of radiant energy absorbed or to facilitate re-radiation of absorbed thermal energy (Vermeij, 1973).

High-zoned littorine species may lie exposed in air for many days or weeks (Rosewater, 1963; Britton, 1992), during which time they are subjected to diurnal heating by either direct absorption of radiant solar energy or transfer of heat absorbed by the substratum. Littorines, including both *Littorina striata* and *Melarhaphe neritoides*, may seek the shelter of crevices, pits or other substratum irregularities to avoid direct radiant solar energy, and further avoid thermal loading from the substratum by minimizing the amount of tissue or shell in contact with it. Frequently, they attach the edge of the aperture to the substratum by a thin mucus strand, withdraw deeply the body into the shell, and hang suspended in air with little or no part of the body in contact with the substratum. In this position the entire shell is ideally positioned to facilitate maximal conductive cooling (Vermeij, 1971b). Both Azorean littorines attach to rock surfaces by means of mucus holdfasts.

Vermeij (1973) has discussed the importance of dissipating absorbed thermal energy by re-radiation and convection. In particular, he has emphasized the importance of increasing total surface area of the shell to facilitate this process. Simply stated, nodulose shells are assumed to be better radiators of absorbed thermal energy than smooth or slightly grooved shells because of the increased surface area provided by nodulosity. Although this seems a reasonable assumption, it has received only limited empirical assessment (Vermeij, 1971a, 1973).

One might expect that species which occupy higher positions on the shore will display an increased incidence of nodulosity, but the general tendency is rarely perfectly expressed (Vermeij, 1973; Britton, 1992). Both nodulose and smoothly sculptured littorine species occupy the littoral fringe of rocky shores in the Caribbean (Britton, 1992), the Azores (Hawkins *et al.*, 1990), Hong Kong (Ohgaki, 1985a) and other shores. Clearly other factors, in addition to an expanded capacity for re-radiation of thermal loads, mediate vertical distributions (McMaon, 1990). Yet, it is a reasonable assumption that nodulosity would be an advantage to any gastropod species occupying a rock surface that, because of its composition, texture, colour, or other characteristic, is easily and excessively heated.

Is there a functional significance to the sequential shell dimorphism of *Littorina striata*? If, as Vermeij (1971a, 1973) has suggested, increased irregularity of the shell surface facilitates conductive re-radiation of absorbed heat, one can hypothesize that nodulose *L. striata* should have an advantage over striate individuals on shores experiencing especially high thermal loadings. If so, they should be more abundant on these shores than the smoother striate shells. The latter should predominate in areas where other factors such as waves or substratum characteristics mediate thermal loading. Thus, this paper tests the hypothesis just stated, considers the effects of desiccation on striate and nodulose *L. striata* and *Melarhaphe neritoides*, and assesses the relationships, if any, between shell morphology, distribution on the shore and desiccation tolerance, especially with respect to *L. striata*.

## Study sites and methods

### Sites

The primary site of this study was the black basaltic rocky shore at Caloura, Ponta da Galera, along the southern coast of São Miguel (Hawkins *et al.*, 1990, Fig. 1). This shore is mostly steeply sloping (45–90°), but interrupted occasionally by narrow, near-horizontal platforms in the upper-intertidal region. Texture of the basalt varies from nearly smooth on the horizontal platforms to extremely vesicular, especially in the upper littoral fringe. This habitat appears to be a very recent, almost unweathered basaltic lava flow.

Additional quantitative samples were taken from the southeastern rocky shore of Ilhéu de Vila Franca (Morton, 1990, Figs 1 and 2). The rocks at this location are dark-coloured, but much less so than the deep black basalt at Caloura. Some rock formations of the littoral fringe are noticeably lighter than those of the eulittoral, being an earthy yellowish colour. Long (2–10 m), broad horizontal platforms fringe the upper intertidal, with vertical cliffs positioned from 1 to 5 m from the seaward edges of the intertidal platforms. Texture of the rocks at the Ilhéu de Vila Franca study site varies from near smooth to moderately vesicular, and they are noticeably more weathered than the fresh basaltic lava at Caloura.

Mid-day air and dry rock-surface temperature profiles were taken at both sites on different, similarly sunny, days to estimate relative thermal loads imposed by the different environments. Air temperature was recorded in shade approximately 1 m above the ground at 30 min intervals during the profile period. Five or more rock-surface temperatures were also recorded at 30 min intervals at both high (littoral fringe) and low (upper eulittoral) positions on the shore.

*Population studies*

*Littorina striata* was abundant on the shore at both Caloura and Ilhéu de Vila Franca, ranging across a vertical distance of as much as 5 m from the upper intertidal to the upper littoral fringe. A preliminary survey of *L. striata* was conducted at Caloura to ascertain whether mean size and/or morphology of the species varied according to position on shore. Approximately 100 individuals were collected from each of three shore levels: (1) upper eulittoral rocks subject to tidal inundation at high tide, (2) the lower littoral fringe subject to intermittent wetting by wave splash but rarely by tidal inundation, and (3) the upper littoral fringe where rocks remained dry during the day for most of the duration of the study. The height (H) of each individual of *L. striata* was measured to the nearest 0.01 mm by vernier caliper, and the presence or absence of nodulose sculpture was recorded. The results of this preliminary study indicated that both size and the morphology of *L. striata* varied with position on the shore; so additional sampling was conducted to assess these differences quantitatively.

Quantitative (0.25 × 0.25 m quadrat) samples of *Littorina striata* were collected at both Caloura and Ilhéu de Vila Franca. The location used for preliminary sampling at Caloura was not resampled quantitatively, but a new site was selected where no previous sampling had been undertaken. Samples were taken from upper eulittoral rocks subject to tidal inundation at high tide (hereafter referred to as low-shore samples) and from littoral fringe rocks wetted only by high waves or atmospheric precipitation (hereafter referred to as high-shore samples). All littorines within the quadrat were collected and preserved in 70% ethanol. They were subsequently identified and the numbers of *L. striata*, shell heights and the presence or absence of nodosity were determined in the manner described for the preliminary survey. The number of individuals of three morphotypes (striate shells, nodulose shells, or an intermediate stage between the two) from each of the quantitative samples within and between shore levels and between the two study sites were compared by Chi-square contingency table analysis to determine the correlation, if any, between morphology and environment. Similar ANOVA comparisons were made with respect to size (height). Finally, the upper and lower shore data obtained during the preliminary survey were combined with pooled data from the quantitative samples to obtain a large-sample frequency distribution of the relationship between the three morphotypes and size.

*Desiccation tolerance*

Desiccation tolerance was determined for striate and nodulose morphs of *Littorina striata* and for *Melarhaphe neritoides*. The former were collected from the black basaltic rocks at Caloura, but *M. neritoides*, being much more abundant at Ilhéu de Vila Franca, were obtained from there. All specimens used to determine body water loss during emersion were collected on 1 August, 1991 and experiments were initiated within 48 h after collection. The heights (H, considered to be the greatest distance from the tip of the spire to the leading edge of the aperture) of 30 specimens of each morph of *L. striata* and of *M. neritoides* were measured to the nearest 0.1 mm with a dial caliper. Prior to emersion, specimens were allowed to hydrate in seawater for 1 h. Specimens which crawled out of water were returned to it during the hydration process. After hydration, specimens were carefully blotted dry and the total wet weight of each was determined to the nearest 0.1 mg using a Mettler analytical balance. The evaporative water loss of each individual was then determined by the method of McMahon & Britton (1985). After the initial wet weights were recorded, all specimens of each species were exposed to air in the laboratory at ambient room temperature (23.2 to 27.2 °C). Each individual was placed in a numbered depression on a plastic tray and each tray was enclosed within a loosely knit nylon cloth bag (mesh size ~ 2 mm) to prevent escape. The trays were carefully monitored for the first few hours until all individuals settled into a state of relative inactivity. The wet weight and behaviour of each individual was recorded twice daily during a fortnight of experimental emersion, after which the experiments were terminated. Decline in wet weight during the period of emersion was assumed to be entirely the result of evaporative water loss.

After the final emersion wet weight of each specimen was recorded, individuals were rehydrated in seawater and their behaviour observed. All specimens extending the foot, righting the shell and crawling within 1 h of rehydration were considered to be alive. The rehydrated snails were blotted free of excess water and dried at 90 °C to constant weight. The total body water weight of each individual was then estimated as the difference between the initial wet weight at the start of the experiment and the final dry weight. Cumulative and successive daily water loss weights for each experimental individual were expressed as percentages of the total body water weight. Tissue water weight was computed for each individual as the difference between tissue wet weight and tissue dry weight. The water loss of each morph of *Littorina striata* and *Melarhaphe neritoides* at each weighing period was determined as the mean cumulative percent of total body water weight lost (i.e., corporal + pallial water). Individual water loss rates were determined as percent of total body water loss per h for periods between adjacent wet weight determinations by subtracting the wet weight of each successive determination from that just preceding it, expressing the result as a percent of total body water weight and dividing that value by the period in hours between determinations. Mean values of water loss rate per h over the duration of emersion provide a measure of the ability of each species to regulate body water loss.

Least squares linear regression analysis of arcsine-transformed cumulative percent of total body water lost determined at each sampling period (wet weight determinations of emersed specimens) versus size (H) indicated that there were significant correlations between size and the evaporative water loss rate. Thus, water loss rates were compared between species by performing a single factor repeated measures ANOVA of arcsine-transformed cumulative percent total body water lost adjusted for a size (H) covariate (Winer, 1971). All statistical comparisons were performed by SYSTAT algorithms (SYSTAT, Inc., Evanston, IL, U.S.A.).

## Results

### Environmental and population studies

Temperature profiles of high- and low-shore rocks at Caloura and Ilhéu de Vila Franca were made on different days, but the general patterns reported here were confirmed by additional spot checks of the differences between air temperature and that of rocks at high- and low-shore positions at both localities. There was little difference in mean rock-surface temperatures between high- and low-shore locations at Ilhéu de Vila Franca when the shore was exposed during lower tides (Fig. 1). Mean low-shore rock-surface temperatures were slightly greater than those from high-shore localities during lower tidal conditions, perhaps because rocks of the lower Ilhéu shore were darker than their higher shore counterparts. When the lower shore at either Ilhéu de Vila Franca or Caloura was washed by waves, mean rock-surface temperatures at both sites plunged below ambient daylight air temperatures.

In contrast, dry littoral fringe rocks at Caloura differed markedly in their capacity to absorb radiant solar energy from either rocks of the upper eulittoral at Caloura or either high- or low-shore positions at Ilhéu de Vila Franca (Fig. 1). Mean high-shore rock-surface temperatures at Caloura were approximately the same as those of the aerially exposed lower shore during the early morning but, as the day progressed, the former rose significantly above the latter. Immediately prior to tidal incurson of the lower Caloura shore, differences in mean rock-surface temperature between high- and low-shore localities reached 3.6 °C. At this time, mean temperatures of low-shore rocks between sites were approximately the same (Caloura, 26.9 °C; Ilhéu, 26.8 °C) but mean temperatures of high-shore rocks between sites were markedly different (Caloura, 30.5; Ilhéu, 26.2). Thus, the dark black basalt of the Caloura littoral fringe attains markedly higher diurnal temperatures than the substrata of the other studied areas.

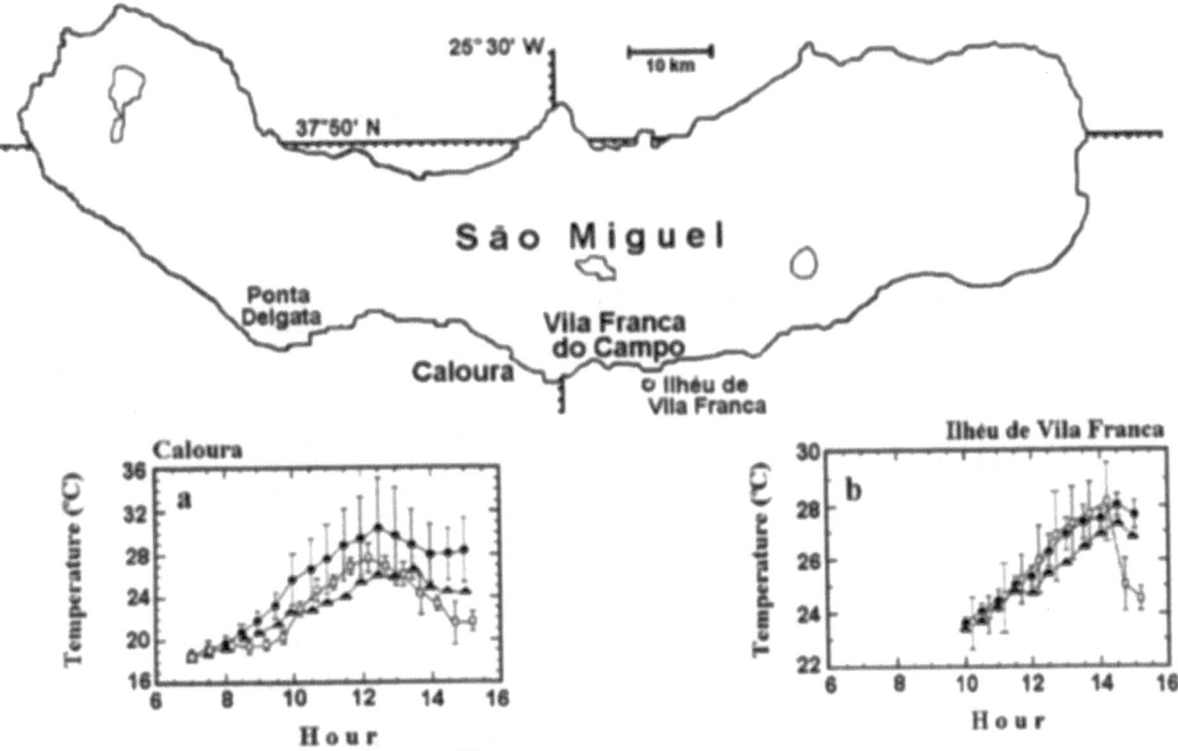

*Fig. 1.* Map indicating the study areas at Caloura and Ilhéu de Vila Franca, São Miguel, Azores, Portugal. Temperature profiles of ambient air temperatures (▲) and rock-surface temperatures at Caloura (a) and Ilhéu de Vila Franca (b), July 1991. ●, mean rock-surface temperatures ($n \geq 5$) at high-shore (littoral fringe) localities which remained dry throughout the profile period; □, mean rock-surface temperatures ($n \geq 5$) at low-shore (upper eulittoral) localities. The latter were subjected to infrequent wave wash at both sites until about 1300 h; thereafter low-shore localities experienced an increasing frequency of wave incursion and eventually tidal inundation. Error bars represent 95% confidence intervals.

The shell height, ornamentation (striate, nodulose or intermediate) and position on shore for 1869 specimens of *Littorina striata* collected from quantitative and non-quantitative samples from Caloura and Ilhéu de Vila Franca were determined during this study. Not all individuals were employed in all analyses (Table 1).

The shells of smaller, younger individuals of *Littorina striata* were usually, but not always, distinctly nodulose (Fig. 2). The smallest individual obtained during the study ($H = 2.58$ mm) had a striate shell lacking any trace of nodosity. Striate individuals comprised an increasing proportion of the individuals in 0.5 mm size classes from 3.0 mm (3.45%) to 5.0 mm (40.14%) (Fig. 2). This trend continued in larger shells and the majority of individuals between 5.5 mm (50.8%) and 7.0 mm (98.2%) were striate, and 100% of individuals $\geq 7.0$ mm had striate shell ornamentation (Fig. 2). Thus, the nodulose ornamentation of many smaller, younger *Littorina striata* is covered over by striate sculpture on the body whorl of individuals

attaining a height of about 7.0 mm. Samples collected during this study contained 99 individuals (5.3%) in which this transition was in progress (Table 1).

Nodulose and striate *Littorina striata* are not uniformly distributed on all shores. Small, nodulose *L. striata* dominated the Caloura littoral fringe (high-shore), represented by 243 individuals (62.3%) of the 390 collected there (Figs 3a, 4a). In contrast, the upper eulittoral (low-shore) at Caloura was dominated by larger, striate *L. striata*, comprising 415 (72.5%) of the 572 individuals collected (Figs 3b, 4b). Thus, the proportions of nodulose, intermediate and striate individuals differed significantly (Pearson Chi-square = 218.5; DF = 2; $P = 0.000$) from high to low positions on this shore.

A similar pattern of distribution was not observed at Ilhéu de Vila Franca. Individuals bearing striate shell ornamentation were dominant and equally represented at high (80.9%) and low-shore (84.5%) positions. Nodulose individuals were a minor component of the population at both levels (Figs 3c and d and 4c and d).

134

*Table 1.* Allocation of the 1869 individuals of *Littorina striata* used in the field and population analyses reported herein.

| Analysis | Excluded | Number included |
|---|---|---|
| Relative proportions of nodulose and striate individuals according to size (H). (Figure 2) | 99 individuals with shell ornamentation intermediate between nodulose and striate | 1770 |
| Relative frequencies of shell morphologies on high and low shore positions at Caloura and Ilhéu de Vila Franca. (Figures 3 and 4) | 203 individuals collected from intermediate positions between the upper eulittoral and the mid-supralittoral fringe | 1687 |
| Quantitative analyses (Tables 2 and 3) | 300 individuals collected non-quantitatively plus 104 individuals from two quantitative samples collected intermediate between the upper eulittoral and the mid-supralittoral fringe. | 1465 |

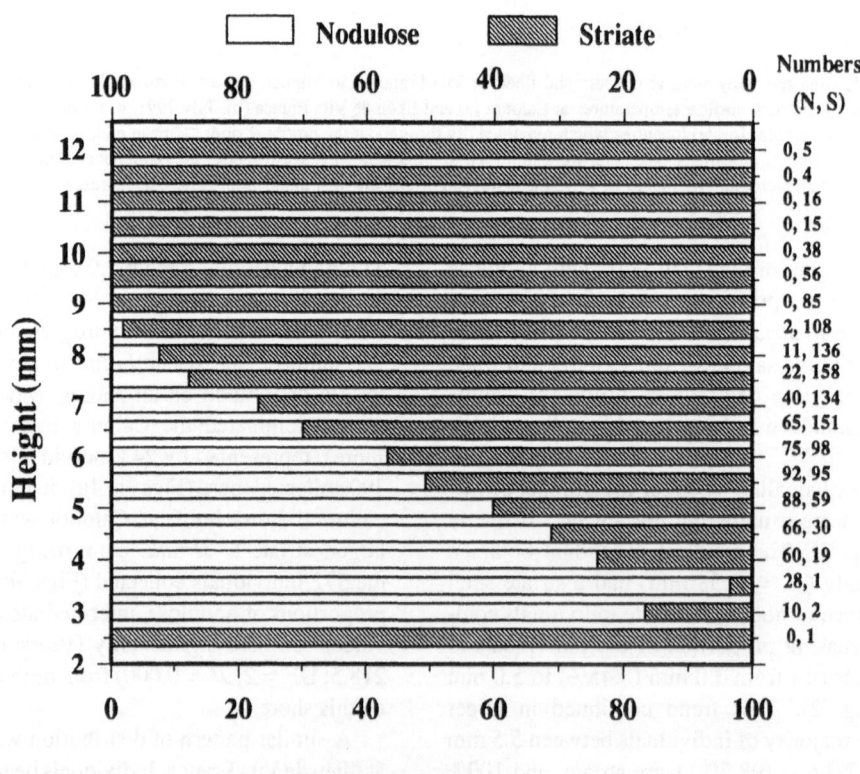

*Fig. 2.* Percent frequency of nodulose and striate individuals of *L. striata* at 0.5 mm size classes. Intermediate forms are excluded from this figure. Actual numbers of nodulose (N) and striate (S) individuals are given for each size class on the right side of the figure.

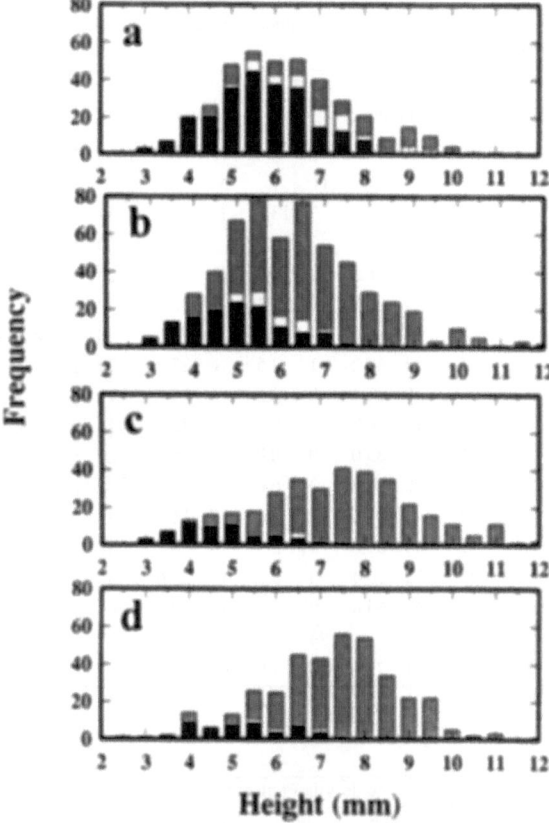

*Fig. 3.* Size-frequency histograms of the distribution of nodulose (filled bars), intermediate (unfilled bars) and striate (cross-hatched bars) *Littorina striata* from Caloura (a high-shore, b low-shore) and Ilhéu de Vila Franca (c high-shore, d low-shore). Note the preponderance of small, nodulose individuals from the Caloura high-shore.

There was no significant difference in the proportion of nodulose, intermediate and striate individuals between high- and low-shore positions (Pearson Chi-square = 2.93; DF = 2; $P = 0.231$).

When only quantitative samples were considered, the overall proportions of nodulose and striate *Littorina striata* at Caloura and Ilhéu de Vila Franca did not change appreciably (Table 2), but these data provided additional information about population structure. The mean densities of *L. striata* at Caloura were less than half that observed at Ilhéu de Vila Franca for both high- and low-shore positions. The mean density of *L. striata* from Caloura high-shore samples was 167.5 ind. m$^{-2}$ (i.e., 6.35 × 26.8, Table 2), and 210.6 ind. m$^{-2}$ from Caloura low-shore samples. In contrast, mean densities at Ilhéu de Vila Franca were 437.5 ind. m$^{-2}$ for high-shore samples and 468.8 ind. m$^{-2}$ for low-shore samples. Note, however, that the elevated

mean density for the Ilhéu de Vila Franca high-shore samples was largely the result of one sample containing more individuals than the other four samples combined, emphasizing that there was clearly variation in density between the quantitative samples.

Nodulose individuals comprised more than 50% of the population for most high-shore quantitative samples at Caloura, but striate individuals equaled or approached 50% of the population in two samples (Table 2, upper left). Similarly, striate shells dominated most low-shore samples at Caloura, but comprised less than 50% of the population in two samples (Table 2, upper right). Striate shells dominated both high- and low-shore samples at Ilhéu de Vila Franca, but the proportion of striate to nodulose individuals varied significantly at both levels (high-shore samples, Pearson Chi-square = 11.3, DF = 4, $P = 0.023$; low-shore samples, Pearson Chi-square = 17.9, DF = 4, $P = 0.001$).

The texture of the substratum and other aspects of microhabitat varied considerably from sample to sample and site to site. Variation ranged from wave-wetted, barnacle-encrusted rocks of the upper eulittoral, to high, smooth, dry, bare rock surfaces in the littoral fringe. These variations are characterized in Table 2. Chi-square contingency table analysis was performed on frequency distributions of striate and nodulose *Littorina striata* from several different microhabitats to determine if these habitat variations contributed to frequency variation observed in the samples. Significant differences between samples collected from high-shore pitted habitats and low-shore wave-swept habitats at both Caloura and Ilhéu de Vila Franca indicate that frequency variations were not the result of microhabitat differences at either site or shore level, but several other contingency table analyses of striate and nodulose frequencies according to habitat were equivocal (Table 3).

*Water loss during emersion*

All specimens of *Littorina striata* and *Melarhaphe neritoides* subjected to 11 days of emersion began active crawling within 10 min of being returned to seawater. Mean cumulative percent of total body water lost (i.e., corporal + pallial water) and percent of total body water h$^{-1}$ during 11 days of emersion for each morph of *L. striata* and *M. neritoides* are presented in Fig. 5. Total percent evaporative water loss by nodulose *L. striata* (17.9%) was significantly greater than that lost by either striate *L. striata* (14.1%) or

*Table 2.* Frequency of striate, nodose and intermediate *Littorina striata* from 0.25 × 0.25 m quadrat samples at Caloura and Ilhéu de Vila Franca. Density (N.m$^{-2}$) = 6.25 × N. Habitat descriptions are provided in the legend.

| Caloura: High Shore | | | | | | | |
|---|---|---|---|---|---|---|---|
| Sample | Striate | | Inter-mediate | | Nodulose | | Total | Habitat |
| | N | % | N | % | N | % | | |
| 1 | 2 | 20.0 | 2 | 20.0 | 6 | 60.0 | 10 | crevice |
| 2 | 1 | 6.3 | 0 | 0.0 | 15 | 93.8 | 16 | crevice |
| 3 | 24 | 49.0 | 9 | 18.4 | 16 | 32.7 | 49 | exposed |
| 4 | 8 | 16.0 | 2 | 4.0 | 40 | 80.0 | 50 | crevice |
| 5 | 8 | 29.6 | 3 | 11.1 | 16 | 59.3 | 27 | pitted |
| 6 | 4 | 23.5 | 1 | 5.9 | 12 | 70.6 | 17 | pitted |
| 7 | 0 | 0.0 | 0 | 0.0 | 16 | 100.0 | 16 | pitted |
| 8 | 2 | 9.5 | 0 | 0.0 | 19 | 90.5 | 21 | pitted |
| 9 | 16 | 50.0 | 2 | 6.3 | 14 | 43.8 | 32 | pitted |
| 10 | 7 | 23.3 | 2 | 6.7 | 21 | 70.0 | 30 | pitted |
| TOTALS: | 72 | | 21 | | 175 | | 268 | |
| MEANS: | 7.2 | 26.9 | 2.1 | 7.8 | 17.5 | 65.3 | 26.8 | |

| Caloura: Low Shore | | | | | | | |
|---|---|---|---|---|---|---|---|
| Sample | Striate | | Inter-mediate | | Nodulose | | Total | Habitat |
| | N | % | N | % | N | % | | |
| 1 | 32 | 59.3 | 7 | 13.0 | 15 | 27.8 | 54 | awash |
| 2 | 16 | 80.0 | 3 | 15.0 | 1 | 5.0 | 20 | awash |
| 3 | 15 | 68.2 | 0 | 0.0 | 7 | 31.8 | 22 | pool |
| 4 | 7 | 36.8 | 4 | 21.1 | 8 | 42.1 | 19 | pool |
| 5 | 18 | 90.0 | 0 | 0.0 | 2 | 10.0 | 20 | awash |
| 6 | 15 | 68.2 | 1 | 4.5 | 6 | 27.3 | 22 | pool |
| 7 | 10 | 62.5 | 2 | 12.5 | 4 | 25.0 | 16 | pool |
| 8 | 10 | 66.7 | 0 | 0.0 | 5 | 33.3 | 15 | exposed |
| 9 | 7 | 35.0 | 0 | 0.0 | 13 | 65.0 | 20 | pool |
| 10 | 40 | 80.0 | 3 | 6.0 | 7 | 14.0 | 50 | crevice |
| 11 | 40 | 81.6 | 2 | 4.1 | 7 | 14.3 | 49 | exposed |
| 12 | 43 | 89.6 | 1 | 2.1 | 4 | 8.3 | 48 | awash |
| 13 | 42 | 76.4 | 0 | 0.0 | 13 | 23.6 | 55 | awash |
| 14 | 43 | 69.4 | 0 | 0.0 | 19 | 30.6 | 62 | awash |
| TOTALS: | 338 | | 23 | | 111 | | 472 | |
| MEANS: | 24.1 | 71.6 | 1.6 | 4.9 | 7.9 | 23.5 | 33.7 | |

| Ilhéu de Vila Franca: High Shore | | | | | | | |
|---|---|---|---|---|---|---|---|
| Sample | Striate | | Inter-mediate | | Nodulose | | Total | Habitat |
| | N | % | N | % | N | % | | |
| 1 | 16 | 72.7 | 1 | 4.5 | 5 | 22.7 | 22 | crevice |
| 2 | 48 | 85.7 | 1 | 1.8 | 7 | 12.5 | 56 | crevice |
| 3 | 29 | 90.6 | 0 | 0.0 | 3 | 9.4 | 32 | pitted |
| 4 | 48 | 94.1 | 0 | 0.0 | 3 | 5.9 | 51 | pitted |
| 5 | 142 | 75.1 | 5 | 2.6 | 42 | 22.2 | 189 | pitted |
| TOTALS: | 283 | | 7 | | 60 | | 350 | |
| MEANS: | 56.6 | 80.9 | 1.4 | 2.0 | 12.0 | 17.1 | 70.0 | |

| Ilhéu de Vila Franca: Low Shore | | | | | | | |
|---|---|---|---|---|---|---|---|
| Sample | Striate | | Inter-mediate | | Nodulose | | Total | Habitat |
| | N | % | N | % | N | % | | |
| 1 | 13 | 92.9 | 0 | 0.0 | 1 | 7.1 | 14 | awash |
| 2 | 105 | 96.3 | 0 | 0.0 | 4 | 3.7 | 109 | awash |
| 3 | 88 | 80.0 | 6 | 5.5 | 16 | 14.5 | 110 | awash |
| 4 | 60 | 83.3 | 2 | 2.8 | 10 | 13.9 | 72 | awash |
| 5 | 51 | 72.9 | 2 | 2.9 | 17 | 24.3 | 70 | awash |
| TOTALS: | 317 | | 10 | | 48 | | 375 | |
| MEANS: | 63.4 | 84.5 | 2.0 | 2.7 | 9.6 | 12.8 | 75.0 | |

Legend:
awash: wetted by waves during sampling
crevice: a linear or deep, completely sheltering, depression.
pitted: many dry, shallow, partially sheltering depressions.
pool: standing water, tidal pools and environs.
exposed: dry rock surfaces lacking sheltering depressions.

*M. neritoides* (13.5%). All test groups experienced maximum body water loss rates during the first day of emersion, the greatest being a mean of 0.31% h$^{-1}$ for the nodulose morph of *L. striata* recorded for the first sample after emersion and the least being a mean of 0.16% h$^{-1}$ for *M. neritoides* recorded for the second sample after emersion. The rates of water loss for each species declined during the experiment, with the greatest changes in rate experienced by *L. striata*. Both morphs experienced high body water loss rates during the first three days of emersion, with the greatest change in rate experienced by the nodulose morph. *M. neritoides* exhibited the least change in water loss rate (i.e., the most consistent prolonged regulation of water loss) during the experiment. Individual variation with respect to percent body water lost generally increased

*Table 3.* Contingency table analysis of frequency distributions of striate and nodulose *Littorina striata* according to various microhabitats. Some samples were not analyzed due to insufficient frequency data. Habitat descriptions are provided in the legend.

| Site | Level | Habitat | Samples (N) | Pearson Chi-square | DF | $P$ |
|---|---|---|---|---|---|---|
| Caloura | High | Crevices | 3 | 1.66 | 2 | 0.435† |
| | | Pitted | 5 | 12.3 | 4 | 0.015* |
| | Low | Awash | 6 | 14.8 | 5 | 0.011* |
| | | Pool | 5 | 8.68 | 4 | 0.069 |
| | | Exposed | 2 | 2.48 | 1 | 0.116† |
| Ilhéu de Vila Franca | High | Crevices | 2 | 1.40 | 1 | 0.236† |
| | | Pitted | 3 | 9.60 | 2 | 0.008* |
| | Low | Awash | 5 | 17.9 | 4 | 0.001* |

* Indicates significance at the 0.05 level.
† More than one-fifth of the frequencies $< 5$ rendering the significance test suspect.

Legend:
awash: wetted by waves during sampling
crevice: a linear or deep, completely sheltering, depression.
pitted: many dry, shallow, partially sheltering depressions.
pool: standing water, tidal pools and environs.
exposed: dry rock surfaces lacking sheltering depressions.

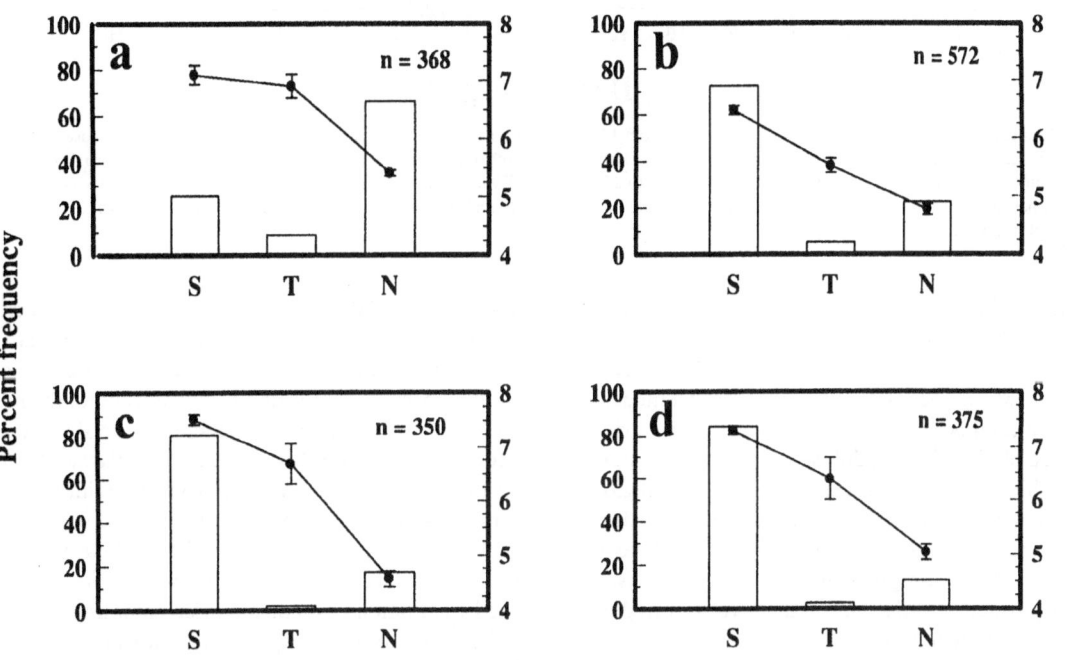

*Fig. 4.* Percent frequency (bars) and mean shell height (•) of striate, intermediate and nodulose *Littorina striata* from the littoral fringe (a and c) and upper eulittoral (b and d) at Caloura (a and b) and Ilhéu de Vila Franca (c and d). Error bars on mean shell length represent the standard error of the mean. The numbers of individuals (n) from each shore are given in the upper right corner of each plot.

with emersion time, as shown by increasing width of error bars for each group tested (Fig. 5).

Each experimental group included individuals which apparently experienced increases in total body water one or more times during the experiment. Suf-

ficient numbers of *Melarhaphe neritoides* experienced increases in total body water between the first two days of emersion to produce a marked decrease in the mean cumulative percent body water lost (Fig. 5). Similar episodes of increased total body water were recorded

138

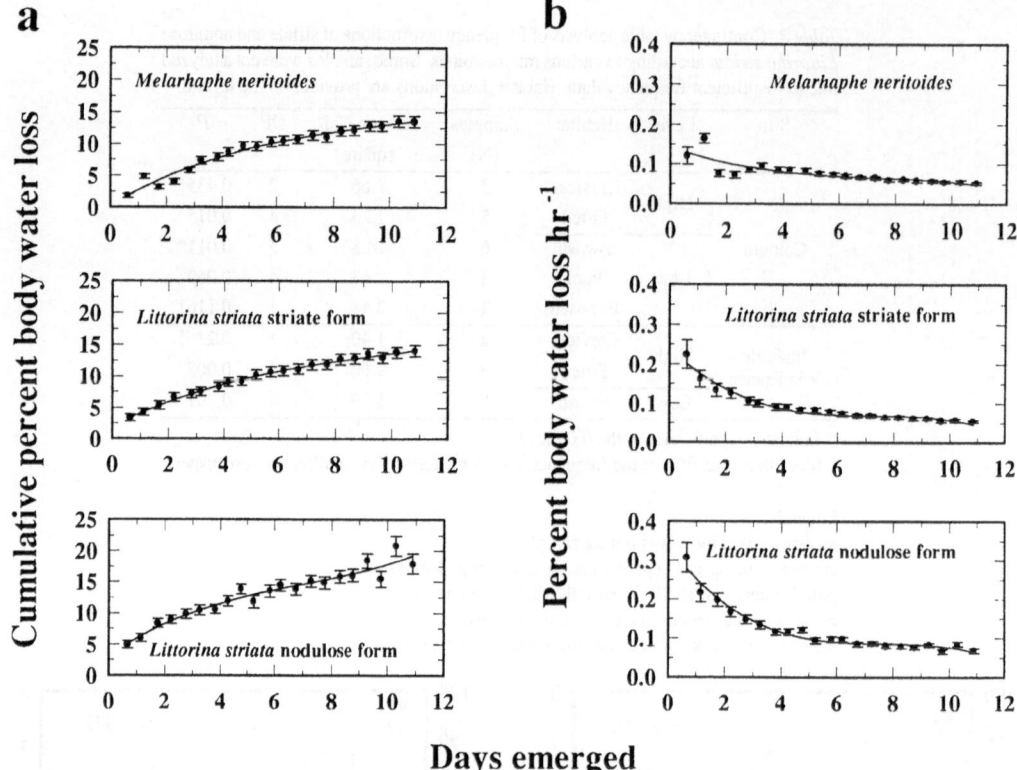

**Days emerged**

*Fig. 5.* Dessication tolerance for *Melarhaphe neritoides* from Ilhéu de Vila Franca, São Miguel, Azores, and striate and nodulose forms of *Littorina striata* from Caloura, São Miguel, Azores. (a) Cumulative percent of total water (= corporal + pallial) lost over approximately 11 days emersion under ambient air temperatures (range = 23.2 to 27.2 °C). (b) Evaporative water loss rates as percent total water (= corporal + pallial) lost per hour for the same duration and under the same conditions. Vertical bars in all figures represent standard error of the mean for samples of 30 individuals.

for nodulose individuals of *Littorina striata* between days 5 and 6, 6 and 7, and especially the last two days of emersion (Fig. 5). Although a few individuals of striate *L. striata* occasionally experienced a modest increase in total body water, it was never to the striking degree observed for nodulose *L. striata* or *M. neritoides*. The only source of water for the emersed gastropods was from water vapour in the atmosphere; during the experiment relative humidity ranged between 44 to 56%. Thus, weight gains reflecting increased total body water of individuals may have occurred as a result of shell hydration.

Water loss rates were compared between the three groups of littorines by means of a single factor repeated measures analysis of covariance (ANCOVA) of arcsine-transformed cumulative percent total body water lost adjusted for the covariate size (H). The analysis indicated a significant difference between species (DF = 2, 70; $MS_{GROUPS}$ = 1441.35; $MS_{ERROR}$ =

214.98; F = 6.70; $P$ = 0.002). Subsequent *a posteriori* Tukey HSD multiple comparisons of cumulative percent evaporative water loss at each of 21 common sampling periods demonstrated that, except for the first and fourth measuring periods, there was no significant difference ($P > 0.05$) between *Melarhaphe neritoides* and the striate form of *Littorina striata* with respect to arcsine-transformed cumulative percent water loss (Table 4). The nodulose form of *L. striata* was significantly different ($P < 0.05$) from both of the other tested populations except for three sampling periods with respect to *Melarhaphe neritoides* and six sampling periods with respect to the striate form of *L. striata* (Table 4). Thus, the nodulose form of *L. striata* lost a significantly greater percentage of total body water during emersion than either the striate form of the same species or *Melarhaphe neritoides*.

The behaviour of each species during emersion was typical of most high zoned littorines (McMahon & Brit-

*Table 4.* Tukey multiple comparisons of arcsine-transformed cumulative percent water loss data. I, intermediate; N, nodulose; S, striate. Significant differences at the 0.05 level are indicated by an asterisk.

| Sample | Comparisons | | |
|---|---|---|---|
| | N–S | N–I | S–I |
| 1 | 0.091 | *0.000 | *0.013 |
| 2 | 0.526 | 0.624 | 0.985 |
| 3 | *0.007 | *0.000 | 0.135 |
| 4 | *0.013 | *0.000 | *0.001 |
| 5 | *0.007 | *0.000 | 0.240 |
| 6 | *0.005 | *0.004 | 0.994 |
| 7 | *0.034 | *0.015 | 0.947 |
| 8 | *0.016 | *0.005 | 0.914 |
| 9 | *0.000 | *0.000 | 0.847 |
| 10 | 0.391 | 0.177 | 0.877 |
| 11 | *0.040 | *0.022 | 0.971 |
| 12 | *0.007 | *0.004 | 0.969 |
| 13 | 0.051 | *0.028 | 0.967 |
| 14 | *0.015 | *0.005 | 0.933 |
| 15 | *0.033 | *0.005 | 0.785 |
| 16 | *0.031 | *0.008 | 0.857 |
| 17 | *0.029 | *0.007 | 0.860 |
| 18 | *0.002 | *0.000 | 0.854 |
| 19 | 0.233 | 0.206 | 0.997 |
| 20 | *0.000 | *0.000 | 0.989 |
| 21 | 0.068 | *0.030 | 0.935 |

ton, 1985, 1991; Britton & McMahon, 1990; Britton, 1992, 1993). Shortly after the experiment commenced, both species initially cemented the edge of the aperture to the substratum by a thin strand of mucus. This fragile junction was broken during the first sampling period. Thereafter, most individuals remained inactive and deeply withdrawn in the shell without reestablishing the mucus connection.

## Discussion

Small, nodulose *Littorina striata* selectively occupy the littoral fringe of Caloura, a shore of black basalt which absorbs high thermal loads from incident solar radiation. A distinctive intraspecific size gradient is evident with larger striate individuals mostly restricted to lower levels of the shore (Figs 3 and 4). Small individuals of *L. striata* are not as numerous at Caloura as in cooler high-shore environs such as at Ilhéu de Vila Franca, but they are much more abundant than either

striate *L. striata* or *Melarhaphe neritoides*. In contrast, larger, striate *L. striata* dominate both high- and low-shore positions on the cooler rocks of Ilhéu de Vila Franca (Figs 3 and 4).

Intraspecific size gradients are common (Vermeij, 1972) and usually follow one of two patterns: (1) shell size tends to increase upshore in high-zoned species, especially in the Littorinidae (Vermeij, 1972; Branch & Branch, 1981; Wells, 1984; Bosch & Moreno, 1987; Chen & Richardson, 1987) and (2) shell size often decreases upshore in mid- to lower-zoned species (Bock & Johnson, 1967; Vermeij, 1972). The latter situation is uncommon among littorinids, but more frequently observed among limpets, trochids, nerites and muricids (Vermeij, 1972). Size gradients related to shore position are absent in some, perhaps many, species of littorines. At least two studies designed to detect intraspecific size gradients in littorine populations found none (James, 1968; Chow, 1975) but McQuaid (1981) reported a vertical size gradient for *Littorina africana*, with smaller individuals occupying higher positions on the shore.

Thus, size distribution of *Littorina striata* on Azorean shores was unusual in two respects. At Caloura, smaller individuals generally occurred higher than larger individuals, in contrast to the situation usually observed among littorine species which size-partition the shore. Also, *L. striata* exhibited a size gradient only upon a shore which experiences high thermal loading from incident solar radiation. As only smaller individuals possess nodulose shells, and shell nodulosity has been implicated in facilitating dissipation of absorbed thermal energy by re-radiation and convection, the link between hot rocks and small nodulose shells seems likely. Before addressing that, however, some other mediating circumstances must be considered.

The distribution of *Littorina striata* at Ilhéu de Vila Franca, with large individuals dominating both the upper and lower shore, is possibly more like the norm for this species rather than the exception. Desiccation tolerance data help to corroborate this assertion. First, the differences in percent total body water loss between large and small *L. striata* were in line with other studies of desiccation tolerance which demonstrate that desiccation rates decrease with increase in body size (Davies, 1969; Chow, 1975). Next, *L. striata* and *Melarhaphe neritoides* are among the Littorinidae most capable of regulating evaporative loss of total body water during emersion (Fig. 6). After 11 days of emersion, they experienced less evaporative water

140

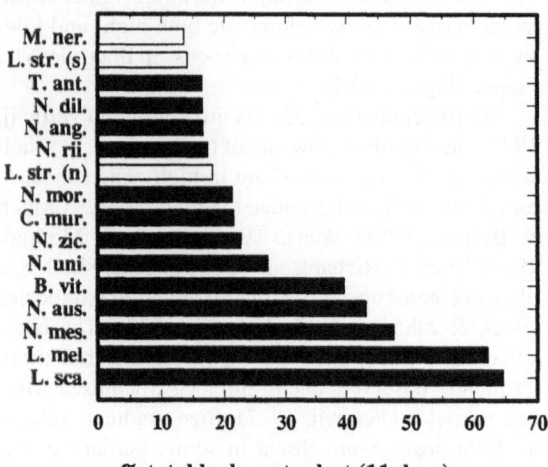

% total body water lost (11 days)

*Fig. 6.* Comparison of evaporative water loss expressed as a percent of total body water lost after 11 days of emersion for 15 species of Littorinidae. All are rocky shore species, except as noted. **B. vit.**, *Bembicium vittatum* (Philippi, 1846), Australia, [A]; **C. mur.**, *Cenchritis muricatus* (Linnaeus, 1758), Caribbean, [B]; **L. stra.**, *Littorina striata* (King & Broderip, 1832), Azores, (s) striate, (n) nodulose, [C]; **L. mel.**, *Littoraria melanostoma* (Gray, 1839), Hong Kong mangrove, [D]; **L. sca.**, *Littoraria scabra* (Linnaeus, 1758), Hong Kong mangrove, [D]; **M. ner.**, *Melarhaphe neritoides* (Linnaeus, 1758), Azores, [C]; **N. ang.**, *Nodilittorina angustior*, Caribbean, [B]; **N. aus.**, *Nodilittorina australis* (Gray, 1826), Australia, [E]; **N. dil.**, *Nodilittorina dilatata* (d'Orbigny, 1842), Caribbean, [B]; **N. mes.**, *Nodilittorina mespillum* (Mühlfeld, 1824), Caribbean, [B]; **N. rii.**, *Nodilittorina riisei* (Mörch, 1876), Caribbean, [B]; **N. uni.**, *Nodilittorina unifasciata* (Gray, 1826), Australia, [A]; **N. zic.**, *Nodilittorina ziczac* (Gmelin, 1791), Caribbean, [B]; **T. ant.**, *Tectarius antonii* (Philippi, 1846), Caribbean, [B]. Sources as follows: [A], McMahon and Britton, 1991; [B], Britton, 1992; [C], this study; [D], McMahon and Britton, 1985; [E], Britton, 1993.

loss (expressed as a percent total body water) than, for example, the tropical western Atlantic high shore species *Cenchritis muricatus*, which is known to be capable of surviving months in air (Vermeij, 1973; Britton, 1992). Thus, despite subtle differences in percent total body water loss between *M. neritoides* and striate and nodulose *L. striata*, all three of these groups are unlikely to find desiccation pressures limiting under most natural conditions, even under the potential temperature extremes of the upper shore of Caloura. Thus, large and small individuals likely range freely from the upper eulittoral to the littoral fringe, as evidenced by populations at Ilhéu de Vila Franca.

While this study was in progress, Grace Vedel and Michael Depledge also investigated heat tolerance of striate and nodose samples of *Littorina striata* by determination of mean heat coma temperatures (HCT) for several replicate samples of each group (McMahon &

Britton, 1985). There was no significant difference in mean HCT for the two groups of *L. striata*, with mean HCT for replicate samples of nodulose *L. striata* ranging from 43.87–44.95 °C and those for striate individuals ranging from 43.79–45.22 °C (Vedel & Depledge, personal communication). These results re-emphasize physiological similarity between large and small *L. striata*, despite their morphological distinctness, and seem to indicate that even the littoral fringe at Caloura is within the tolerance of both. One must, however, use the HCT findings with care. Thermal limits determined in water (as in HCT determinations) for species rarely submerged in nature provide comparative benchmarks, but do not aid understanding of thermal relations in the natural environment (Garrity, 1984). This is not to imply that thermal tolerances are less in air, but only that they have not been determined.

With these facts in mind, there seems to be no obvious physiological constraint to inhibit movements of either striate or nodose *Littorina striata* on most shores from the upper eulittoral to the upper littoral fringe. Indeed, occasional large individuals occupied the upper shore at Caloura and were dominant there at Ilhéu de Vila Franca. Yet, Hawkins *et al.* (1990) found *L. striata* exerting the greatest grazing pressure in the upper eulittoral, and this study confirms that the highest population densities on both studied shores was focussed here. Intertidal species usually occupy higher levels on the shore for one of two reasons, either to minimize intra- or interspecific competition (Branch & Branch, 1981) or to avoid predation from predators restricted to the eulittoral zone (Vermeij, 1972; McQuaid, 1982). Either or both of these reasons is sufficient to account for the distribution of *L. striata* in the littoral fringe. Upon occupying the upper shore, however, individuals may voluntarily abandon it, especially for reproductive or spawning activities. For example, Ohgaki (1985b) found the mean size of *Nodilittorina exigua* to be larger on higher levels of the shore in all seasons except summer, when the upper shore apparently attracted smaller individuals. In fact, the necessity to spawn planktonic eggs produced a downward migration or larger, sexually mature *N. exigua* during the summer, with smaller immature individuals remaining in abundance on the higher shore (Ohgaki, 1988). If that were the case with *L. striata*, one should expect similar numbers of small individuals in the littoral fringe at both Caloura and Ilhéu de Vila Franca. As this was not observed, it is reasonable to assume that some factor other than a spawning migration was responsible for the prevalence of small, nodulose individuals on the higher rocks at Caloura.

Having considered some additional mediating circumstances, we can reconsider the role of nodulose texture for facilitating the dissipation of absorbed thermal energy by re-radiation and convection (Vermeij, 1971a, 1973). Small *Littorina striata* with nodulose shells should be more efficient in dissipating thermal energy than larger individuals with striate shells. This advantage possibly contributes to producing higher densities of smaller individuals on the Caloura littoral fringe, especially when one considers the posturing behaviour which renders re-radiation of absorbed thermal energy most effective.

Posture strongly influences heat regulation in most high shore littorines (Vermeij, 1971b; Garrity, 1984). Heat is more readily absorbed from the substratum if a portion of the body or shell is in contact with it (Vermeij, 1971a). Thus many species of littoral fringe littorines, including *Littorina striata*, reduce physical contact with the substratum to little more than a strand of mucus attached to the aperture lip, with most of the body of the animal suspended in air free of the substratum. In this position, absorbed thermal energy is most readily re-radiated to the atmosphere by convection. Small, slender shapes posture more effectively attached to mucus holdfasts, whereas it becomes increasingly difficult to suspend larger, more globose shapes completely in air without some portion of the shell touching the substratum. That is the case with large *L. striata* in the natural environment. At both Caloura and Ilhéu de Vila Franca, smaller individuals were more frequently observed attached by mucus holdfasts with the body suspended in air than were larger individuals. The latter, although also attached by mucus, usually had some, and often much, of the shell in contact with surrounding rock surfaces, from which they would undoubtedly absorb considerable heat, especially on shores like Caloura which also readily absorb it.

If, as Wolcott (1973) has hypothesized, intertidal gastropods tend to occupy shore zones just below the limits of their tolerance to various environmental factors and if small, nodulose individuals of a species are better equipped to survive thermal limits, it is reasonable to assume that they will be more abundant in environments experiencing diurnal substratum temperatures elevated higher than normal. It is unlikely that the larger individuals experience mortality when they venture into these environments, but simply readjust their position on the shore to a level where substratum temperatures are more compatible with the physiological demands imposed by their morphology.

The question remains, however, as to the selective forces which may have fostered nodulose ornamentation in small individuals of *Littorina striata*. This is a species mostly restricted to the volcanic islands of the mid-Atlantic ridge. Although it possibly prefers shore conditions such as those at Ilhéu de Vila Franca, it is probably frequently dispersed to geologically young, dark basaltic rocks such as those at Caloura, where substratum temperatures potentially approach the limits of physiological tolerance. Large adult individuals are, as I have shown, more poorly adapted to live in the littoral fringe of such shores than smaller individuals which often also possess an additional morphological adaptation, nodulosity, which further increases their fitness under these conditions. With smaller individuals living higher on the shore, there is less intraspecific competition for resources at the level of maximal population size, usually the upper eulittoral, and perhaps at least some individuals are protected from eulittoral predators. Thus, small, young *L. striata* seem ideally suited to colonize dark basaltic rocks typical of most mid-oceanic Atlantic islands, and more so than many other high-shore littorine species, the larger individuals of which typically seek higher ground.

## Acknowledgements

This study was conducted during the Second International Workshop of Malacology of the Azores, organized by Dr António M. de Frias Martins. I extend my most sincere gratitude to Dr Frias Martins for the opportunity to visit the Azores, the hospitality and assistance provided during my visit, and permission to submit this report as a contributed paper to the Fourth International Littorinid Symposium. Thanks also to Ms. Regina Tristão da Cunha for provision of various and sundry laboratory supplies required during the field studies. I also thank Thierry Backeljau, Robert Bullock and Brian Morton for the benefit of helpful discussions related to this project.

## References

Bock, C. E. & R. E. Johnson, 1967. The role of behavior in determining the intertidal zonation of *Littorina planaxis* Philippi, 1847, and *Littorina scutulata* Gould, 1849. Veliger 10: 42–54.

Bosch, M. & I. Moreno, 1986. Spatial distribution of *Littorina neritoides* L. 1758 (Mollusca: Gastropoda) in the supralittoral zone of the Balearic Islands, Spain. Cah. Biol. mar. 27: 53–62.

Branch, G. M. & M. L. Branch, 1981. Experimental analysis of intraspecific competition in an intertidal gastropod, *Littorina unifasciata*. Aust. J. mar. Freshwat. Res. 32: 573–589.

Britton, J. C., 1992. Evaporative water loss, behavior during emersion, and upper thermal tolerance limits in seven species of eulittoral fringe Littorinidae (Mollusca: Gastropoda) from Jamaica. In J. Grahame, P. J. Mill & D. G. Reid (eds), Proceedings of the 3rd International Symposium on Littorinid Biology. The Malocological Society of London, London: 69–83.

Britton, J. C., 1993. The effects of experimental protocol and behavior on evaporative water loss during emersion in three species of rocky shore gastropods. In F. E. Wells, D. I. Walker, H. Kirkman & R. Lethbridge (eds), Proceedngs of the Fifth International Marine Biological Workshop: The Marine Flora and Fauna of Rottnest, Western Australia. Western Australian Museum, Perth, 2: 601–619.

Britton, J. C. & R. F. McMahon, 1990. The relationship between vertical distribution, evaporative water loss rate, behaviour, and some morphometric parameters in four species of rocky intertidal gastropods from Hong Kong. In B. Morton (ed.), Proceedings of the Second International Marine Biological Workshop: The Marine Flora and Fauna of Hong Kong and Southern China, Hong Kong, 1986. Hong Kong University Press, Hong Kong, II, 3: 1153–1171.

Chen, Y. S. & A. M. M. Richardson, 1987. Factors affecting the size structure of two populations of the intertidal periwinkle, *Nodilittorina unifasciata* Gray, 1839, in the Derwent River, Tasmania, Australia. J. moll. Stud. 53: 69–78.

Chow, V., 1975. The importance of size in the intertidal distribution of *Littorina scutulata*. Veliger 18: 69–78.

Davies, P. S., 1969. Physiological ecology of *Patella*, III. Desiccation effects. J. mar. biol. Ass. U.K. 49: 291–304.

Garrity, S. D., 1984. Some adaptations of gastropods to physical stress on a tropical rocky shore. Ecology 65: 559–574.

Hawkins, S. J., L. P. Burnay, A. I. Neto, R. Tristão da Cunha & A. M. Frias Martins, 1990. A description of the zonation patterns of molluscs and other important biota on the south coast of São Miguel, Azores. In A. M. de Frias Martins (ed.), The Marine Fauna and Flora of the Azores (Proceedings of the First International Workshop of Malacology, São Miguel, Azores, 1988). Açoreana, Supplement: 21–28.

James, B. L., 1968. The characters and distribution of the subspecies and varieties of *Littorina saxatilis* (Olivi 1792) in Britain. Cah. Biol. mar. 9: 143–165.

McMahon, R. F., 1990. Thermal tolerance, evaporative water loss, air-oxygen consumption and zonation of intertidal prosobranchs: a new synthesis. Hydrobiologia 193 (Dev. Hydrobiol. 56): 241–260.

McMahon, R. F. & J. C. Britton, 1985. The relationship between vertical distribution, thermal tolerance, evaporative water loss rate, and behaviour on emergence in six species of mangrove gastropods from Hong Kong. In B. Morton & D. Dudgeon (eds), Proceedings of the Second International Workshop on the Malacofauna of Hong Kong and Southern China, 1983. Hong Kong University Press, Hong Kong, 2: 563–582.

McMahon, R. F. & J. C. Britton, 1991. The relationship between vertical distribution, thermal tolerance, evaporative water loss

rate, behaviour, and morphometrics in six species of rocky shore gastropods from Princess Royal Harbour, Western Australia. In F. E. Wells, D. I. Walker, H. Kirkman & R. Lethbridge (eds), Proceedings of the Third International Marine Biological Workshop: The Marine Flora and Fauna of Albany, Western Australia. Western Australian Museum, Perth, 2: 675–692.

McQuaid, C. D., 1981. The establishment and maintenance of vertical size gradients in populations of *Littorina africana knysnaensis* (Philippi) on an exposed rocky shore. J. exp. mar. Biol. Ecol. 54: 77–89.

McQuaid, C. D., 1982. The influence of desiccation and predation on vertical size gradients in populations of the gastropod *Oxystele variegata* (Anton) on an exposed rocky shore. Oecologia 53: 123–127.

Morton, B., 1990. The intertidal ecology of Ilhéu de Vila Franca – a drowned volcanic crater in the Azores. In A. M. de Frias Martins (ed.), The Marine Fauna and Flora of the Azores (Proceedings of the First International Workshop of Malacology, São Miguel, Azores, 1988). Açoreana, Supplement: 3–20.

Newell, R. C., 1979. Biology of intertidal animals. Marine Ecological Surveys, Ltd., Faversham, Kent, 781 pp.

Ohgaki, S. I., 1985a. Distribution of the family Littorinidae (Gastropoda) on Hong Kong rocky shores. In: B. Morton & D. Dudgeon (eds), Proceedings of the Second International Workshop on the Malacofauna of Hong Kong and Southern China, 1983. Hong Kong University Press, Hong Kong, 2: 457–464.

Ohgaki, S. I., 1985b. Vertical variation in size structure and density of the littoral fringe periwinkle *Nodilittorina exigua*. Venus (Jap. J. Malacol.) 44: 260–269.

Ohgaki, S. I., 1988. Vertical migration and spawning in *Nodilittorina exigua* (Gastropoda: Littorinidae). J. Ethol. 6: 33–38.

Porter, W. P. & D. M. Gates, 1969. Thermodynamic equilibria of animals with environment. Ecol. Monogr. 39: 227–244.

Reid, D. G., 1989. The comparative morphology, phylogeny and evolution of the gastropod family Littorinidae. Phil. Trans. r. Soc., Lond., B., Biol. Sci. 324: 1–110.

Rosewater, J., 1963. Resistance to desiccation in dormancy by *Tectarius muricatus*. Nautilus, 76: 111.

Rosewater, J., 1981. The family Littorinidae in tropical West Africa. Atlantide Rep. 13: 7–48.

Underwood, A. J., 1979. The ecology of intertidal gastropods. Adv. Mar. Biol. 16: 111–210.

Vermeij, G. J., 1971a. Temperature relationships of some tropical Pacific intertidal gastropods. Mar. Biol. 10: 308–314.

Vermeij, G. J., 1971b. Substratum relationships of some tropical Pacific intertidal gastropods. Mar. Biol. 10: 315–320.

Vermeij, G. J., 1972. Intraspecific shore-level size gradients in intertidal molluscs. Ecology 53: 693–700.

Vermeij, G. J., 1973. Morphological patterns in high-intertidal gastropods: adaptive strategies and their limitations. Mar. Biol. 20: 319–346.

Wells, F. E., 1984. Population characteristics of the periwinkle, *Nodilittorina unifasciata*, on a vertical rock cliff in Western Australia. Nautilus 98: 102–107.

Winer, B. J., 1971. Statistical principles in experimental design, 2nd edn. McGraw-Hill, New York, 907 pp.

Wolcott, T. G., 1973. Physiological ecology and intertidal zonation in limpets (*Acmaea*): a critical look at 'limiting factors'. Biol. Bull. 145: 389–422.

*Hydrobiologia* **309**: 143–150, 1995.
*P. J. Mill & C. D. McQuaid (eds), Advances in Littorinid Biology.*
©1995 *Kluwer Academic Publishers.*

# Maintenance of zonation patterns in two species of flat periwinkle, *Littorina obtusata* and *L. mariae*

Gray A. Williams

*Department of Zoology, University of Bristol, Woodland Road, Bristol BS8 1UG, UK*
[1]*Present address: The Swire Institute of Marine Science and Department of Ecology and Biodiversity, The University of Hong Kong, Hong Kong*

*Key words:* epiphytes, behaviour, rocky shore, *Littorina obtusata, Littorina mariae*

## Abstract

The zonation patterns of *Littorina obtusata* (L.) and *Littorina mariae* Sacchi et Rastelli were shown to be quite distinct on a sheltered rocky shore. *L. obtusata* was found at all the heights sampled; it reached peak numbers at mid shore on the alga *Ascophyllum nodosum* L. (Le Jol). There was no difference in the tidal height occupied by adults or juveniles; or in the mean size of *L. obtusata* along the vertical gradient of the shore. In contrast *L. mariae* occurred exclusively low on the shore, on *Fucus serratus* L. Translocation of the two species within their respective levels resulted in random movement after 4 days, although initial movements after 1 and 2 days were sometimes directional. Animals transplanted to the normal level of the other species showed directional movement towards their home zone; this was most pronounced after 4 days. There was no difference in the distance moved by the two species, although the distance moved did vary with tidal height, both species moving further at mid shore than low shore. Distances moved by littorinids at replicate areas in the low shore were similar but those at mid shore did vary. There was an interaction between the species and the different tidal heights which revealed that transplanted species moved further than translocated species at the same tidal level. However, this was only significant in the case of *L. mariae*. It is suggested that the close relationship between the winkles and their host algae may direct the homing behaviour of displaced individuals.

## Introduction

*Littorina obtusata* (L.) and *Littorina mariae* Sacchi et Rastelli are closely related, epiphytic intertidal gastropods (see Williams, 1990 for review of their ecologies), which were only distinguished in 1966 (Sacchi & Rastelli, 1966). The behavioural responses of flat winkles, *L. obtusata sensu lato*, to varying environmental cues such as light and gravity had been investigated before 1966 (Janssen, 1960; Charles, 1961; Evans, 1965). The most important taxis noted was the attraction to fucoid algae. Flat winkles were attracted to algae from up to a metre away (Barkmann, 1955; Van Dongen, 1956) and observed to form zones in tidal tanks only in the presence of fucoids (Evans, 1965; Thompson, 1968; Underwood, 1972). The results, however, were confusing and equivocal, perhaps due to inadver-

tant inclusion of *L. mariae* with *L. obtusata*. This was clarified by Guiterman (1970) who recognized both species and described their behaviour: *L. obtusata* is negatively phototactic and therefore retreats into the algal mass when exposed to light, whereas *L. mariae*, exhibiting less strong negative phototaxis, actively crawls over the surface of the algae when exposed.

The zonation patterns of the two species are known to differ and are also related to the zonation of their host algal species. On sheltered shores, *L. obtusata* inhabits all shore levels but is dominant on the alga *Ascophyllum nodosum* L. (Le Jol) whereas *L. mariae* is restricted to low shore living on *Fucus serratus* L. (Watson & Norton, 1987; Williams, 1990, 1994). The degree of overlap varies and is related to wave exposure (Reimchen, 1974; Goodwin, 1975), possibly through variation in the quality and abundance of fucoid algae

(Williams, 1990, 1994). This paper describes the zonation patterns of the two species on a sheltered rocky shore and tests the hypothesis that the vertical distribution may be maintained through differential patterns of behaviour.

## Materials and methods

All the work was conducted at Sawdern Point (British National Grid Reference SM 888032), a sheltered rocky shore in Dyfed, west Wales. This shore has luxuriant stands of algae, especially *Ascophyllum nodosum* and *Fucus serratus*. The initial zonation study was carried out in October 1985 and the movement experiments in March 1987. Monthly population surveys showed that the zonation patterns of the two species did not vary significantly during the year (Williams, 1987, 1992). The density of the winkles was recorded at one metre vertical intervals up the shore. At each height ten $50 \times 50$ cm quadrats were sampled. These were positioned, using a random number table, in a 30 m horizontal band at sites where the substratum was sufficiently homogeneous to give a satisfactory replication of tidal height and algal cover. Percentage algal cover was scored using a double strung 100-point quadrat (Jones *et al.*, 1980). The algae were then systematically searched, using 'Tru-Touch' medical gloves, to which small individuals stick, until no more winkles were found. All damaged air-bladders were searched for juveniles (see Reichmen, 1974; Goodwin, 1975). The number and species of the winkles found were noted and the length (value 'a' of Goodwin & Fish, 1977) of each winkle above 2.5 mm was measured to within $\pm 0.05$ mm, using vernier calipers. Those below 2.5 mm were assigned to the smallest size class.

For the behavioural experiments, adult winkles, as defined by thickening of the aperture lip (Goodwin & Fish, 1977), were selected: *L. obtusata* (size range 14–17 mm) from mid shore (4.4 m above Chart Datum, C.D.); and *L. mariae* (size range 8–12 mm) from low shore (1.4 m above C.D.). The winkles collected were numbered using 'Micromarkers' (W. H. Brady and Co., Banbury) which were stuck to the shell using 'Super Glue' (Loctite). The experimental design involved four treatments: two of these used *L. obtusata* and *L. mariae* within their home ranges (4.4 m above C.D. for *L. obtusata* and 1.4 m above C.D. for *L. mariae*). These treatments investigated the movement of the two species at their respective tidal heights. The second two treatments were of the winkles moved to the shore

level (and as a result host algae) appropriate to the other species: *L. obtusata* transplanted to low shore on *F. serratus*, and *L. mariae* transplanted to mid shore on *Ascophyllum*. These treatments investigated the effect of displacement from their normal zone on the movement of the winkles. Treatments were duplicated at different parts of the shore. Four sites were chosen which had similar algal cover and uniform slopes; two at low shore, with a cover of *F. serratus* (sites L1 and L2); and two at mid shore with a cover of *Ascophyllum* (sites M1 and M2). The approximate direction of the slope of the shore was measured by surveying the experimental areas. In each replicate 30 individuals of each species were used. All the winkles used in the experiments were handled in the same way prior to replacement onto the weed. Any possible disturbance effects were therefore similar for all the treatments, although there was no control for the effect of handling on behaviour (see Chapman, 1986; Chapman & Underwood, 1992a). The animals were released by placing them at 5 cm intervals in a grid which was orientated to two fixed points – bolts screwed into the substrate. Therefore the exact position of each individual was known at the start of the experiment. The line between these points was parallel to the slope of the shore, and the bearing of this line relative to true North was determined. All calculations of movement were made from these reference bolts using coordinate geometry (following the methods of Underwood, 1977).

The two components of movement are the distance travelled and the direction of this movement. The distance moved was calculated using coordinate geometry and the direction was scored as angular bearings relative to North. The original measurements were analysed by a BASIC computer programme which calculated the distance moved by an individual from the initial release point to the next point and the direction of this movement. From these data the mean angle and the mean resultant vector (a measure of the variability around the mean direction, see Underwood & Chapman, 1985) were calculated following the methods of Zar (1974). The significance of the direction was examined using Rayleigh's Test (Zar, 1974), a significant result indicating that movement was not random but directional. The distance moved was analysed by a Three Factor ANOVA with the different sites nested within the shore heights. Significant differences between the means were identified using SNK tests (Zar, 1974).

*Table 1.* Pearson product moment correlation coefficients between winkle numbers and algal cover (arcsin transformed) found at all the heights sampled ($n=50$)

| | F. spiralis | F. vesiculosis | A. nodosum | F. serratus |
|---|---|---|---|---|
| L. obtusata | −0.349* | 0.347* | 0.697** | −0.552** |
| L. mariae | | | −0.521** | 0.884** |

*$P<0.05$, **$P<0.001$; N.S. = Not significant

Movement was estimated during emersion periods and was scored after one, two and four days. This technique only measures the direction and distance of final displacement from one period to another and does not account for any complexity in the path of the winkles (see Chapman & Underwood, 1992a). Due to the nature of the substrate, and the tendency of *L. obtusata* to crawl into the weed mass, extensive searching was necessary. The movement of each winkle was scored to the nearest 5 cm because searching often slightly displaced the individuals.

## Results

### Vertical zonation

The heights sampled represented the entire vertical range which the winkles could potentially inhabit as extension upshore, or downshore, was not possible due to the absence of suitable substrate or algae. *L. obtusata* was found at all sample heights (Fig. 1) but its density varied with height on the shore (Kruskal Wallis test: $\chi^2 = 44.32$, $P \leq 0.001$; see Fig. 1). The peak density was at mid shore ($\simeq 4.4$ m above C.D.) and decreased very rapidly towards both high and low shore. Furthermore, the peak density was on *Ascophyllum nodosum* (high positive correlation, Table 1) and the species was negatively correlated with *Fucus spiralis* (L.) and *F. serratus* (Table 1). There were no significant differences in the size of *L. obtusata* with height on the shore, indicating that both adults and juveniles inhabit the same shore levels (One Way ANOVA, $F = 1.01$; 4,38 df; $P>0.05$). In contrast, *L. mariae* was found almost exclusively at the two lowest sites on *F. serratus*, and was positively correlated with this alga and negatively correlated with *Ascophyllum* (Table 1).

### Direction of movement

The directions of movement of *L. obtusata* translocated at mid shore (the control) were variable. In one of the replicate areas, M1, *L. obtusata* showed random movement for all three days but at site M2 directional movement occurred on days one and two, this was constant for both days at 248 °, a slightly downshore direction. However, after four days this control sample showed random movement, in a similar fashion to M1. In contrast, both replicates of *L. mariae* transplanted to mid shore moved in a downward direction for all three days (Table 2, Fig. 2).

At low shore the translocated animals, *L. mariae*, showed a random direction of movement for all the days recorded, whereas transplanted *L. obtusata* tended to move upshore. At L2 this movement was significant for all three days, while at L1 it was random after one day but was upshore on the two remaining sampling occasions.

Thus, after four days all of the transplanted animals moved in the direction of their home zone whereas all control animals (i.e. at their normal shore levels) exhibited random movement.

### Distance moved

Animals at mid shore tended to move further than those at low shore for all the days recorded (Tables 2 and 3). There was no significant difference in the distances moved by the two species for the duration of the experiment (Table 3). There was, however, an interaction between the factors 'Species' and 'Height' on all three days, suggesting different movement patterns for the two species at mid and low shore. Further analysis (SNK Tests, Table 3) revealed that *L. mariae* transplanted to mid shore moved further than any of the other groups. For days one and two *L. obtusata* at mid shore also moved further than either species at low shore, which moved similar distances. During the entire period of the experiment, transplanted animals moved further than translocated animals at the same tidal level although this was only significant in the case of *L. mariae*. There was also a difference in the distances moved by the winkles between the replicate areas for the first two observation periods. Initially winkles at site M1 moved significantly shorter distances than those at M2 (which moved the furthest distance of all the treatments). After four days, however, replicate sites at mid shore as well as those at low shore did

146

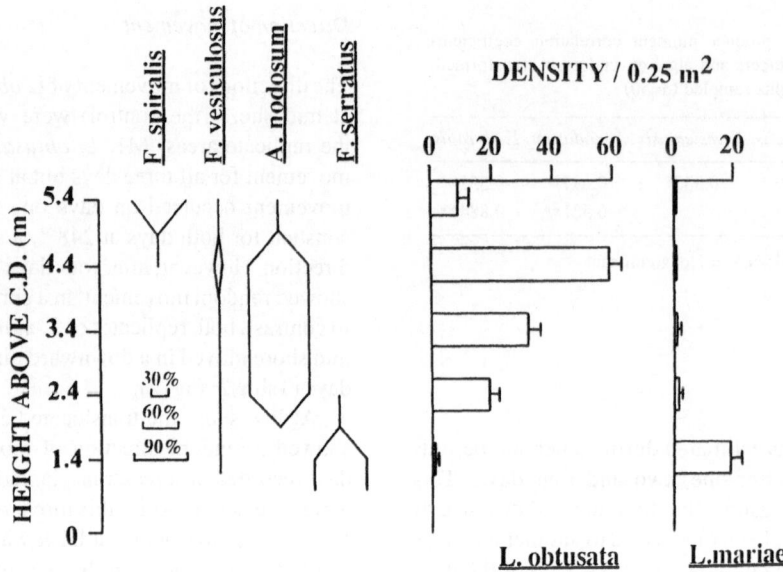

*Fig. 1.* Zonation pattern of fucoid algae cover (pattern illustrated by kite diagrams) and flat winkle species (mean density $+$S.D./0.25 m$^2$) (after Williams, 1994).

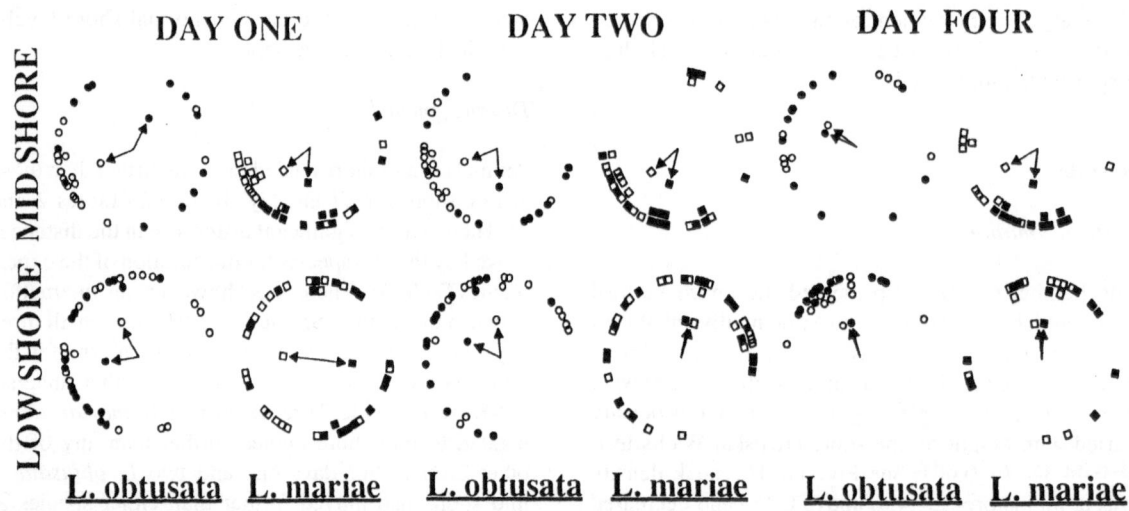

*Fig. 2.* Directions of the movement of individual littorinids. ○, ●, individual *L. obtusata*; □, ■, individual *L. mariae*. The top diagrams illustrate mid shore replicates (○, □, M1; ●, ■, M2); the bottom diagrams illustrate low shore replicates (○, □, L1; ●, ■, L2). Arrows and associated symbols indicate mean directions of movement.

not differ in the distances moved by the experimental animals.

**Discussion**

Previous workers have noted a difference in tidal height occupied by *Littorina obtusata* and *L. mariae* (Sacchi & Rastelli, 1966; Sacchi, 1969; Reimchen, 1974;

Goodwin, 1975; Watson & Norton, 1987). In the present study *L. obtusata* was found at all heights examined on the shore, reaching a maximum density of about 60/0.25 m$^2$ at mid-shore. The density of winkles was far greater than previously recorded (Watson & Norton, 1987). The present study also showed that there was no difference in the zonation patterns of juveniles and adults. The range of distribution of *L. mariae* was more restricted than that of

*Table 2.* Sample sizes (n), mean distances (cm±S.D.) and directions (mean angle, °N, and mean resultant vector – the higher the value of the mean resultant vector, the lower the variability about the mean angle) moved by experimental animals after one, two and four days. Significant results (*=$P<0.05$; **$P<0.01$; ***$P<0.001$; N.S. = Not significant) indicate a departure from a random direction of movement as designated by Rayleigh's Test.

| Treatment | | Day one | | | | Day two | | | | Day four | | | |
| | | Distance | Direction | | Sig. | | Distance | Direction | | Sig. | | Distance | Direction | | Sig. |
| | n | Mean±S.D. (cm) | Mean resultant vector | Mean angle °N | | n | Mean±S.D. (cm) | Mean resultant vector | Mean angle °N | | n | Mean±S.D. (cm) | Mean resultant vector | Mean angle °N | |
|---|---|---|---|---|---|---|---|---|---|---|---|---|---|---|---|
| **Midshore** | | | | | | | | | | | | | | | |
| M1 *L. obtusata* | 12 | 24.1±13.7 | 0.147 | 28 | N.S. | 12 | 31.2±15.8 | 0.349 | 178 | N.S. | 12 | 48.9±29.4 | 0.426 | 293 | N.S. |
| *L. mariae* | 15 | 79.2±29.6 | 0.713 | 185 | *** | 20 | 106.6±44.4 | 0.548 | 193 | ** | 13 | 123.4±39.7 | 0.932 | 188 | *** |
| M2 *L. obtusata* | 16 | 120.1±56.8 | 0.688 | 248 | *** | 17 | 119.7±53.9 | 0.826 | 248 | *** | 9 | 92.5±43.0 | 0.358 | 306 | N.S. |
| *L. mariae* | 15 | 120.2±64.4 | 0.757 | 233 | *** | 18 | 135.4±68.8 | 0.487 | 228 | * | 10 | 128.6±89.3 | 0.713 | 240 | ** |
| **Low shore** | | | | | | | | | | | | | | | |
| L1 *L. obtusata* | 19 | 28.1±16.4 | 0.635 | 262 | *** | 17 | 46.0±41.3 | 0.624 | 297 | *** | 14 | 83.6±51.3 | 0.916 | 344 | *** |
| *L. mariae* | 17 | 24.7±14.6 | 0.256 | 97 | N.S. | 18 | 36.9±19.2 | 0.365 | 11 | N.S. | 11 | 66.9±56.3 | 0.331 | 3 | N.S. |
| L2 *L. obtusata* | 22 | 44.5±23.6 | 0.342 | 330 | N.S. | 16 | 56.5±33.4 | 0.532 | 355 | * | 12 | 80.6±33.9 | 0.865 | 338 | *** |
| *L. mariae* | 19 | 25.7±10.5 | 0.334 | 275 | N.S. | 19 | 33.2±21.7 | 0.351 | 17 | N.S. | 11 | 49.9±26.1 | 0.491 | 357 | N.S. |

Direction of shore slope
M1 Upshore approx. 360°; Downshore approx. 180°; M2 Upshore approx. 360°; Downshore approx. 220°
L1 Upshore approx. 360°; Downshore approx. 180°; L2 Upshore approx. 360°; Downshore approx. 240°

*Table 3.* Three factor analysis of variance (with the factor Area nested within the factor Height) to investigate the distance moved by the experimental animals after one, two and four days. Data were log $e$ transformed prior to analysis (Height = Mid shore versus low shore; Species = *L. obtusata* versus *L. mariae*; Area = M1, M2, L1 and L2; * = $P<0.05$; ** = $P<0.001$; N.S. = Not significant). Outcomes of SNK Tests to compare cell means are indicated.

| Source of variation | Day one | | | | Day two | | | | Day four | | | |
| | d.f. | MS | F | Sig | d.f. | MS | F | Sig | d.f. | MS | F | Sig |
|---|---|---|---|---|---|---|---|---|---|---|---|---|
| Height (HT) | 1 | 28.24 | 60.62 | ** | 1 | 25.71 | 50.02 | ** | 1 | 2.30 | 4.49 | * |
| Species (SP) | 1 | 0.69 | 1.48 | NS | 1 | 1.21 | 2.34 | NS | 1 | 0.34 | 0.66 | NS |
| HT×SP | 1 | 5.55 | 11.91 | ** | 1 | 7.33 | 14.25 | ** | 1 | 7.55 | 14.72 | ** |
| Area | 2 | 7.56 | 16.23 | ** | 2 | 5.46 | 10.63 | ** | 2 | 1.12 | 2.19 | NS |
| Error | 130 | 0.47 | | | 136 | 0.51 | | | 86 | 0.51 | | |
| SNK Tests | | | | | | | | | | | | |
| Height | Mid>Low | | | | Mid>Low | | | | Mid>Low | | | |
| HT×SP | M*L. ma*>M*L. ob*>L*L. ob*>L*L. ma* | | | | M*L. ma*>M*L. ob*>L*L. ob*>L*L. ma* | | | | M*L. ma*>L*L. ob*>M*L. ob*>L*L. ma* | | | |
| Area | M2=L2=L1>M1 | | | | M2=L1=L2>M1 | | | | NS | | | |

*L. obtusata*, being confined to the lowest height sampled, and there was only a slight overlap between the species (Fig. 1). *L. obtusata* and *L. mariae* are evidently sharply zoned on this sheltered shore and, as they are both mobile species it is likely that their zonation is effected by behavioural responses. At mid shore, *L. obtusata* will experience greater desiccation stress and temperature range than *L. mariae* at low shore. *L. obtusata* shows behavioural adaptations to counteract these problems, crawling into the centre of the algal mass (as the alga dries) and remaining relatively dormant until immersed (Guiterman, 1970). Laboratory experiments have shown *L. obtusata* to be more tolerant of acute desiccation and temperature fluctuations than is *L. mariae* (Sacchi, 1969;1972a, b).

When transplanted to the shore level normally occupied by the other species, both species moved towards their normal zones. The two species, when translocated within their own zone, showed random movement after an initial settling period. Initial movement may have been the result of disturbance factors such as the effect of displacement and marking, a phenomenon which has been noted by other workers (Petraitis, 1982; Underwood & Chapman, 1985; Chapman, 1986; Chapman & Underwood, 1992a and b).

The random direction of movement of translocated individuals after four days illustrated a passive maintenance of shore zone by the two species. Similar random movements have been recorded for *Nerita atramentosa* Reeve and *Bembicium nanum* (Lamarck) (Underwood, 1977); *L. littorea* (Alexander, 1960; Gendron, 1977; Petraitis, 1982) and *L. africana knysnaensis* (Philippi) (McQuaid, 1981) at their respective tidal levels. Very few authors have examined the movement of translocated animals at their respective tidal heights when investigating movement patterns of displaced animals (see Chapman, 1986 for review). The displaced, recaptured winkles in this experiment all showed directional movement towards their home zones after four days. Such maintenance of shore level in displaced gastropods has been noted by many authors: e.g. *L. littorea* (Alexander, 1960; Gendron, 1977; Petraitis, 1982); *Nodilittorina unifasciata* (Gray) (Chen & Richardson, 1987); *Tegula funebralis* (Adams) (Byers & Mitton, 1981; Doering & Phillips, 1983); *L. africana knysnaensis* (McQuaid, 1981); *Nerita* spp (Bovbjerg, 1984); *Gibbula umbilicalis* (Da Costa) (Thain *et al.*, 1985) and *L. scutulata* Gould and *L. planaxis* (Philippi) (Bock & Johnson, 1967).

Winkles at mid shore moved further than those at low shore, this was especially true of *L. mariae*. Even after four days winkles displaced from their home zones appeared to move further than the controls at the same tidal height. Although this was significant only in the case of *L. mariae*, it appeared to represent a consistent trend for both species. The greater distance moved by transplanted *L. mariae* may be explained by the fact that *L. mariae* are generally more active than *L. obtusata*. When the winkles are emersed, *L. mariae* continue to crawl over the surface of the algae while *L. obtusata* retreat into the algal mass. This difference in behaviour, combined with the strongly directional movement of *L. mariae*, will result in greater distances

being moved by *L. mariae* than *L. obtusata* at the mid-shore level.

Relating the factors responsible for these behavioural patterns purely to vertical, tidal height is confounded by the presence of different algae at these heights. The greater distance moved by both species at mid shore as compared to low shore may be due to the ease with which the winkles could move over the substratum, a factor noted for *L. unifasciata* moving over rocky substratum of varying topographies (Underwood & Chapman, 1989). At mid shore this would be maximal when the algae were emersed, as the *Ascophyllum* fronds would cover approximately 90% of the rock surface and form a regular homogeneous carpet. Movement in the horizontal plane would have been prevented at high water when *Ascophyllum* floats vertically. When emersed the fronds may settle randomly, which may account for the random distribution of *L. obtusata* at this shore level. At low shore the substratum was *Fucus serratus*, which is more irregular in distribution than *Ascophyllum*. When emersed, the stipes of *F. serratus* hold the fronds partly off the rock surface and effectively make movement between plants more difficult for the winkles than for those on *Ascophyllum*. The differences in movement between the sites (M1 and M2) at mid shore may be due to variation in algal cover, general topography or in time of sampling, because one site would be sampled before the other on the ebbing tide. Variation in distances of movement of *L. unifasciata* have also been recorded on different spatial and temporal scales (Underwood & Chapman, 1989; Chapman & Underwood, 1992a).

The host algae release secondary chemicals which influence the behaviour of *L. obtusata* and *L. mariae* (Norton *et al.*, 1990; Norton & Manley, 1990; Williams & Seed, 1992). In the laboratory, *L. obtusata* is positively attracted to exudates of *Ascophyllum* whereas *L. mariae* is repelled by this alga but attracted to *F. serratus*. It has been suggested that the winkles use these chemical cues to orientate themselves (Watson & Norton, 1987; Norton & Manley, 1990), although the exact mechanisms of this attraction in the field are still a subject of debate. The close relationship between the littorinids and their host algae supports the role of algal chemical cues reinforcing the littorinid distribution patterns and aiding recovery of the host when displaced (Williams & Seed, 1992).

In conclusion, a number of factors (see reviews by Underwood & Chapman, 1985 and Chapman & Underwood, 1992b), may combine to provide the stimuli to which *L. obtusata* and *L. mariae* respond by directional movement. The physiological tolerances of the species will result in selection for the maintenance of an optimum shore level in evolutionary time, but the physical presence of the host algae of the winkles is probably the overriding factor responsible for orientation *in situ* (Barkman, 1955; Van Dongen, 1956; Underwood, 1972, 1979).

## Acknowledgments

This work was supported by a N.E.R.C Studentship. I am grateful to Dr C. Little for his continual help and encouragement. The BASIC programme was written by Dr J. Rayner and G. Smith (Bristol University). Dr S. J. Hawkins (Port Erin Marine Laboratory), Professor A. J. Southward (Plymouth Marine Laboratory) and Dr D. G. Reid (British Museum) kindly commented on the manuscript. I would also like to thank Dr and Mrs R. Crump, the staff at Orielton Field Studies Centre and the many people who helped with the fieldwork. Presentation of this work was supported by a Hong Kong University Conference grant.

## References

Alexander, G. D., 1960. Directional movement of the intertidal snail, *Littorina littorea*. Biol. Bull. 119: 301–302.

Barkman, J. J., 1955. On the distribution and ecology of *Littorina obtusata* (L.) and its subspecific units. Arch. Neerl. Zool. 11: 22–86.

Bock, C. E. & R. E. Johnson, 1967. The role of behaviour in determining the intertidal zonation of *Littorina planaxis* Philippi 1847, and *Littorina scutulata* Gould, 1849. Veliger. 10: 42–54.

Bovbjerg, R. V., 1984. Habitat selection in two intertidal snails, genus *Nerita*. Bull. mar. Sci. 34: 185–196.

Byers, B. A. & J. B. Mitton, 1981. Habitat choice in the intertidal snail *Tegula funebralis*. Mar. Biol. 65: 149–154.

Chapman, M. G., 1986. Assessment of some controls in experimental transplants of intertidal gastropods. J. exp. mar. Biol. Ecol. 103: 181–201.

Chapman, M. G. & A. J. Underwood, 1992a. Experimental designs for the analyses of movements by molluscs. In J. Grahame, P. J. Mill & D. G. Reid (eds), Proceedings of the 3rd International Symposium on Littorinid Biology. Malacological Society of London, London: 169–180.

Chapman, M. G. & A. J. Underwood, 1992b. Foraging behaviour of marine benthic grazers. In D. M. John, S. J. Hawkins & J. H. Price (eds), Plant-animal interactions in the marine benthos. Systematics Association Special Volume No 46. Clarendon Press, Oxford: 289–317.

Charles, G. H., 1961. The mechanism of orientation of freely moving *Littorina littoralis* (L.) to polarized light. J. exp. Biol. 38: 203–212.

Chen, Y. S. & A. M. M. Richardson, 1987. Factors affecting the size structure of two populations of the intertidal periwinkle, *Nodilittorina unifasciata* (Gray, 1839), in the Derwent River, Tasmania. J. moll. Stud. 53: 69–78.

Doering, P. H. & D. W. Phillips, 1983. Maintenance of the shore-level size gradient in the marine snail *Tegula funebralis* (A. Adams): importance of behavioural responses to light and sea star predators. J. exp. mar. Biol. Ecol. 67: 159–173.

Evans, F., 1965. The effect of light on zonation of the four periwinkles, *Littorina littorea* (L); *L. obtusata* (L); *L. saxatilis* (Olivi) and *Melaraphe neritoides* (L) in an experimental tidal tank. Neth. J. Sea Res. 2: 556–565.

Gendron, R. P., 1977. Habitat selection and migratory behaviour of the intertidal gastropod, *Littorina littorea* (L.). J. anim Ecol. 46: 79–92.

Goodwin, B. J., 1975. Studies on the biology of *Littorina obtusata* and *L. mariae* (Mollusca Gastropoda). Unpublished Ph. D. thesis. University College of Wales, Aberystwyth, U.K.

Goodwin, B. J. & J. D. Fish, 1977. Inter- and intraspecific variation in *Littorina obtusata* and *L. mariae* (Gastropoda: Prosobranchiata). J. moll. Stud. 43: 241–254.

Guiterman, J. D., 1970. The population biology of *Littorina obtusata* (L.) (Gastropoda: Prosobranchiata). Unpublished Ph.D. thesis. University College of Wales, Bangor, U.K.

Janssen, C. R., 1960. The influence of temperature on geotaxis and phototaxis in *Littorina obtusata* (L). Arch. Neerl. Zool. 3: 500–510.

Jones, W. E., S. Bennell, C. Beveridge, B. McConnell, S. Mack-Smith & J. Mitchell, 1980. Methods of data collection and processing in rocky intertidal monitoring. In: Price, J. H., D. E. Irvine & W. F. Farnham (eds), The shore environment. Vol 1: Methods. Academic Press, Lond.: 137–170.

McQuaid, C. D., 1981. The establishment and maintenance of vertical size gradients in populations of *Littorina africana knysnaensis* (Philippi) on an exposed rocky shore. J. exp. mar. Biol. Ecol. 54: 77–89.

Norton, T. A., S. J. Hawkins, N. L. Manley, G. A. Williams & D. C. Watson, 1990. Scraping a living: a review of littorinid grazing. Hydrobiologia 193 (Dev. Hydrobiol. 56): 117–138.

Norton, T. A. & N. L. Manley, 1990. The characteristics of algae in relation to their vulnerability to grazing snails. In R. N. Hughes (ed.), Behavioural mechanisms of food selection. NATO ASI Series Vol G20, Springer-Verlag, Berlin and Heidelberg: 461–478.

Petraitis, P., 1982. Occurrence of random and directional movements in the periwinkle *Littorina littorea*. J. exp. mar. Biol. Ecol. 59: 207–218.

Reimchen, T. E., 1974. Studies on the biology and colour polymorphism of two sibling species of marine gastropod (*Littorina*). Unpublished Ph. D. thesis. University of Liverpool, U.K.

Sacchi, C. F., 1969. Recherches sur l'écologie comparée de *Littorina obtusata* (L.) et de *L. mariae* Sacchi et Rast. (Gastropoda, Prosobranchia) en Galice et en Bretagne. Invest. pesq. 33: 381–414.

Sacchi C. F., 1972a. Recherches sur la valence thermique du couple d'espèces intertidales *Littorina obtusata* (L.) et *L. mariae* (Sacchi et Rast) (Gasteropoda, Prosobranchia). Fifth Marine Biology Symposium: 209–215.

Sacchi C. F., 1972b. Recherches sur l'écologie comparée de *Littorina obtusata* (L.) et de *L. mariae* Sacchi et Rast. (Gasteropoda, Prosobranchia). II. Recherches sur la valence thermique. Boll. Pesca. Piscic. Idrobiol. 27: 105–137.

Sacchi, C. F. & M. L. Rastelli, 1966. *Littorina mariae* nov. sp. Les différences morphologiques et écologiques entre 'nain' et 'normaux' chez l''espèce' *L. obtusata* (L.) (Gastr. Prosobr) et leur signification adaptative et évolutive. Atti della. Soc. Ital. Sci. nat. 105: 351–370.

150

Thain, V. M., J. F. Thain & J. A. Kitching, 1985. Return of the prosobranch *Gibbula umbilicalis* (Da Costa) to the littoral region after displacement to the shallow sublittoral. J. moll. Stud. 51: 205–210.

Thompson, T. E., 1968. Experiments with molluscs on the shore and in a laboratory tide model. School Science Review. 149: 97–102.

Underwood, A. J., 1972. Tide-model analysis of the zonation of intertidal Prosobranchs. I. Four species of *Littorina* (L.). J. exp. mar. Biol. Ecol. 9: 239–225.

Underwood, A. J., 1977. Movement of intertidal gastropods. J. exp. mar. Biol. Ecol. 26: 191–201.

Underwood, A. J., 1979. The ecology of intertidal gastropods. Adv. mar. Biol. 16: 111–210.

Underwood, A. J. & M. G. Chapman, 1985. Multifactorial analysis of directions of movement of animals. J. exp. mar. Biol. Ecol. 91: 17–43.

Underwood, A. J. & M. G. Chapman, 1989. Experimental analyses of the influences of topography of the substratum on movement and density of an intertidal snail, *Littorina unifasciata*. J. exp. mar. Biol. Ecol. 134: 175–196.

Van Dongen, A., 1956. The preference of *Littorina obtusata* for Fucaceae. Arch. Neerl. Zool. 11: 373–386.

Watson, D. C. & T. A. Norton, 1987. The habitat and feeding preferences of *Littorina obtusata* and *L. mariae* Sacchi et Rastelli. J. exp. mar. Biol. Ecol. 112: 61–72.

Williams, G. A., 1987. Niche partitioning in *Littorina obtusata* and *L. mariae*. Unpublished Ph. D. thesis, University of Bristol, U.K.

Williams, G. A., 1990. The comparative ecologies of the flat periwinkles, *Littorina obtusata* (L.) and *L. mariae* Sacchi et Rastelli. Fld. Stud. 7: 469–482.

Williams, G. A., 1992. The effect of predation on the life histories of *Littorina obtusata* and *L. mariae*. J. mar. biol. Ass. U.K. 72: 403–416.

Williams, G. A., 1994. Variation in populations of *Littorina obtusata* L. and *L. mariae* Sacchi et Rastelli in the Severn Estuary. Biol. J. linn. Soc. 51: 189–198.

Williams, G. A. & R. Seed, 1992. Interactions between macrofaunal epiphytes and their host algae. In D. M. John, S. J. Hawkins & J. H. Price (eds), Plant-animal interactions in the marine benthos. Systematics Association Special Volume No 46. Clarendon Press, Oxford: 189–211.

Zar, J. H., 1974. Biostatistical Analysis. Prentice Hall, New Jersey, U.S.A., 620 pp.

*Hydrobiologia* **309**: 151–159, 1995.
*P. J. Mill & C. D. McQuaid (eds), Advances in Littorinid Biology.*
©1995 *Kluwer Academic Publishers.*

# Seasonal migration promoting assortative mating in *Littorina brevicula* on a boulder shore in Japan

Yoshitake Takada
*Amakusa Marine Biological Laboratory, Kyushu University, Reihoku, Amakusa, Kumamoto, 863-25, Japan*

*Key words:* assortative mating, migration behaviour, habitat selection, reproductive season, boulder shore, *Littorina brevicula*

### Abstract

*Littorina brevicula* Philippi is one of the most common snails found in the upper intertidal zone of Japan. In Amakusa, some of the population of *L. brevicula* migrate to the lower zone in the winter, while the rest stay in the upper zone. Thus, during the winter, which is its reproductive season, the population of *L. brevicula* divides into two sub-populations. This leads to a hypothesis that the migration pattern in winter is genetically controlled and this behavioural dimorphism is maintained by reproductive isolation between the two sub-populations. In order to test this hypothesis, the following three points were investigated: (1) whether the same snails migrate in a similar way every winter, (2) whether there is a significant tidal level preference in snails, and (3) whether reproductive isolation occurs between the two sub-populations. The results showed (1) the migration behaviour of each snail was consistent over two successive winters, i.e. the same group of snails migrated downward every winter and the same group of snails stayed in the upper zone every winter, (2) transplanted snails moved toward the original zones where they were caught, suggesting that the snails actively selected their tidal zone in winter, and (3) most of the snails copulated within each sub-population. Therefore, reproductive isolation between the two sub-populations was considered to be established to some extent by the dimorphic migration behaviour. In conclusion, the migratory behaviour of *L. brevicula* is determined separately for each individual and might be genetically controlled, and the behavioural dimorphism may be maintained by partial reproductive isolation between the two sub-populations.

### Introduction

Differentiation both within shell character and isozymes of local populations of littorinid snails are frequently found (e.g. Ward, 1990; Grahame & Mill, 1992), but the mechanisms behind such differentiation are sometimes not clear. Recently Johannesson *et al.* (1993) suggested that genetic and morphological differentiation of *Littorina saxatilis* on a microgeographic cline were maintained by a partial barrier to gene flow due to assortative mating.

*Littorina brevicula* Philippi is one of the most common snails found in the intertidal zone of Japan. The reproductive season of *L. brevicula* in northern (Asamushi) and central (Shirahama) parts of Japan is from February to April (Kojima, 1957; Ohgaki, 1981). *L. brevicula* produces pelagic egg capsules, each containing one egg (Kojima, 1957). In these parts of Japan,

*L. brevicula* has a yearly migration behaviour. In summer, it occurs in the upper intertidal zone and in winter it migrates downward to the lower zone. In the following spring, it again migrates upward (Abe, 1935; Kojima, 1959; Luckens, 1970). Kojima (1959) considered this downward migration in winter to be related to reproduction because of the coincidence of reproduction and downward migration of mature snails. However, in Amakusa, southern Japan, *L. brevicula* shows a dimorphism in its migratory behaviour during the reproductive season (Takada, 1992). One part of the population remains on the upper shore, while another part moves to the lower shore. This partial migration in the population contrasts to the report of Kojima, in which all mature snails migrated downward. As a result of this partial migration in Amakusa, the population of *L. brevicula* divides into two sub-populations in winter; one in the upper-intertidal zone and the other

152

in the mid-intertidal zone. As mid-winter is the main copulation season, the two sub-populations are probably more or less reproductively isolated from each other.

The above facts lead to a hypothesis that the migratory behaviour of *L. brevicula* is genetically controlled and the dimorphic behaviour is maintained by reproductive isolation between the two sub-populations. If this migratory behaviour is individually consistent, that is, if the same individuals return to the same place each year, assortative mating is a likely result.

In order to test this hypothesis, I first investigated the consistency of the migratory behaviour. If the downward migration occurs at random, some snails will stay in the upper zone even though they had migrated downward in the previous winter. On the other hand, if the migration behaviour is an inherent character of some snails, then the same snails will migrate downward every year. Secondly, I studied the tidal zone preference. The hypothesis was that if the snails have an inherent preference for a certain tidal zone, then transplanted snails will tend to return to their original tidal zone. Thirdly, I studied the schedule of migration and copulation. If the snails start to copulate before the beginning of the downward migration of the sub-population, reproductive isolation will not occur.

**Material and methods**

*Littorina brevicula* is distributed from Vladivostok and southern Kurile Island to Hong Kong (Reid, 1992; Tatarenkov, 1992). Amakusa lies in the center of its geographic range. The field experiments were carried out on a moderately exposed boulder shore on the eastern side of Magarizaki spit, Amakusa Shimo-shima Island, west Kyushu, Japan (32°31′N, 130°02′E). The mean tidal difference at this site is 3 m. The seawater temperature ranges from 13 °C in winter to 27 °C in summer. The entire intertidal zone is covered by one or two layers of boulders (average diameter < 50 cm). The boulders are stable because wave action in this area is weak. While active, *L. brevicula* forages on microalgae growing on the upper surface of the boulders; while inactive, they rest underneath the boulders. Because the shore has only a gentle slope, *L. brevicula* do not change their tidal zone during their daily feeding excursion.

Fifteen stations (St) were established on the shore at 2.5 m intervals (Fig. 1) so as to cover the entire vertical

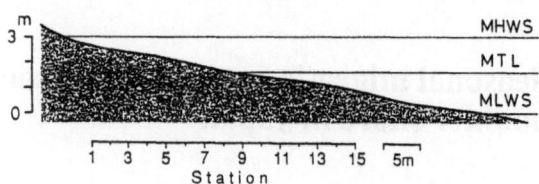

*Fig. 1.* Profile of the shore and 15 stations (St).

distribution zone of *L. brevicula*. St1 (station 1) was located 1.0 m above the MTL and St15 0.7 m below the MTL. In summer, all *L. brevicula* occurred in the upper zone (around St4). In winter, the upper sub-population occurred around St4 (193 snails m$^{-2}$) and the lower sub-population occurred around St9 to St13 (12 and 45 snails m$^{-2}$, respectively, Takada, 1992). The two sub-populations were not completely discrete, because very few snails occurred at St6 (5.5 snails m$^{-2}$, Takada, 1992), which could be regarded as the boundary of the sub-populations.

*Consistency of migration behaviour*

In order to study the consistency of migration behaviour in *L. brevicula*, I carried out two mass marking experiments. In the early spring, snails from the upper (St4) and lower (St7–9) sub-populations were marked with different colours using spray paint, and released in their original tidal zones. In late spring, the snails in the lower sub-populaton migrated upward and mixed with the upper sub-population. However, in the following winter, after the downward migration, snails were recaptured from the same area where they were released in the previous year. The null hypothesis is that the snails migrate randomly. In order to test this, proportions of recaptured to newly captured snails in the upper and lower zones were compared using $\chi^2$-tests. The experiment was repeated twice, (1) from 18 March, 1991 to 20 February, 1992, and (2) from 3 March, 1992 to 24 February, 1993.

*Tidal level preference*

In order to investigate the tidal preference of *L. brevicula* in winter, a transplant experiment was carried out from 19 December, 1992 to 16 January, 1993. Snails were caught randomly from the upper (adjacent to St4) and lower (adjacent to St9) zones. Each sample was divided into two groups (control and experimental). Snails of the experimental groups were transplanted reciprocally between the upper (St4) and lower (St9) zones. Control groups contained snails caught from the same tide level. Each group contained 20 snails ranging from 11.6 mm to 15.6 mm in shell width (diameter perpendicular to columella, Table 1). Snails were individually marked by fixing a plastic string of 15 cm length to the apex of the shell. The snails could easily be located from this string even if they were resting on the under side of the boulders. Even though this marking method might have hindered the movement of the snails to a small extent, it ensured a high rate of recapture. The position of each snail was marked on a grid with 0.1 m accuracy. Data were collected 7, 14, 21 and 28 days after release.

The mean distance and direction from the released points were compared. Distances from the released point $(Xm)$ were log-transformed $(Log_{10}[10X + 1])$ before testing using a two-way ANOVA. Factors were origin of the snails' (upper or lower) sub-population and treatment (control or experiment). The data for different days were not independent of each other because snails were observed repeatedly. For the comparison of mean direction between the control and experiment groups, mean vector and angular deviation were calculated and a Rayleigh-test was used to test the randomness of the direction of movement (Batschelet, 1981). For snails which moved short distances, an error in recording the position causes a serious error in estimating direction. Hence, only snails which moved more than 0.2 m from the released point were used for the analysis of directional data.

## Schedule of migration and copulation

During copulation, while the male *L. brevicula* was on the female shell inserting his penis into the mantle cavity of the female, the female stopped moving and clung to the boulder. However, in some pairs, the male did not insert his penis into the female, even though he was on her shell. Such pairs were not regarded as copulating pairs.

In order to study the schedule of migration and copulation, the occurrence of copulating pairs at five tide levels (St4, St7, St9, St11, and St13) was observed bi-weekly at spring tides from 3 November, 1991 to 6 April, 1992. From May to October, activity was very low and copulation did not occur (pers. obs.). At each tide level, five quadrats (1 m × 1 m) were set and the number of active snails and copulating pairs in each quadrat were counted. To confirm the insertion of the penis, I picked up the male in each copulating pair.

Copulation was observed within 30 min of emersion from the night ebbing tide, because the snails became inactive and rested under the boulders 30 min after they were emersed. The mean number of active individuals and the ratio of copulating snails to active individuals were calculated. The observations were carried out for three days at each spring tide, except for the first and second spring tide in November 1991.

## Results

### Consistency of migration behaviour

In 1992, the marked snails from the lower sub-population were significantly more numerous in this zone than in the upper zone (Table 2a). Very few marked snails from the upper sub-population were recaptured in this year, and this resulted in such low power of the test that no significant result could be obtained even if there had been an actual difference. In the second experiment (Table 2b), marked low shore snails were most common in the lower zone and this time marked upper shore snails were most common among the snails recaptured in the upper shore. This result confirmed the expected migration pattern of the snails, i.e. the same snails migrate downward each winter, while the other snails stay in the upper zone.

### Tide level selection

Table 1 gives mean shell width and number of snails used; shell width of upper and lower sub-populations were not statistically different (Student's t-test, $df = 78 t = 0.117 p = 0.91$). After 28 days, out of 40 snails marked and released in the upper zone, one was found dead and one snail had disappeared. Of 40 snails in the lower zone two died and two disappeared. Mortality rates after 28 days in the upper and lower zones did not differ statistically (Fisher's exact probability test, $p = 0.62$).

Figure 2 shows the mean distance moved after 7, 14, 21, and 28 days from release. The mean distance steadily increased with time. Two-way ANOVA (Table 3) showed that the sub-population from which a snail came did not affect the distance, while treatment showed a significant effect on all observation days after day 7. Thus, experimentally transplanted snails moved significantly longer distances than did control snails. Interaction between origin and treatment was significant only on day 7. The record maximum distance

*Table 1*. Experimental design and number of snails recaptured for the tide level preference experiment. In each treatment, 20 snails were released on 19 December, 1992. Numbers of snails which moved $\geq 0.2$ m are in parentheses. L, lower zone; U, upper zone.

| Treatment | | Shell width (mm, mean ± SE) | Number of snails recaptured (day) | | | |
|---|---|---|---|---|---|---|
| | | | 7 | 14 | 21 | 28 |
| 1. Control | U–U | $13.61 \pm 0.13$ | 18 (8) | 19 (14) | 19 (14) | 19 (17) |
| 2. Experiment | L–U | $13.43 \pm 0.19$ | 19 (12) | 20 (15) | 20 (16) | 19 (18) |
| 3. Experiment | U–L | $13.59 \pm 0.21$ | 19 (14) | 19 (15) | 19 (18) | 19 (19) |
| 4. Control | L–L | $13.73 \pm 0.15$ | 20 (9) | 20 (14) | 17 (14) | 17 (16) |

*Table 2*. Results of mass marking experiments to determine the consistency of migration behaviour in *Littorina brevicula*. The null hypothesis is that the marked snails migrate randomly. Frequencies of recaptured snails in the upper and lower zones were compared using $\chi^2$-tests.

| | | Sub-populations | | Total number of snails captured |
|---|---|---|---|---|
| | | Upper (St4) | Lower (St7–9) | |
| (a) | **First experiment** | | | |
| | No. snails released (March, 1991) | c. 200 | c. 300 | |
| | No. snails recaptured (February, 1992) | | | |
| | in the upper zone | 3 | 1 | 581 |
| | in the lower zone | 0 | 39 | 625 |
| | sum | 3 | 40 | |
| | $\chi^2$ (d$f$ = 1) | 4.23 | 33.42 | |
| | | ns | $p < 0.001$ | |
| (b) | **Second experiment** | | | |
| | No. snails released (March, 1992) | 556 | 500 | |
| | No. snails recaptured (February, 1993) | | | |
| | in the upper zone | 22 | 3 | 600 |
| | in the lower zone | 3 | 34 | 1045 |
| | sum | 25 | 37 | |
| | $\chi^2$ (d$f$ = 1) | 28.64 | 12.85 | |
| | | $p < 0.001$ | $p < 0.001$ | |

moved in 28 days was 10.1 m; this was observed in an experimental snail transplanted from the lower zone to the upper zone. This snail moved downward and returned almost to its original zone. The distance between the upper zone and the lower zone was 12.5 m. Thus most of the experimental snails did not return to their original tidal zone despite an increased migration rate.

Figure 3 shows the direction of movement. In the upper zone, the snails in the control group (upper sub-population) moved upwards (days 7 and 14) and then at random (days 21 and 28). Direction, represented by the length of the mean resultant vector, was higher on day 7 than on day 14. Movement of the experimental snails (transplanted from the lower sub-population) was not random and the mean direction was always downward, which was the direction of the original

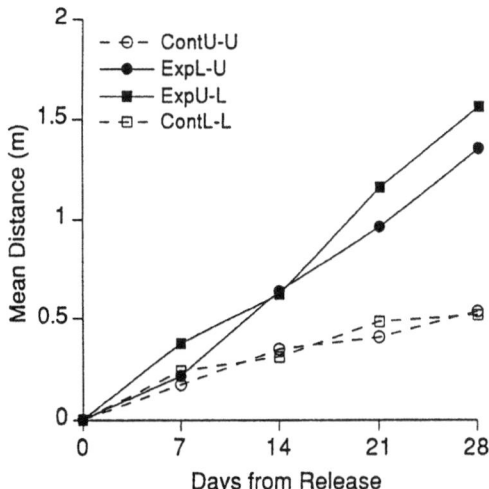

*Fig. 2.* Mean distances moved by the four groups of snails in the transplant experiment. 20 snails in each group were released in the upper zone (Control U–U and Experimental L–U) and in the lower zone (Experimental U–L and Control L–L). Numbers of snails recaptured on days 7, 14, 21, and 28 after release are listed in Table 1. Mean values were calculated for log-transformed distances. L, lower zone; U, upper zone.

*Table 3.* Two-way ANOVAs to compare origin (upper and lower sub-populations) and treatment (control and experimental groups) effects on distance moved by *Littorina brevicula* over the four observation days. Distances from the released point ($Xm$) were log-transformed ($\text{Log}_{10}[10X + 1]$).

| Factor | df | MS | F | p |
|---|---|---|---|---|
| *Day 7* | | | | |
| Origin | 1 | 0.183 | 0.405 | 0.527 |
| Treatment | 1 | 1.094 | 2.422 | 0.124 |
| O × T | 1 | 1.903 | 4.211 | 0.044 |
| Error | 72 | 0.452 | | |
| | | | | |
| *Day 14* | | | | |
| Origin | 1 | 0.022 | 0.028 | 0.867 |
| Treatment | 1 | 5.471 | 6.914 | 0.010 |
| O × T | 1 | 0.061 | 0.077 | 0.782 |
| Error | 74 | 0.791 | | |
| | | | | |
| *Day 21* | | | | |
| Origin | 1 | 0.005 | 0.006 | 0.938 |
| Treatment | 1 | 10.627 | 13.705 | 0.0004 |
| O × T | 1 | 0.449 | 0.579 | 0.449 |
| Error | 71 | 0.775 | | |
| | | | | |
| *Day 28* | | | | |
| Origin | 1 | 0.121 | 0.185 | 0.669 |
| Treatment | 1 | 15.160 | 23.086 | < 0.0001 |
| O × T | 1 | 0.051 | 0.077 | 0.782 |
| Error | 70 | 0.657 | | |

tidal level. In the lower zone, the snails in the control group (lower sub-population) moved at random at first (days 7 and 14) and then laterally (days 21 and 28). In the experimental group (snails transplanted from the upper sub-population), the null hypothesis of random movement could not be rejected on days 14 and 21, but there was a significant upward migration on days 7 and 28. In summary, the experimental groups in both the lower and the upper zones showed a tendency to move towards their original tidal zones.

### Schedule of migration and copulation

Figure 4 shows numbers of active snails and copulating pairs. In November, active snails occurred only in the upper zone ($137.4 \pm 49.8$ snails m$^{-2}$, mean $\pm$ sd; $n = 5$). At the end of the year, some snails migrated downward gradually and a low shore maximum was reached in January and February. By the end of March most snails had returned to the upper zone.

The frequency of copulating pairs was generally higher in the lower sub-population (St7–13) than in the upper sub-population (St4). At St4, copulation occurred from November to March. In the lower zone (St7–13), snails began to copulate as they reached the specific tide level. Maximum frequency

of copulation ($5.1 \pm 5.8\%$, mean $\pm$ sd, $n = 15$) was observed in the lower sub-population (St11) in February, 1992. From December to February, *L. brevicula* copulated assortatively within the upper and the lower sub-populations. However, in November and March, copulation occurred before the population divided and after the population united. Therefore, cross mating between the upper and the lower sub-populations may have occurred at the beginning and at the end of the mating season.

### Discussion

In this paper, I have treated the two sub-populations as behavioural dimorphs within a single species, while someone might suspect that these were two sibling species with different migration patterns. Up to now, I have detected no characters which differ between the

156

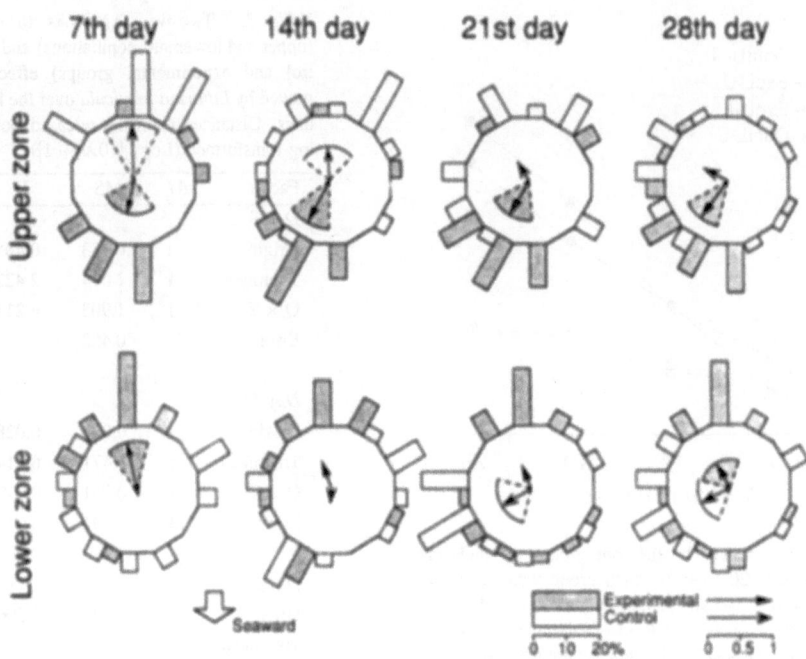

*Fig. 3.* Direction of movement of four groups of *Littorina brevicula* 7, 14, 21 and 28 days after release. Bars indicate per cent snails moving in each direction. The mean direction ($\rightarrow$), the length of the mean resultant vector and the 95% confidence limits are indicated. No confidence limits were calculated for random movement (Rayleigh-test $P < 0.05$). Only snails which moved $\geq 0.2$ m from the release point were used for the calculation (see Table 1).

two sub-populations other than the seasonal migration behaviour. They show the same adult size (Table 1) and no difference was detected in a multivariate analysis of shell shape (Takada, unpublished data). Recently, preliminary data from allozyme electrophoresis also rejected the possibility of two sibling species (Zaslavskaya & Takada, unpublished data).

*Tide level selection*

Since the design of this transplant experiment included only 'translocated control' and 'transplanted experimental' (*sensu* Chapman & Underwood, 1992), a generalization of the results is limited. It did not consider the effect of handling disturbance and encounter to a new position (Chapman & Underwood, 1992). Handling disturbances caused directional movement in *Littorina littorea* (Petraitis, 1982) and affected the response to substratum topography in *Nodilittorina unifasciata* (Chapman, 1986). In this study, significant interaction between treatment and origin in the two-way ANOVA and directional movement (Cont U–U and Exp U–L) on day 7 were probably due to handling effects during marking and releasing. However,

the mean distances showed a linear increase with days after release (Fig. 2), and the effect of handling disturbance on the movement of *N. unifasciata* decreased with time (Underwood & Chapman, 1989). Hence, I believe that the results of the later observations were not influenced by the initial handling disturbance.

Although the samples from different days were not independent, the transplanted snails moved in the direction of their original tidal zone. This indicates that upper and lower snails have different tide level preferences, and that the snails differentiated between upward and downward directions on the shore. Abe (1935) reported that *L. brevicula* formed aggregations on the side of rocks where the waves did not strike directly, suggesting that wave movement is used for the directional stimulus, as in *L. littorea* (Gendron, 1977).

*Reproductive isolation between two sub-populations*

As in other species of *Littorina* (e.g. Saur, 1990), male *L. brevicula* sometimes form mating pairs with other males in the laboratory (Takada, pers. obs.). In

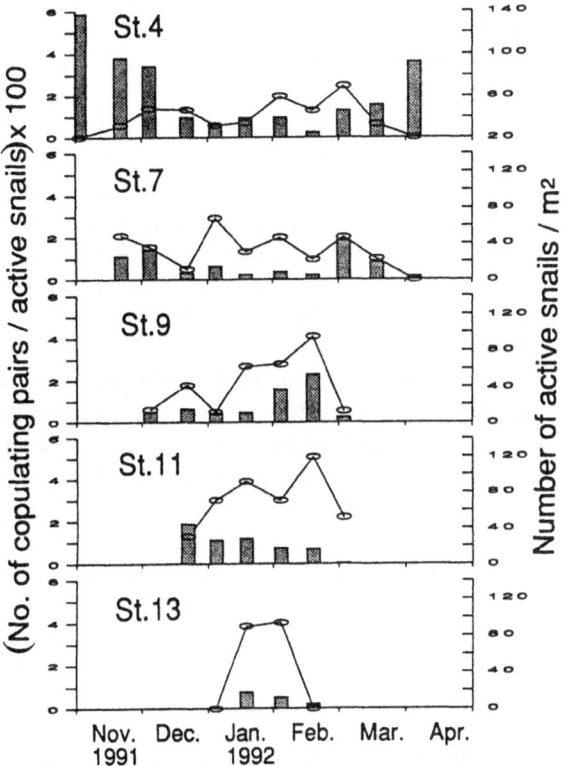

*Fig. 4.* Bi-weekly fluctuations of the mean number of active *Littorina brevicula* per 1 m² (bars) and the copulation ratio (ellipses). The copulation ratio was obtained by dividing the number of copulating pairs by the number of active snails. The graphs are arranged from the upper zone (St4) to the lower zone (St13).

this study, the mating activity of the population was expressed as the frequency of copulating pairs in which the male inserted his penis into the snail beneath him. Male-male pairs were probably counted during the observation on the shore. However, the duration of the copulation of male-male pairs is shorter than normal male-female pairs in *L. keenae* (Gibson, 1965) and *L. littorea* (Saur, 1990). Probably, males of *Littorina* judge the partner's sex after insertion of the penis. The interrupted mating attempts of male-male pairs mean that such pairings will be observed less frequently on the shore, even if encouter rates are random. Although the actual relation between the insertion of the male's penis and the transfer of sperm is uncertain, the frequency of copulating pairs can be used as an indicator of the rate of gene mixing within and between the two sub-populations.

In the upper zone, copulation of *L. brevicula* continued for four months, from late November to March.

Spawning occurred at the end of the copulation season, i.e. from late February to March (Takada, unpublished data). Snails started to copulate well before the gonad became mature. Furthermore, the long copulation period leads to the question of whether the sperm inseminated early in the copulation season could fertilize eggs late in the season. Sperm are thought to keep their activity for a long period. Female *L. brevicula* spawned their eggs one week after copulation in the laboratory (Kojima, 1957). In *L. saxatilis*, isolated females continue to give birth to offspring more than one year after isolation (Johannesson, pers. comm.). However, female littorinids may be able to digest a certain amount of the sperm and the nurse cells in the vesicula seminalis (Fretter & Graham, 1962). Therefore, if the period in the vesicula seminalis of the female becomes long, the probability of digestion will become greater. Furthermore, during the four month copulation season, females most probably copulated several times. If there is any relation between the order of sperm insemination and the priority of the sperm used for fertilization, the degree of gene mixing between the two sub-populations could be modified in relation to the direct estimate of the copulation frequency. To date, there are no data on sperm activity, sperm digestion or sperm competition in *L. brevicula*, and it is impossible to evaluate the importance of these factors in reproductive isolation between the two sub-populations.

In general, each snail showed consistent migration behaviour over the two successive winters. This character is not a phenotypic response to early growth in different environments because all snails spend their juvenile period in the upper zone (Takada, 1992). Downward migration leads to a reduction of local density of *L. brevicula*. However, density-dependent migration is not plausible because the snails migrated downward even in a low density population adjacent to the study area (Takada, pers. obs.). It seems most likely that the different migration behaviours of the upper and lower sub-populations are inherited.

From January to February, the middle of the copulation season, 7.5% of the total population migrates vertically between the upper and the lower zones. (Takada, unpublished data), which suggests a little gene flow between the two sub-populations. As copulation and downward migration start simultaneous, in November, and copulation continues until the end of the upward migration in March, snails of the upper and lower sub-populations may copulate with each other. Hence, reproductive isolation between the upper and the lower sub-populations may not be complete.

158

## *Factors maintaining the dimorphic migration behaviour*

One of the important findings in this study is that *L. brevicula* copulates assortatively depending on position on the shore. The dimorphism in migration behaviour during the reproductive season results in partial reproductive isolation between the two sub-populations, and this reproductive isolation promotes the dimorphic behaviour. Thus, once reproductive isolation is established to some extent, the genetic diversity may be maintained, assuming perfect isolation and neutral characters (Rice, 1984). Two main factors for gene mixing in littorinid snails are migration of adults and dispersal of planktonic larvae. Selection and migration models for genetic differentiation in species without planktonic dispersion were developed by Boulding (1990) and Johannesson & Sundberg (1992). In the case of *L. brevicula*, because some gene mixing between the two sub-populations is expected during copulation and during the planktonic stage, disruptive selection is needed to maintain differentiation.

Unfortunately, I cannot find any plausible selection pressures on this population. In winter, survival rates of the two sub-populations were very high (Table 1). Growth rate in winter and egg number per female are not different between the two sub-populations (Takada, unpublished data). Downward migration during the reproductive season has been regarded as an adaptive behaviour in order to deposit egg masses in the damper, lower zone (Hannaford Ellis, 1985), or to increase the chance of spawning during immersion (Ohgaki, 1988). Since *L. brevicula* releases planktonic eggs and both sub-populations are below MHWN, these two models do not explain the separation into two sub-populations. Parasitic infection by trematodes affected migration behaviour in *L. littorea* (Lambert & Farley, 1968) and in *Ilyanassa obsoleta* (Curtis, 1990), but parasitism does not seem to be a major factor in the migration of *L. brevicula* because of the low infection rate (< 1%; Takada, unpublished data). A possible explanation is that the two sub-populations are descendants from geographically different populations under different selection pressures. That is, disruptive selection acts over a large geographic area and Amakusa is a hybridizing zone for two populations. In Asamushi, northern Japan, all the adults migrated downward (Kojima, 1959), suggesting that the lower sub-population in Amakusa shows a similar migration behaviour to the northern population. On the other hand, the upper sub-population may be a descendant

of an unknown southern population which does not migrate downward because of high predation pressure on the low shore (Lubchenco *et al.*, 1984; Menge *et al.*, 1985).

Generally, species with planktonic stages are believed to have low diversity between local populations (Johannesson, 1988; Behrens Yamada, 1989). Relatively large genetic differences between local populations of *L. brevicula* (Tatarenkov, 1992) may perhaps be explained by local differences in selection direction or the type of assortative mating suggested in this study.

In this study, I have pointed out the possibility of reproductive isolation between two sub-populations, but I did not investigate direct evidence of genetic differentiation or selective pressures against those snails with intermediate migration behaviour. It is necessary to study why there are a few snails with intermediate migration behaviour and why the dimorphism has been established.

## Acknowledgements

I thank Prof. T. Kikuchi for giving me the opportunity to study this subject. I also thank M. Tanaka, K. Mori, K. Nandakumar and I. Yosho for valuable suggestions during the course of this study.

## References

Abe, N., 1935. The colony of the Littorina: *Littorivaga brevicula* (Philippi). Sci. rep. Tohoku Imp. Univ. Fourth Ser. (Biol.) 9: 279–296.

Batschelet, E., 1981. Circular statistics in biology. Academic Press, London, 371 pp.

Behrens Yamada, S., 1989. Are direct developers more locally adapted than planktonic developers? Mar. Biol. 103: 403–411.

Boulding, E. G., 1990. Are the opposing selection pressures on exposed and protected shores sufficient to maintain genetic differentiation between gastropod population with high intermigration rates? Hydrobiologia 193 (Dev. Hydrobiol. 56): 41–52.

Chapman, M. G. & A. J. Underwood, 1992. Experimental designs for analyses of movements by molluscs. In J. Grahame, P. J. Mill & D. G. Reid (eds), Proceedings of the 3rd International Symposium on Littorinid Biology. The Malacological Society of London, London: 169–180.

Chapman, M. G., 1986. Assessment of some controls in experimental transplants of intertidal gastropods. J. exp. mar. Biol. Ecol. 103: 181–201.

Curtis, L. A., 1990. Parasitism and the movement of intertidal gastropod individuals. Biol. Bull. 179: 105–112.

Fretter, V. & A. Graham, 1962. British prosobranch molluscs. Ray Society, London, 755 pp.

Gendron, R. P., 1977. Habitat selection and migratory behavior of the intertidal gastropod *Littorina littorea* (L.). J. anim. Ecol. 46: 79–92.

Gibson, D. G., 1965. Mating behavior in *Littorina planaxis* Philippi (Gastropoda: Prosobranchiata). Veliger 7: 134–139.

Grahame, J. & P. J. Mill, 1992. Local and regional variation in shell shape of rough periwinkles in southern Britain. In J. Grahame, P. J. Mill & D. G. Reid (eds), Proceedings of the 3rd International Symposium on Littorinid Biology. The Malacological Society of London, London: 99–106.

Hannaford Ellis, C. J., 1985. The breeding migration of *Littorina arcana* Hannaford Ellis, 1978 (Prosobranchia: Littorinidae). Zool. J. linn. Soc. 84: 91–96.

Johannesson, K., 1988. The paradox of Rockall: why is a brooding gastropod (*Littorina saxatilis*) more widespread than one having a planktonic larval dispersal stage (*L. littorea*)? Mar. Biol. 99: 507–513.

Johannesson, K. & P. Sundberg, 1992. Speciation in *Littorina saxatilis* (Olivi)? – a one-dimensional selection-migration model. In J. Grahame, P. J. Mill & D. G. Reid (eds), Proceedings of the 3rd International Symposium on Littorinid Biology. The Malacological Society of London, London: 1–8.

Johannesson, K., B. Johannesson & E. R. Alvarez, 1993. Morphological differentiation and genetic cohesiveness over a microenvironmental gradient in the marine snail *Littorina saxatilis*. Evolution 47: 1770–1787.

Kojima, K., 1957. On the breeding of a periwinkle, *Littorivaga brevicula* (Philippi). Bull. biol. Stn Asamushi 8: 59–62.

Kojima, K., 1959. The relation between seasonal migration and spawning of a periwinkle, *Littorina brevicula* (Philippi). Bull. biol. Stn Asamushi 9: 183–186.

Lambert, T. C. & J. Farley, 1968. The effect of parasitism by the trematode *Cryptocotyle lingua* (Creplin) on zonation and winter migration of the common periwinkle, *Littorina littorea* (L.). Can. J. Zool. 46: 1139–1147.

Lubchenco, J., B. A. Menge, S. D. Garrity, P. Lubchenco, L. R. Ashkenas, S. D. Gaines, R. Emlet, J. Lucas & S. Strauss, 1984. Structure, persistence, and role of consumers in a tropical rocky intertidal community (Taboguilla Island, Bay of Panama). J. exp. mar. Biol. Ecol. 78: 23–73.

Luckens, P. A., 1970. Seasonal distributional variation within a limited shore area at Asamushi. Sci. Rep. Tohoku Univ. Ser. IV (Biol.) 35: 161–170.

Menge, B. A., J. Lubchenco, S. D. Gaines & L. R. Ashkenas, 1985, Diversity, heterogeneity and consumer pressure in a tropical rocky intertidal community. Oecologia (Berlin) 71: 394–405.

Ohgaki, S., 1981. Spawning activity in *Nodilittorina exigua* and *Peasiella roepstorffiana* (Littorinidae, Gastropoda). Publ. Seto Mar. Biol. Lab. 26: 437–446.

Ohgaki, S., 1988. Vertical migration and spawning in *Nodilittorina exigua* (Gastropoda: Littorinidae). J. Ethol. 6: 33–38.

Petraitis, P. S.: 1982. Occurrence of random and directional movements in the periwinkle, *Littorina littorea* (L.). J. exp. mar. Biol. Ecol. 59: 207–217.

Reid, D. G., 1992. The gastropod family Littorinidae in Hong Kong. In B. Morton (ed.), The marine flora and fauna of Hong Kong and Southern China III. Proceedings of the Fourth International Marine Biological Workshop: The Marine Flora and Fauna of Hong Kong and Southern China. Hong Kong University Press: 187–210.

Rice, W. R.: 1984. Disruptive selection on habitat preference and evolution of reproductive isolation. Evolution 38: 1251–1260.

Saur, M., 1990. Mate discrimination in *Littorina littorea* (L.) and *L. saxatilis* (Olivi) (Mollusca: Prosobranchia). Hydrobiologia 193 (Dev. Hydrobiol. 56): 261–270.

Takada, Y., 1992. The migration and growth of *Littorina brevicula* on a boulder shore in Amakusa, Japan. In J. Grahame, P. J. Mill & D. G. Reid (eds), Proceedings of the 3rd International Symposium on Littorinid Biology. The Malacological Society of London, London: 277–279.

Tatarenkov, A. N., 1992. Allozyme variation in *Littorina brevicula* (Philippi) from Peter The Great Bay (Sea of Japan). In J. Grahame, P. J. Mill & D. G. Reid (eds), Proceedings of the Third International Symposium on Littorinid Biology. The Malacological Society of London: 25–30.

Underwood, A. J. & M. G. Chapman, 1989. Experimental analyses of the influences of topography of the substratum on movements and density of an intertidal snail, *Littorina unifasciata*. J. exp. mar. Biol. Ecol. 134: 175–196.

Ward, R. D., 1990. Biochemical genetic variation in the genus *Littorina* (Prosobranchia: Mollusca). Hydrobiologia 193 (Dev. Hydrobiol. 56): 53–69.

*Hydrobiologia* **309**: 161–166, 1995.
*P. J. Mill & C. D. McQuaid (eds), Advances in Littorinid Biology.*
©1995 *Kluwer Academic Publishers.*

# Extreme morphological diversity between populations of *Littorina obtusata* (L.) from Iceland and the UK

R. I. Lewis* & Gray A. Williams#
*Port Erin Marine Laboratory, The University of Liverpool, Port Erin, Isle of Man, UK*
*Present addresses: *School of Biological Sciences, University College of Swansea, Singleton Park, Swansea, SA2 8PP, UK; #The Swire Institute of Marine Science and Department of Ecology and Biodiversity, The University of Hong Kong, Hong Kong*

*Key words:* Taxonomy, morphological variation, allozyme variation, *Littorina palliata*, dispersal

## Abstract

Marine gastropods which do not disperse larvae in the plankton exhibit a relatively high degree of interpopulation morphological variability. This phenomenon has been the root of considerable taxonomic confusion, particularly in the Littorinidae. In the present study, specimens of *Littorina obtusata* from the UK were compared morphologically and genetically with two samples of a northern high-spired form from Iceland which has been referred to by some workers as *Littorina palliata*. A multivariate discriminant function analysis based on three measured shell dimensions clearly separated the three samples, correctly predicting the origin of shells on the basis of morphology alone in 96% of cases. However, genetic analysis revealed that the most distant relationship, based on allozyme data at 13 loci, was surprisingly close (Nei's $I = 0.983$) providing no evidence to suggest that *L. palliata* is not conspecific with *L. obtusata*.

## Introduction

Marine gastropods can be divided into two groups on the basis of whether or not they disperse larvae in the plankton. Attention has been drawn to this functional division by various authors (e.g. Berger, 1973; Vermeij, 1982; Behrens Yamada, 1987, 1989), since it is likely to have resulted in profound differences between the groups in response to two major evolutionary influences. Firstly, taxa without widely dispersing larvae (such as *Littorina obtusata* (L.)) will presumably experience relatively low rates of interpopulation migration. This facilitates relatively high levels of genetic divergence between populations through higher rates of genetic drift (Maynard-Smith, 1989). Secondly, populations of locally recruiting taxa have the opportunity to evolve heterogeneously in response to spatially heterogeneous selective agents. This possibility is much reduced in broadcast spawners, since parents and progeny are less likely to experience the same set of selective phenomena. The most significant consequence of this functional division is that locally recruiting taxa are likely to exhibit a relatively high degree of interpopulation morphological variability. Indeed some authors have cited the comparatively high degree of interpopulation morphological variation in such taxa as evidence for heritability of shell shape characters (Vermeij, 1982; Behrens Yamada, 1987, 1989). Experimental evidence that shell characters show at least some degree of heritability has been provided by a number of authors (Newkirk & Doyle, 1975; Janson, 1982; Boulding & Hay, 1993).

Traditionally, taxonomy has been based upon morphological discrimination. Marine gastropods which do not disperse progeny in the plankton have presented taxonomists with particular problems due to their relatively high degrees of inter-population morphological variation. The literature concerning the Littorinidae presents many examples of this particular problem. *L. saxatilis* (Olivi) for example (often conservatively referred to as the *L. 'saxatilis'* complex) is still the subject of considerable debate and revision (see review by Raffaelli, 1982; and papers in this volume).

In the present study, we examined specimens of *L. obtusata* taken from the Isle of Man (UK), and compared them with morphologically different samples taken from the west coast of Iceland. The taxonomy of this northern form has been the subject of some debate (see reviews by Dautzenberg & Fischer, 1914; Colman, 1932; Knudsen, 1949). Some authors have referred to northern high-spired forms as *L. palliata* (Say) (e.g. Thorson, 1941; Hubendick & Warén, 1975), whilst others have considered them as variants of *L. obtusata* (e.g. Colman, 1932; Knudsen, 1949; Seeley, 1986).

The analyses in the present study combine a multivariate morphological analysis of measured shell dimensions with a genetic analysis based on comparisons of allozyme frequencies between samples. Such genetic analyses have become invaluable additions to the tools of the taxonomist, because they have the potential to provide data on gene flow and relatedness that for taxonomic purposes can be considered to be largely independent of significant environmental influence (for general principles and techniques see Ferguson, 1980; Richardson *et al.*, 1986). Indeed, genetic analyses have often been applied with useful effect to particular taxonomic questions in the Littorinidae (see review by Ward, 1990).

**Materials and methods**

During the summer of 1989, samples of high-spired *L. palliata* forms were collected from Grótta, Seltjarnarnes Cape, Iceland (64°10′ N, 22°03′ W), and S.W. Borgarnes, Iceland (64°33′ N, 21°53′ W). At the latter site, animals were collected from the alga *Ascophyllum nodosum* from a sheltered rocky shore near the head of the fjord. Low-spired *L. obtusata* forms were collected from St. Michael's Island, Derbyhaven, off the Isle of Man, UK. This sample was also taken from *Ascophyllum nodosum* on a sheltered rocky shore, on the north side of the island (54°05′ N, 4°33′ W). All samples were transported live and kept in running seawater for no longer than two weeks prior to genetic and morphological analysis.

*Morphological analysis*

Morphological differences in shell shape between sites were investigated using a multivariate discriminant function analysis (from SPSS 6.0 for Windows; Norušis, 1992). Measurements of the three shell

dimensions length (a), aperture width (b), and height (c) (after Goodwin & Fish, 1977) were used as input variables for the analysis. This analysis maximises the discriminatory power of the data by combining variables into a series of uncorrelated linear functions which maximise the between group variance of function scores. The predictive power of the analysis is then assessed by the *a posteriori* success with which individuals are assigned to the correct group based solely on measured variables. Specific procedures for the discriminant function analysis follow those of Lewis & Thorpe (1994).

*Genetic analysis*

Genetic analysis was carried out using standard techniques for horizontal starch gel electrophoresis of allozymes (Harris & Hopkinson, 1978; Ferguson, 1980; Richardson *et al.*, 1986). Specific procedures and reaction conditions follow those of Lewis & Thorpe (1994). The protein homogenate was prepared by mechanically homogenising approximately 3 mm$^3$ of fresh foot tissue in 50 $\mu$l of distilled water. Stain recipes were taken from Harris & Hopkinson (1978) and Shaw & Prasad (1970). Results are given for a total of thirteen putative enzyme coding loci, detected by staining for the following eight enzymes; glucose phosphate isomerase (E.C. No. 5.3.1.9), mannose phosphate isomerase (E.C. No. 5.3.1.8.), esterase (E.C. No. 3.1.1.1), adenylate kinase (E.C. No. 2.7.4.3), isocitrate dehydrogenase (E.C. No. 1.1.1.42), leucine aminopeptidase (E.C. No. 3.4.11.-), malate dehydrogenase (E.C. No. 1.1.1.37) and nucleoside phosphorylase (E.C. 2.4.2.1). Electromorphs were labelled alphabetically in order of anodal migration. Consistency of scoring was maintained by the use of standard 'marker' organisms of known genotype. Allele frequencies were used to calculate Nei's genetic identity (Nei, 1972) for each locus.

**Results**

Morphological discrimination between samples, based on the three measured shell characters, was highly significant (Table 1, $P < 0.001$). This finding is not unexpected, because the geographically extreme samples are clearly quite distinct in morphology (Fig. 1). Two discriminant functions were generated. Exclusion of the first function from the analysis still resulted in highly significant discrimination (Table 1, $P < 0.001$), indi-

Table 1. Summary statistics for discriminant functions based on shell dimensions.

| Function | Eigenvalue | % variance | Cumulative % variance | Canonical correlation | After function excluded | Wilks' lambda | Chi-square | d.f. | P |
|----------|-----------|-----------|----------------------|----------------------|------------------------|--------------|-----------|------|---|
|          |           |           |                      |                      | 0                      | 0.075        | 113.81    | 6    | <0.001 |
| 1        | 7.433     | 92.82     | 92.82                | 0.939                | 1                      | 0.635        | 20.00     | 2    | <0.001 |
| 2        | 0.575     | 7.18      | 100.00               | 0.604                |                        |              |           |      |   |

Table 2. Summary statistics for shell dimensions. Wilks' Lambda, $F$ ratios and probability ($P$) values refer to One-Way ANOVA on shell measurements using sample location as the grouping factor.

| Variable | { One-Way ANOVA on shell dimensions} | | | Function 1 coefficients | Correlations between shell dimensions and function 1 discriminant scores |
|----------|------------------|----------|--------|------------------------|----------------------------------------|
|          | Wilks' lambda    | $F$-ratio | $P$    |                        |                                        |
| Length   (a) | 0.211        | 84.381   | <0.001 | 1.269                  | 0.681                                  |
| Aperture (b) | 0.177        | 104.879  | <0.001 | 1.555                  | 0.779                                  |
| Height   (c) | 0.309        | 50.333   | <0.001 | −2.144                 | 0.502                                  |

Table 3. Classification Results (confusion matrix) showing percentage of cases (individuals) from each sample classified *a posteriori* to the correct home site (the leading diagonal) on the basis of shell shape (mean correct classification = 95.83%). The expected correct classification by chance is 33.3%.

| Actual sample | Predicted Sample | | | No. of cases |
|---------------|--------|-----------|------|------|
|               | GRÓTTA | BORGARNES | IOM  |      |
| GRÓTTA        | 92.3   | 0.0       | 7.7  | (13) |
| BORGARNES     | 10.0   | 90.0      | 0.0  | (10) |
| IOM           | 0.0    | 0.0       | 100  | (25) |

Table 4. Allelic frequencies at 13 loci for all individuals scored at the three sampled sites.

| LOCUS | ALLELE | IOM | GRÓTTA | BORGARNES |
|-------|--------|-------|--------|-----------|
| *Pgi* | a | 0.025 | 0.000 | 0.000 |
|       | b | 0.625 | 0.565 | 0.429 |
|       | c | 0.338 | 0.435 | 0.571 |
|       | d | 0.013 | 0.000 | 0.000 |
| *Mpi* | a | 0.613 | 0.305 | 0.429 |
|       | b | 0.338 | 0.695 | 0.571 |
| *Est 1* | a | 1.000 | 1.000 | 1.000 |
| *Est 2* | a | 1.000 | 1.000 | 1.000 |
| *Est 3* | a | 1.000 | 1.000 | 1.000 |
| *Ak 1* | a | 1.000 | 1.000 | 1.000 |
| *Ak 2* | a | 1.000 | 1.000 | 1.000 |
| *Idh* | a | 1.000 | 1.000 | 1.000 |
| *Lap 1* | a | 1.000 | 1.000 | 1.000 |
| *Lap 2* | a | 1.000 | 1.000 | 1.000 |
| *Mdh 1* | a | 1.000 | 1.000 | 1.000 |
| *Mdh 2* | a | 1.000 | 1.000 | 1.000 |
| *Np* | a | 1.000 | 1.000 | 1.000 |
| n |  | 40 | 77 | 21 |

cating that the second function offers an important contribution to the analysis. A plot of the function scores, however, shows that most of the separation between groups is apparent in the plane of function 1 (Fig. 2). This is also demonstrated by the relatively high Eigenvalue (between groups sum of squares/within groups sum of squares for function scores) and the percentage of total variance (92.82%) for function 1 (Table 1).

The distribution of each measured variable is highly heterogeneous between samples (Table 2, $P<0.001$). Aperture width shows the greatest correlation with discriminant scores for function 1 (Table 2), suggesting that it may be the most influential of the variables. All three variables, however, have correlations of a simi-

lar order of magnitude, indicating that shell length and height are by no means unimportant in contributing to group discrimination.

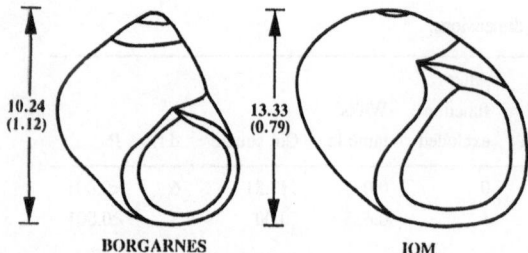

*Fig. 1.* Mean heights (standard deviations in brackets) and appearances of typical shells from the most northerly (Borgarnes) and most southerly (IOM) samples.

*Table 5.* Nei's (1972) Genetic Identity (*I*) values between samples, based on *I* values averaged over all loci.

|  | GRÓTTA | BORGARNES |
|---|---|---|
| BORGARNES | 0.995 | |
| IOM | 0.983 | 0.986 |

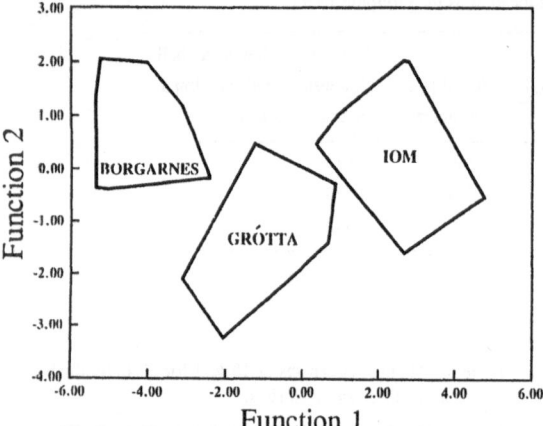

*Fig. 2.* Plot of discriminant function scores for both functions based on analysis of three measured shell dimensions (each polygon encloses all points from one site).

The confusion matrix (Table 3) shows that, on average, the analysis would correctly predict the origin of an individual from one of the three sites purely on the basis of the three measured shell parameters in 95.83% of cases. This compares very favourably with a chance correct classification rate of 33.33%. Borgarnes samples are the most northerly, and would only be incorrectly assigned in 10% of cases to the more southerly Iceland site. Manx samples would always be correctly assigned on the basis of morphology.

The results of the genetic analysis (Table 4) contrast sharply with the clear distinction between samples offered by statistical, and even by visual, analysis of morphology. Eleven out of the thirteen loci scored are monomorphic for the same allele. Although the two polymorphic loci show differences in allele frequencies between sites, these are not of a scale normally associated with taxonomic differences. This is borne out by the overall genetic identity (*I*) values (Table 5), the lowest of which is 0.983, between the Grotta and

Isle of Man samples. The highest *I* value is 0.995, between the two Icelandic samples.

## Discussion

The results of the two approaches in the present study serve to illustrate the development of approaches to taxonomy. The advent of Darwinian evolutionary models based on heritable variability has led to the recognition that considerable variation may exist within taxa. Prior to the general acceptance of these models, such observed variability undoubtedly resulted in many cases where single taxa were given more than one specific name (Oldroyd, 1980). This situation was probably exacerbated by relatively poor communications and a lack of standardised taxonomic criteria in the 18th and 19th centuries, during which time many of the most prolific taxonomists were active. Modern genetic techniques have usefully assisted taxonomic studies, not only in confirming suspected synonymy (Moyse *et al.*, 1982), but also in uncovering unexpected taxonomic divisions in situations of morphological crypsis between taxa (Lindstrom & Cole, 1992).

In the present study, the three samples are clearly quite different in morphology. The most northerly sample from Borgarnes in Iceland is thin shelled and high spired compared to the sample from the Isle of Man (Fig. 1). All three measured shell characters are significantly heterogeneous, and all three contribute greatly to discrimination between samples. The very high rate of correct prediction of origin (95.83%) based on shell characteristics measured for an average of only 16 individuals per sample is indicative not only of the power of multivariate approaches to the study of variability, but also of the high degree of morphological variability found in *L. obtusata* (see also Williams, 1994). Indeed the observation of high degrees of interpopulation morphological variation in marine gastropods which – in common with *L. obtusata* – do not disperse larvae in the plankton, compared with those species which do, is a relatively common finding (Vermeij, 1982; Behrens Yamada, 1987, 1989).

Many factors have been shown to influence shell shape determination in littoral gastropods, including population density (Kemp & Bertness, 1984; Boulding & Hay, 1993), wave exposure (Etter, 1988; Gibbs, 1993), predation (Vermeij, 1982; Johanesson, 1986; Seeley, 1986), heritability (Newkirk & Doyle, 1975; Janson, 1982; Boulding & Hay, 1993), and water-borne stimuli (Appleton & Palmer, 1988). In the case of *L. obtusata*, crab predation has been shown to be of particular influence. Seeley (1986) for example found experimentally that crabs have lower rates of successful attack against low spired shells, and that distribution of relative spire height between *L. obtusata* populations correlated with crab abundance. Local dispersal of progeny in *L. obtusata* would be expected to facilitate rapid response to local selective agents. Since such agents are likely to vary geographically, local dispersal would also be expected to facilitate relatively high degrees of interpopulation variability such as those found in the present study. Given this situation, it is not surprising that the taxonomy of the northern form has been the subject of some debate (Colman, 1932; Knudsen, 1949). The high genetic identity values between samples in the present study, however, (minimum $I = 0.983$) are in the upper range of values generally considered to be representative of conspecific groups (see review by Thorpe (1982)). In a review on biochemical genetic variation in the Littorinidae, Ward (1990) presents a UPGMA tree of relationships between populations of *L. arcana* Hannaford Ellis, *L. saxatilis*, and *L. nigrolineata* Gray, whereby the most distantly related intraspecific clusters correspond approximately to the genetic identity values 0.931, 0.944, and 0.959 respectively. Genetic identity values averaged from values given by Ward (1990) for 14 pairs of sibling and 11 pairs of non-sibling species are 0.808 and 0.432 respectively (all values converted from Nei's (1972) genetic distance values). Thus, in view of these comparisons, we have no evidence to suggest that the northern high spired *L. palliata* form is not synonymous with *L. obtusata*. Seeley (1986) came to a similar conclusion comparing samples of *L. obtusata* differing in relative spire heights on the American atlantic coast, finding a genetic identity of 0.997 (converted from a genetic distance ($D$) value of 0.003). Our genetic results support the results of Seeley (1986) and the views of Colman (1932) and Knudsen (1949) – that the northern high spired *L. palliata* form

should be regarded as being conspecific with *L. obtusata*.

One of the striking aspects of the present study is the clear morphological distinction between the two Icelandic samples, despite their geographic proximity. It is of interest to note that both samples were taken from a region found by Thorson (1941) to accommodate both *L. obtusata* and *L. palliata* forms, and that he failed to find *L. obtusata* on northern coasts of Iceland, or *L. palliata* forms on southern coasts. Thus it appears that the coasts of Iceland could provide systems of particular interest to investigators of the determinants of shell shape in the Littorinidae, since influential factors may vary considerably over relatively small geographic distances.

## Acknowledgements

The authors are most grateful to Dr Sigmar Steingrimsson and Dr David Reid for collection of samples from Iceland, and to Dr Tim Brailsford for assistance with analysis.

## References

Appleton, R. D. & A.R. Palmer, 1988. Water-borne stimuli released by predatory crabs and damaged prey induce more predator resistant shells in a marine gastropod. Proc. natn. Acad. Sci. U.S.A. 85: 4387–4391.

Behrens Yamada, S., 1987. Geographic variation in growth rates of *Littorina littorea*, and *L. saxatilis*. Mar. Biol. 96: 529–534.

Behrens Yamada, S., 1989. Are direct developers more locally adapted than planktonic developers? Mar. Biol. 103: 403–411.

Berger, E. M., 1973. Gene-enzyme variation in three sympatric species of *Littorina*. Biol. Bull. 145: 83–90.

Boulding, E. G. & T.K. Hay, 1993. Quantitative genetics of shell form of an intertidal snail: constraints on short-term response to selection. Evolution 47: 576–592.

Colman, J., 1932. A statistical test of the species concept in Littorina. Biol. Bull. 62: 223–243.

Dautzenberg, P. H. & H. Fischer, 1914. Etude sur le *Littorina obtusata* et ses variations. J. Conch. 62: 87–128.

Etter, R. J., 1988. Asymmetrical developmental plasticity in an intertidal snail. Evolution 42: 322–334.

Ferguson, A., 1980. Biochemical systematics and evolution. Blackie, Glasgow.

Gibbs, P. E., 1993. Phenotypic changes in the progeny of *Nucella lapillus* (Gastropoda) transplanted from an exposed shore to sheltered inlets. J. moll. Stud. 59: 187–194.

Goodwin, B. J.& J. D. Fish, 1977. Inter- and intraspecific variation in *Littorina obtusata* and *L. mariae* (Gastropoda: Prosobranchia). J. Moll. Stud. 43: 241–251.

Harris, H., & D. A. Hopkinson, 1978. Handbook of electrophoresis in human genetics. North-Holland, Amsterdam.

166

Hubendick, B., & A. Warén, 1969-76. Framgälade Snäckor fra Svenska Västkusten. Göteborg Naturhistoriska Museum. 36–43.

Janson, K., 1982. Genetic and environmental effects on the growth rate of *Littorina saxatilis*. Mar. Biol. 69: 73–78.

Johannesson, B., 1986. Shell morphology of *Littorina saxatilis* Olivi: the relative importance of physical factors and predation. J. exp. mar. Biol. Ecol. 102: 183–195.

Kemp, P. & M. D. Bertness, 1984. Snail shape and growth rates: evidence for plastic shell allometry in *Littorina Littorea*. Proc. natn. Acad. Sci. U.S.A. 81: 811–813.

Knudsen, J., 1949. Geographical variation of *Littorina obtusata* (L.) in the North-Atlantic. Vidensk. Meddr. Dansk. Naturh. Foren. III: 247–255.

Lewis, R. I., & J. P. Thorpe, 1994. Temporal Stability of Gene Frequencies within Genetically Heterogeneous Populations of the Queen Scallop *Aequipecten (Chlamys) opercularis* (L.). Mar. Biol.121: 117–126.

Lindstrom, S. C. & K. M. Cole, 1992. The *Porphyra lanceolata – P. pseudolanceolata* (Bangiales, Rhodophyta) complex unmasked – recognition of new species based on isozymes, morphology, chromosomes and distributions. Phycologia 31: 431–448.

Maynard-Smith, J., 1989. Evolutionary genetics. Oxford University Press, Oxford.

Moyse, J., J. P. Thorpe, & E. Al-Hamadani, 1982. The status of *Littorina aestuarii* Jeffreys. An approach using morphology and biochemical genetics. J. Conch. 31: 7–15.

Nei, M., 1972. Genetic distance between populations. Am. Nat. 106: 283–292.

Newkirk, G. F., & R. W. Doyle, 1975. Genetic Analysis of Shell Shape Variation in *Littorina saxatilis* on an Environmental Cline. Mar. Biol. 30: 227–237.

Norušis, M. J., 1992. SPSS for Windows Professional Statistics Release 5. SPSS Inc., Michigan.

Oldroyd, D. R., 1980. Darwinian Impacts, 2nd edn. The Open University Press, Milton Keynes.

Raffaelli, D., 1982. Recent ecological research of some European species of Littorina. J. moll. Stud. 48, 342–354.

Richardson, B. J., P. R. Baverstock & M. Adams, 1986. Allozyme Electrophoresis. Academic Press, Orlando.

Seeley, R. H., 1986. Intense natural selection caused a rapid morphological transition in a living marine snail. Proc. natn. Acad. Sci. U.S.A. 83: 6897–6901.

Shaw, P. R. & R. Prasad, 1970. Starch gel electrophoresis of enzymes – a compilation of recipes. Biochem. Genet. 4: 297–320.

Thorpe, J. P., 1982. The molecular clock hypothesis: Biochemical evolution, genetic differentiation and systematics. Ann. Rev. Ecol. Syst. 13: 139–68.

Thorson, G., 1941. Marine *Gastropoda Prosobranchiata*. In: The Zoology of Iceland, vol 4., (part 60): 30–33.

Vermeij, G. J., 1982. Phenotypic evolution in a poorly dispersing snail after arrival of a predator. Nature 299: 349–350.

Ward, R. D., 1990. Biochemical variation in the genus *Littorina* (Prosobranchia: Mollusca). Hydrobiologia 193 (Dev. Hydrobiol. 56): 53–69.

Williams, G. A., 1994. Variation in populations of *Littorina obtusata* and *L. mariae* (Gastropoda: Prosobranchia) along the Severn Estuary. Biol. J. Linn. Soc. 51: 189–198.

*Hydrobiologia* **309**: 167–172, 1995.
*P. J. Mill & C. D. McQuaid (eds), Advances in Littorinid Biology.*
©*1995 Kluwer Academic Publishers.*

# Frequency- and density-dependent sexual selection in natural populations of Galician *Littorina saxatilis* Olivi

Emilio Rolán-Alvarez[1]*, Kerstin Johannesson[2] & Anette Ekendahl[2]
[1] *Unidad de Genética, Biológicas Módulo A 201, Universidad Autónoma de Madrid (Cantoblanco), 28049 Madrid, Spain*
[2] *Tjärnö Marine Biological Laboratory, S-452 96 Strömstad, Sweden*
*Present address: Canovas del Castillo 22, 6 ° S, 36202 Vigo, Spain*

*Key words:* reproductive isolation, mating components, assortative mating, sexual selection, fitness estimate

## Abstract

Galician exposed shore populations of the direct developing periwinkle *Littorina saxatilis* are strikingly polymorphic, with an ornamented and banded upper shore form and a smooth and unbanded lower shore form. Intermediates between the two pure forms occur in a narrow mid shore zone together with the parental forms. We have previously shown that the two pure forms share the same gene pool but that mating between them is non-random. This is due to a non-random microdistribution in the zone of overlap, and also to assortative mating. In this study we present data which show that intermediate (hybrid) females mate less often than pure females in micropatches dominated by either of the pure forms, but not in micropatches in which the two pure forms are equally common. Thus, sexual fitness in intermediate females depends on the frequency of both pure morphs. Furthermore, sexual selection against intermediate females also varies with the densities of snails within each micro patch. The biological mechanisms which may explain this particular reduction of female hybrid fitness are discussed.

Assortative mating between the pure morphs is sometimes almost complete, while both morphs do not mate the intermediates assortatively. In the light of this, sexual selection against intermediate females may contribute considerably to restrict gene flow between the pure forms.

## Introduction

Mating behaviour is of central importance to the development of prezygotic reproductive isolation, and thus to speciation and evolution (Spieth & Ringo, 1983; Coyne, 1992). For descriptive purposes, mating behaviour can be partitioned into three components: sexual isolation (assortative mating or random mating component), male sexual selection and female sexual selection (Merrel, 1950; Spieth & Ringo, 1983; Coyne, 1992). These components are frequently used to study mate choice and sexual selection within species but may also be applied to situations where incipient species exist (Spieth & Ringo, 1983). There are some biological and theoretical justifications for dividing mating behaviour into these three components. They are observed independently, often differing between sexes, which suggests that they may

be caused by different mechanisms (O'Donald, 1980; Spieth & Ringo, 1983; Santos *et al.*, 1986; Johannesson *et al.*, in press).

Sexual selection is defined as the selection component that can be observed during mating within species or between incipient species (O'Donald, 1980; Spieth & Ringo, 1983; Endler, 1986). Theoretical studies have suggested the possibility that sexual selection contributes to reproductive isolation (Gilbert & Starmer, 1985; Marin, 1991), although sexual isolation is the most efficient component of mating behaviour contributing to reproductive isolation (Spieth & Ringo, 1983; Gilbert & Starmer, 1985; Marin, 1991). Johannesson *et al.* (in press) describe a micro-scale hybrid zone between one upper and one lower shore form of *Littorina saxatilis* (Olivi), over which gene flow is restricted mainly due to sexual isolation between the two pure morphs. In this paper we show further analyses of sexual selection in the same populations, using the same data set as Johannesson et al., (in press).

Frequency- or density-dependent selective mechanisms can be powerful forces in maintaining genetic polymorphism in natural populations, especially when the selective mechanisms originate from intraspecific competition or predation (Allen, 1988; Antonovics & Kareiva, 1988; Christiansen, 1988; Clarke et al., 1988). A negative frequency-dependent male sexual selection has been found in many species and is known as 'the rare male effect' (O'Donald, 1980; Knoppien, 1985; Partridge, 1988). The rare male effect is not explained by male-male competion models, and genetic models based on female choice fit the observed data better (O'Donald, 1980; O'Donald & Majerus, 1988; Partridge, 1988). The mating success of females is usually unrelated to their frequency (Partridge, 1988). In this study, however, we report sexual selection in female Littorina saxatilis which is both frequency and density dependent.

An important statistical problem in studies of natural selection is that many estimators used to infer fitness values have been shown to be frequency-dependent (Knoppien, 1985; Partridge, 1988). Moreover, many sexual fitness estimators confound the three components of mating behaviour: male and female sexual selection and sexual isolation (Spieth & Ringo, 1983; Knoppien, 1985). However, this is not the case for the cross product estimator, used in our studies. It is a statistically frequency-independent estimator and has been suggested one of the best for polymorphic characters (Cook, 1971; Spieth & Ringo, 1983; Knoppien, 1985; Partridge, 1988).

In exposed rocky shores in Galicia, Spain, two conspecific morphs of L. saxatilis, one adapted to a high shore barnacle zone and the other to a low shore mussel zone, overlap and hybridize in a narrow and patchy mid shore zone (see Johannesson et al., 1993 and in press). Both morphs are genetically very similar over a scale of several kilometres in that the two are indistinguishable on the basis of allozyme markers (Johannesson et al., 1993). Furthermore, fertile intermediates (hybrids) can be observed both in nature and in the laboratory. However, a partial reduction in gene flow between pure morphs can be detected at a scale of a few metres, and we have suggested that this may be explained by disruptive selection with incipient reproductive isolation between both pure morphs (op. cit.). In fact two different mechanisms contributed to the partial prezygotic reproductive isolation: a non-random microdistribution of the different morphs in the patchy mid shore, and a behavioural assortative mating between pure morphs. A third mechanism contributing to the reproductive isolation was also suggested: sexual selection against intermediate females. The following, more detailed analyses support this later suggestion. Moreover, we have found that the fitnesses of the intermediate females are both frequency- and density-dependent.

## Materials and methods

A full description of sampling areas and methods, variables studied and statistical tests are given in Johannesson et al. (in press).

We collected pairs of copulating snails and surrounding non-copulating individuals from two different areas (100 m apart) and three vertical habitats in each area (high, mid and low shore). On both the low and the mid shore more than one morph was present, but all snails collected from high shore sites were of the same morph, i.e. ridged and banded, and hence the high shore data have not been considered in this paper. The non-copulating snails were collected within small areas. These were 9 cm in diameter in low shore sites, but 11 cm in diameter in mid shore sites in order to obtain reasonable samples of snails for morph analysis in these lower density areas. In a few cases, where the snail density was very low, the collection area was increased for the above reason. The effect of the larger patches was to produce a slight overestimate of the mean number of snails ('density') in those patches. All mating pairs and surrounding snails were allotted corresponding numbers. Thus, we could reanalyse mating pairs from mid shore depending on morph frequency or number of surrounding individuals within the patches. In one analysis, samples were divided into three groups, each being defined by the frequency of ridged and banded morphs in a patch: ridged and banded patches (>66.6% of ridged and banded morph), mixed patches (33.3–66.6%) and smooth and unbanded patches (<33.3%). In a second analysis, we divided the whole sample into three different groups taking into account the number of adult snails in the patches (<10 snails, 10–15 snails and >15 snails, respectively). The sex of each individual was also noted, and male/male and male/juvenile pairings were excluded from the analyses.

We used a pseudo-probability chi-square contingency test (Zaykin & Pudovkin, 1993) to estimate the sexual selection component for males and females of each morph, and the cross product estimator to quantify the relative sexual advantage of each morph (sexual fitness estimate). The latter gives the fitness of one morph relative to the chosen reference morph, which in this study was the intermediate morph (see Johannesson *et al.*, in press). The significance of each fitness estimator was obtained by bootstrapping (Rolan-Alvarez, 1993). Yule's *V* was used to estimate the sexual isolation component (Gilbert & Starmer, 1985).

## Results

The frequencies of the three morphs and the frequency of mating pairs in each sample have been shown elsewhere (Johannesson *et al.*, in press.). The percentage of intermediate snails, the mean number of snails surrounding mating pairs, the number of patches studied, and the total number of snails observed are shown in Table 1 for mid and low shore samples and the different subsamples from the mid shore. Pooled mid and pooled low shore samples showed significant differences in the morph frequency ($\chi^2 = 31.8$, 1 d*f*; $p < 0.001$) and in the mean number of snails in the patches ('density') ($T = 4.46$, 321 d*f*; $p < 0.001$). However, the low shore patches were smaller than the mid shore ones and thus the apparent difference in the number of snails found within the patches may be an artefact. However, there were no differences in patch size among subsamples from the mid shore. Among the mid shore samples, only ridged and banded patches and mixed patches showed different frequencies of intermediates ($\chi^2 = 6.6$, 1 d*f*, $p < 0.05$). On the other hand, smooth and unbanded patches had different numbers of snails compared to ridged and banded patches ($T = 2.6$, 106 d*f*, $p < 0.05$) and mixed patches ($T = 2.4$, 119 d*f*, $p < 0.05$), while ridged and banded patches had the same number of snails ('density') as mixed patches. Comparing patches with different 'densities', the frequency of intermediate animals was smaller in high 'density' patches than in both low density patches ($\chi^2 = 8.7$, 1 d*f*, $p < 0.001$) and intermediate density patches ($\chi^2 = 14.2$, 1 d*f*, $p < 0.001$). The average number of snails per patch was different among the three density patches (Table 1).

The results of the sexual fitness estimates for the three morphs of males and females are shown in Table 2 for mid and low shore samples and grouped subsamples. In pooled mid shore samples, ridged and banded females were significantly fitter than the intermediate females, while smooth and unbanded females were

nearly so. In pooled low shore samples, the fitness values of ridged and banded, and smooth and unbanded females tended to be larger than the intermediate females (although the differences were not significant). Unfortunately, the statistical power of this test is quite low if the morphs involved occurred at low frequencies, which was the case for the ridged and banded and the intermediates in these samples. In summary, pooled mid shore samples and pooled low shore samples showed similar trends, with a sexual advantage in both pure morphs of females with respect to intermediate females (although this was only significant in mid shore samples).

In the frequency-dependent patches we found an interesting pattern among the females. The fitness of pure morphs tended to be greater than that of intermediates in ridged and banded patches and in smooth and unbanded patches (although this was significant only in the latter), while the fitness of females of both pure morphs and intermediates were similar in mixed patches (Table 2). This frequency-dependent fitness pattern is difficult to explain simply by the frequency of the different morphs (see Table 1 and 2). However, fitness of intermediate females depended on the frequency of both pure morphs simultaneously, with greatest fitness of hybrid females occurring when both pure morphs showed similar and intermediate frequencies. Although smooth and unbanded patches showed the highest number of snails ('density') of the subsamples, high density patches showed similar fitness estimates between pure and intermediate females (see Table 2). Thus, there was not a simple density-dependent pattern.

The index of sexual isolation between the pure morphs is also presented for the different samples and subsamples (Table 2). Sexual isolation is a mating component separate from sexual selection and it is not surprising to find that most Yule V estimates were large and highly significant, showing similar values (0.66–1) in all frequency- and density-dependent patches. Noticeably, there was no assortative mating between pure morphs in low shore samples, suggesting that the degree of sexual isolation is affected by the extremely low frequency of ridged and banded snails in these samples (1.3%).

## Discussion

We have previously concluded that mating behaviour contributes to the partial reproductive isolation

Table 1. Frequencies of the intermediate morph, mean numbers of snails in a patch ('densities') with standard errors in parentheses, the number of patches, and the total number of snails in the sampled patches. Each patch surrounded one mating pair. Only patches from mid and low shore are shown as no intermediates were found on the high shore. Mid shore patches are divided into frequency-dependent patches and density-dependent patches.

| Patches | Frequency of intermediates | 'Density' in patches | No. of patches | No. of snails |
|---|---|---|---|---|
| **Mid shore** | | | | |
| Pooled | 12.0% | 15.0 (0.50) | 165 | 2470 |
| Frequency dependent | | | | |
| Ridges and banded | 8.8% | 13.1 (0.78) | 44 | 577 |
| Mixed | 14.0% | 13.6 (0.71) | 57 | 774 |
| Smooth and unbanded | 12.2% | 17.5 (0.91) | 64 | 1119 |
| Density dependent | | | | |
| Low density | 14.2% | 8.5 (0.27) | 56 | 478 |
| Intermediate density | 14.7% | 15.1 (0.33) | 60 | 909 |
| High density | 8.7% | 20.2 (0.73) | 49 | 1083 |
| **Low shore** | | | | |
| Pooled | 6.2% | 10.4 (0.54) | 158 | 1637 |

Table 2. The degree of sexual isolation (Yule's V; where 0 is complete random mating and 1 is complete assortative mating), cross-product estimates of fitnesses for the pure morphs (relative to hybrid fitness = 1) together with bootstrap estimated probabilities (Johannesson et al., in press), and the results of pseudo-probability chi-square tests (Zaykin & Pudovkin, 1993) of male and female sexual selection within each patch. Mid shore patches are divided into frequency-dependent patches and density-dependent patches. I, intermediate; SU, smooth and unbanded; RB, ridged and banded.

| Patches | Sexual isolation Yule's V | Relative sexual fitness of males | | | | Relative sexual fitness of females | | | |
|---|---|---|---|---|---|---|---|---|---|
| | | RB | I | SU | $\chi^2$ | RB | I | SU | $\chi^2$ |
| **Mid shore** | | | | | | | | | |
| Pooled | 0.77*** | 1.34 | 1 | 0.95 | ns | 2.54* | 1 | 2.03? | * |
| Frequency dependent | | | | | | | | | |
| Ridged and banded | 0.85** | – | – | – | ns | 3.12 | 1 | 2.77 | ns |
| Mixed | 0.70*** | 1.02 | 1 | 1.09 | ns | 1.54 | 1 | 1.08 | ns |
| Smooth and unbanded | 0.66*** | 1.04 | 1 | 0.72 | ns | 9.49** | 1 | 4.51* | * |
| Density dependent | | | | | | | | | |
| Low density | 0.81*** | 1.95 | 1 | 2.81 | ns | – | – | – | ? |
| Intermediate density | 0.60* | 1.07 | 1 | 0.59 | ns | 5.18* | 1 | 4.82* | ns |
| High density | 10*** | 1.04 | 1 | 0.73 | ns | 1.04 | 1 | 0.86 | ns |
| **Low shore** | | | | | | | | | |
| Pooled | −0.01 | – | – | – | ? | 13 | 1 | 2.06 | ns |

?, $0.05 > p < 0.10$; *, $p < 0.05$; **, $p < 0.01$; ***, $p < 0.001$.

observed between the high and low shore morphs of Galician *L. saxatilis* (Johannesson *et al.*, in press). We will now consider more specifically the nature of the sexual selection component.

Although mating between pure morphs is largely assortative, we found no sexual isolation between any of the pure morphs and the intermediates (Johannesson *et al.*, in press). This suggests that the gene flow between the pure morphs occurs mostly via the intermediates or hybrids. Thus, a reduction of sexual fitness of intermediate females, as described in this study, further enhances reproductive isolation between the pure morphs.

The fitness of intermediate females during copulations was related to the morph frequency and the number of snails within each patch ('density'). Usually, there is no effective difference between frequency- and density-dependence, as population composition tends to vary with population density (Christiansen, 1988). This is the case in our study because different morph compositions and 'densities' can be observed in the different microhabitats of the exposed Galician *L. saxatilis* populations (Johannesson *et al.*, 1993; Johannesson *et al.*, in press.; Table 1). Among females from the mid shore, intermediates showed a significantly lower sexual fitness than both pure morphs in smooth and unbanded patches, and probably also in ridged and banded patches, although we were not able to show the latter statistically. In mixed patches, however, intermediate fitnesses were similar to those of pure morphs. There was also a tendency towards low sexual fitness of intermediate females in the low shore, which had a high frequency of the smooth and unbanded morph.

There are several mechanisms that may explain this phenomenon. There are many behavioural models that may produce frequency-dependence (O'Donald, 1980; O'Donald & Majerus, 1988; Partridge, 1988). One possibility would be a behavioural mechanism related to the known size differences between morphs (see Johannesson *et al.*, 1993; Johannesson *et al.*, in press). Size has been suggested as being of fundamental importance in the local adaptation of *L. saxatilis* populations (Sundberg, 1988) and indeed, large *L. littorea* females have some mating advantages over small ones (Erlandsson & Johannesson, in press). If, for example, males had a preference for different female sizes or there was sexual competition among different sizes of females, this might result in different fitness values among morphs might occur. Although ridged and banded females had higher fitnesses than other morphs, the smallest smooth and unbanded

females also showed higher fitnesses than the hybrids of intermediate size. Thus, this mechanism alone cannot fully explain the main sexual selection pattern in females. However, some contribution cannot be excluded because ridged and banded females always showed higher fitness than smooth and unbanded ones (although these differences were not significant).

Other possibilities are behavioural mechanisms directly related to morph frequencies. These mechanisms may be originated by male choice for female morphs or sexual female competition between morphs. In fact, female choice with fixed preferences for different mating types has been suggested as the most probable explanation of the rare-male effect. On the other hand, competition between different forms may explain some natural frequency-dependent polymorphisms in insects, amphibians and salmon (Partridge, 1988). At present we have no conclusive evidence to support any of these alternatives for the *L. saxatilis* hybrid zone.

However, the existence of a frequency-dependent mechanism cannot fully explain the observed density-dependent fitness pattern. There are significant differences in 'density' between the different groups of patches from the mid shore (Table 1). We can explain the density-dependent sexual fitness of intermediate females as an artefact of this phenomenon, but the same reasoning can apply to the frequency-dependent pattern. Johannesson *et al.* (in press) found a non-random microdistribution of the different morphs in mid shore areas. This suggests that different microhabitats may exist, with strong habitat preferences for each morph. Habitat choice in *L. saxatilis* would be related to the sexual selection component if, for example, the intermediate morph was well adapted to mixed patches but showed some general disadvantages, with respect to both pure morphs, in smooth and unbanded patches, and perhaps also in ridged and banded patches. The fitnesses of the different morphs would depend therefore on the habitat, and so they would be indirectly frequency- and density-dependent because each habitat shows particular morph frequencies and densities. In this case, we would expect to find other habitat-dependent fitness components (viability, fecundity, etc.) for the morphs involved.

The low sexual fitness of intermediate females is interesting from an evolutionary point of view, whatever the mechanism involved, because it suggests a decline in frequency of hybrids in the contact area. An equilibrium would eventually be achieved with a very narrow hybrid zone (as observed here). At this point, the system would be able to evolve, decreasing the hybrids fitness in the zone and thus completing the reproductive isolation into speciation. It should be emphasized that this mechanism possesses some theoretical advantages as a mechanism for reinforcement, because it can be initiated not only by mate choice but also by sexual competition (O'Donald & Majerus, 1988; Partridge, 1988). The theoretical problems of many reinforcement models are associated with the difficulty of the further evolution of assortative mating genes (Butlin, 1987 and 1989). New experiments will be necessary to confirm the fitness pattern found in this study and to distinguish the underlying mechanisms that can produce it.

## Acknowledgements

We thank María Cruz Alvarez for room and facilities during the preparation of preliminary drafts and Phill Mason for correcting a later version. E. R-A. was supported with grants from the Xunta de Galicia and the Universidad de Santiago de Compostela, and K.J. and A.E. had financial support from the Swedish Natural Research Council and the Collianders Foundation.

## References

Allen, J. A., 1988. Frequency-dependent selection by predators. Phil. Trans. r. Soc., Lond. B 319: 485–503.

Antonovics, J. & P. Kareiva, 1988. Frequency-dependent selection and competition: empirical approaches. Phil. Trans. r. Soc., Lond. B 319: 601–613.

Butlin, R., 1987. Speciation by reinforcement. Trends in Ecology and Evolution 2: 8–13.

Butlin, R., 1989. Reinforcement of premating isolation. In D. Otte & J. A. Endler (eds), Speciation and its consequences. Sinauer, Sunderland, MA USA: 158–179.

Christiansen, F. B., 1988. Frequency dependence and competition. Phil. Trans. r. Soc., Lond. B 319: 587–600.

Clarke, B. C., F. R. S. Shelton, P. R. & G. S. Mani, 1988. Frequency-dependent selection, metrical characters and molecular evolution. Phil. Trans. r. Soc., Lond. B 319: 631–640.

Cook, L. M., 1971. Coefficients of natural selection. Hutchinson University library, London.

Coyne, J. A., 1992. Genetics and speciation. Nature 355: 511–515.

Endler, J. A., 1986. Natural selection in the wild. Princenton University Press, Princenton.

Erlandsson, J. & K. Johannesson, (in press). Sexual selection on female size in a marine snail, *Littorina littorea* (L.). J. exp. mar. Biol. Ecol.

Gilbert, D. G. & W. T. Starmer, 1985. Statistics of sexual isolation. Evolution 39: 1380–1383.

Johannesson, K., B. Johannesson & E. Rolán-Alvarez, 1993. Morphological differentiation and genetic cohesiveness over a microenvironmental gradient in the marine snail *Littorina saxatilis*. Evolution 47: 1770–1787.

Johannesson, K., E. Rolán-Alvarez & A. Ekendahl, (in press). Incipient reproductive isolation between two sympatric morphs of the intertidal snail *Littorina saxatilis*. Evolution.

Knoppien, P., 1985. Rare male mating advantage: a review. Biol. Rev. 60: 81–117.

Marin, I., 1991. Sexual isolation in *Drosophila*, I. Theoretical models for multiple-choice experiments. J. theor. Biol. 152: 271-284.

Merrel, D. J. 1950. Measurement of sexual isolation and selective mating. Evolution 4: 326–331.

O'Donald, P., 1980. Genetic models of sexual selection. Cambridge University Press, London.

O'Donald, P. & M. E. N. Majerus, 1988. Frequency-dependent sexual selection. Phil. Trans. r. Soc., Lond. B 319: 571–586.

Partridge, L., 1988. The rare-male effect: what is its evolutionary significance? Phil. Trans. r. Soc., Lond. B 319: 525–539.

Rolán-Alvarez, E. 1993. Estructura genética y selección sexual en poblaciones naturales de dos especies gemelas del género *Littorina*. Ph. D. Thesis, University of Santiago, Spain.

Santos, M., R. Tarrio, C. Zapata & G. Alvarez. 1986. Sexual selection on chromosomal polymorphism in *Drosophila subobscura*. Heredity 57: 161–169.

Saur, M. 1990. Mate discrimination in *Littorina littorea* (L.) and *L. saxatilis* (Olivi) (Mollusca: Prosobranchia). Hydrobiologia 193 (Dev. Hydrobiol. 56): 261–270.

Spieth, H. T. & J. M. Ringo, 1983. Mating behavior and sexual isolation in *Drosophila*. In M. Ashburner, H. L. Carson & J. N. Thompson (eds), The genetics and biology of Drosophila, 3c. Academic Press, London: 224–284.

Sundberg, P. 1988. Microgeographic variation in shell characters of *L. saxatilis* Olivi- a question mainly of size? Biol. J. linn. Soc. 35: 169–184.

Zaykin, D. V. & A. I. Pudovkin, 1993. Two programs to estimate significance of $\chi^2$ values using pseudo-probability tests. J. Hered. 84: 152.

*Hydrobiologia* **309**: 173–180, 1995.
*P. J. Mill & C. D. McQuaid (eds), Advances in Littorinid Biology.*
©1995 *Kluwer Academic Publishers.*

# Dispersal and population expansion in a direct developing marine snail (*Littorina saxatilis*) following a severe population bottleneck

Kerstin Johannesson & Bo Johannesson
*Tjärnö Marine Biological Laboratory, S-452 96 Strömstad, Sweden*

*Key words:* cross water dispersal, migration, *Chrysochromulina polylepis* bloom, rafting

## Abstract

Most marine benthic invertebrate species have planktonic larvae, and in species in which juveniles and adults have low vagility a larva is obviously an efficient way of active dispersal. A minority of benthic invertebrate species develop without any pelagic phase at all. A largely unsolved question is how and at what rate do these species disperse. We have addressed this question using the marine littoral snail *Littorina saxatilis* (Olivi) as an example of a species that completely lacks larval dispersal.

In the Koster archipelago (north part of the Swedish west coast), *L. saxatilis* occupies rocky island habitats of different sizes, from large islands to small intertidal skerries (islets). In 1988 an extremely dense bloom of a toxin-producing flagellate killed more than 99% of this snail species in this area. Populations of larger islands were reduced, often to less than 1%, but were restored over 2–4 yr. In contrast, populations of small intertidal skerries were completely wiped out and thus could not increase by local recruitment. Four years later, however, four of 33 skerries (12%) were successfully recolonized with relatively dense populations, and another five had received a few founder individuals. These results indicate recruitment through founder individuals, and are rough estimates of dispersal rate in a snail species that lacks a pelagic developmental stage.

## Introduction

We know little about how intertidal invertebrates that lack a planktonic dispersal stage colonize island habitats. However, remote islands are often colonized by species lacking long-lived pelagic larvae (Highsmith, 1985; Johannesson, 1988). Rafting is, perhaps, the most likely mechanism for over-sea dispersal by directly developing benthic invertebrates (Gerlach, 1977; Highsmith, 1985; O Foighil, 1989; Jokiel, 1990). Alternative mechanisms that may be possible for very small benthic invertebrates are flying with birds (Malone, 1965), floating in the surface layer (Highsmith, 1985) and drifting by the aid of mucous threads (Sigurdsson *et al.*, 1976; Martel & Chia, 1991) .

If our knowledge of the mechanisms of dispersal for direct developing benthic invertebrates may be considered rudimentary, even less is known about the actual rates of dispersal across barriers of open water. Since estimates of dispersal rates indicate levels of gene flow, these are crucial to the understanding of the dynamics and evolution of populations and species. Furthermore, knowledge of dispersal rates and dispersal distances are valuable when interpreting biogeographic patterns of species. In this study we have recorded the rate at which the snail *Littorina saxatilis* (Olivi) recolonized island habitats in a Swedish archipelago. As we assume that a recolonization event is possible by the migration of a low number of founder individuals, the rate at which new populations were established reflects the dispersal rate. In another study (Johannesson & Johannesson, in prep.) we will compare the dispersal rate estimates with gene flow estimates obtained from isozyme data.

In May and June 1988, a bloom of the toxin-producing prymnesiophycean flagellate *Chrysochromulina polylepis* killed high numbers of a range of invertebrates, vertebrates and macro algae along the Danish, Swedish, and Norwegian coasts of Skagerrak, Kattegat, the Belt and the Sound (Rosenberg *et al.*, 1988). *Littorina saxatilis* was among those species that suffered severe losses during the bloom. On the

Swedish west coast, this species inhabits exposed and semi-exposed rocky and boulder shores from the low water level up to the top of the splash zone (e.g. Janson, 1982). It is also found more or less submerged in coastal lagoons (Janson & Ward, 1985). Juveniles and adults have low vagility (Janson, 1983) and the embryos have a direct development inside the females brood pouch. Generation time is around six months (Fretter & Graham, 1980).

During May and June 1988 over 99% of the individuals of rocky and boulder shore populations of *L. saxatilis* were killed by the flagellate, while the lagoon populations were not affected, probably due to the flagellate being most common in the outer coastal areas (Rosenberg *et al.*, 1988). The nearly complete elimination of this else very common intertidal gastropod gave us an excellent opportunity to study recolonization rates. Detailed records of the distribution of this species in a 20 $km^2$ archipelago in the north part of the Swedish west coast had been made during the years 1979 to 1987 (e.g. Janson & Sundberg, 1983; Janson, 1987); this includes estimates of the sizes of a number of populations occupying small intertidal rocky skerries (islets). All of these skerry populations were eliminated by the flagellate bloom. An important point is that this species of snail is confined to the littoral zone, and therefore we believe that the skerries could only be recolonized through cross water dispersal.

The aim of this study was to estimate the pattern and approximate rate of cross water dispersal of a species lacking the possibility of pelagic spreading, in an archipelago of small and relatively close (<1 km) island habitats.

### Description of sites, materials and methods

The studied archipelago west of Strömstad on the Swedish west coast (58° 50′ N 11° 5′ E) has hundreds of island habitats from small intertidal skerries or islets (meters across) to islands of different sizes (up to a few kilometers across). The whole area is rocky (granite and gneiss), and rocky cliffs and boulder shores dominate while sandy beaches are rare. Thus practically all shores are potential habitats of *Littorina saxatilis* (Olivi). The species is, however, much more dense among moderately to extremely exposed cliffs than among protected cliffs within this area, but in protected boulder shores it is found in high densities

*Table 1.* Description of the four classes of rocky skerries surveyed in the summers of 1984 and 1985.

| Category of skerry (N) | Small & protected (5) | Large & protected (36) | Small & exposed (26) | Large & exposed (38) |
|---|---|---|---|---|
| **Size above mean water ($m^2$)** | | | | |
| Mean | 0.75 | 15.6 | 0.54 | 18.9 |
| Range | 0.25–1.5 | 2–100 | 0.25–1.5 | 2–450 |
| **Height above mean water (m)** | | | | |
| Mean | 0.3 | 0.4 | 0.2 | 0.4 |
| Range | 0.1–0.4 | 0.1–1.5 | 0–0.5 | 0.2–0.9 |
| **Distance to nearest island (m)** | | | | |
| Mean | 8 | 13 | 15 | 104 |
| Range | 3–15 | 2–50 | 2–80 | 2–1000 |

The area is almost atidal (maximum tidal range is 0.3 m) but the water level changes irregularly within a range of approximately 1.5 m due to variation in air pressure and direction of the wind (Johannesson, 1989).

The distribution of *L. saxatilis* in this area is well documented. It has been studied on the islands of Saltö and Ursholmen (Janson & Sundberg, 1983; Johannesson & Johannesson, 1990), both of which were used for detailed studies of recolonization after 1988. Furthermore, we have unpublished observations of distributions and densities on a number of islands during the period 1979 to 1987, including the sizes of 105 populations inhabiting intertidal skerries. The density of these skerry populations was recorded during the summers of 1984 and 1985 using a roughly logarithmic scale, that is, numbers below 10 were indicated exactly but otherwise numbers were given as tens (<100), hundreds (<1000), or thousands and more (>1000). For each skerry we also noted maximum height over mean water level, area over mean water level, distance to nearest island (Table 1) and the presence or absence of *Ascophyllum nodosum*, a macro alga that is an indicator of low energy shores (Lewis, 1972). We grouped the skerries into four categories: (1) small & protected (<2 $m^2$, *Ascophyllum* present), (2) large & protected (≥2 $m^2$, *Ascophyllum* present), (3) small & exposed (<2 $m^2$, *Ascophyllum* absent), (4) large & exposed (≥2 $m^2$, *Ascophyllum* absent).

The bloom of the toxin-producing flagellate, *Chrysochromulina polylepis*, started in early May 1988 and ended in June 1988 (Rosenberg *et al.*, 1988). Dur-

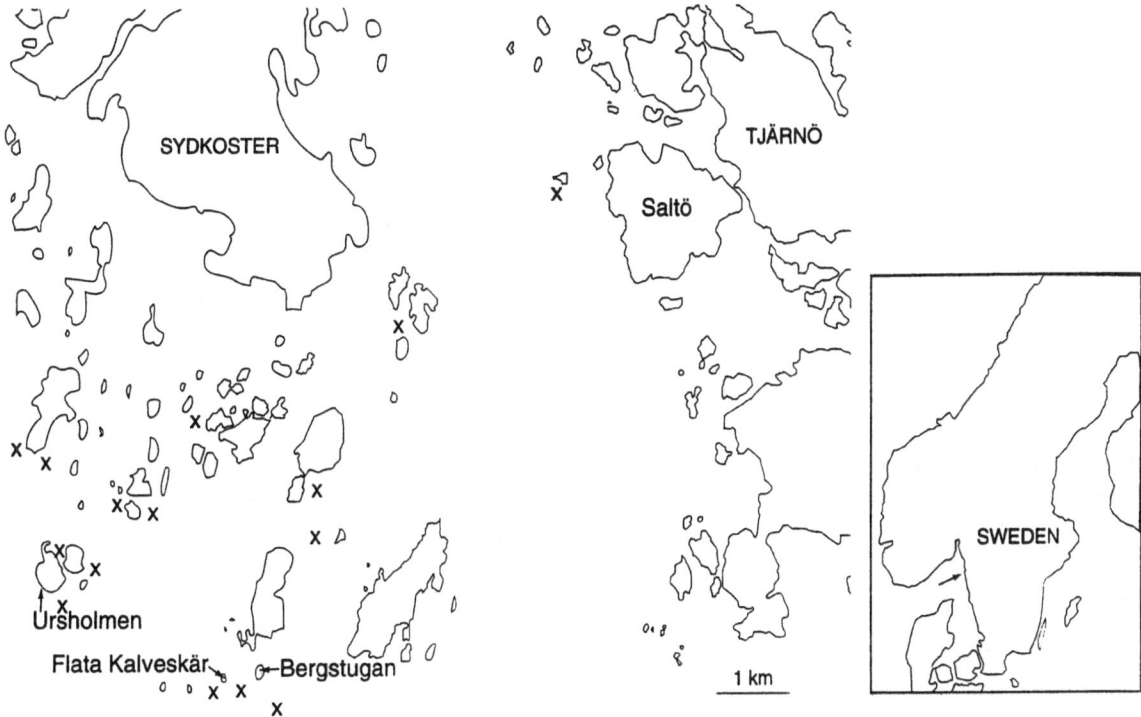

*Fig. 1.* Map of the study archipelago on the Swedish west coast. The four islands are indicated by their names and the positions of the 35 intertidal skerries are indicated by crosses (each cross shows one or several skerries).

ing most of this period there were easterly winds and a relatively low water level due to high air pressure.

Four islands (Saltö, Flata Kalveskär, Bergstugan and Ursholmen) and 35 skerries were visited during the summers of 1988, 1989, 1990 and 1992 (Fig. 1). We also made casual observations of recolonization rates at a number of islands in the neighbourhood of the skerries.

## Results

### Population expansion on islands

Very large parts of the populations of *Littorina saxatilis* disappeared during the bloom in 1988 and as a result the shore zone which *L. saxatilis* usually inhabits was overgrown by ephemeral algae by the end of the summer. Grazed patches indicated the presence of surviving snails, while ungrazed pieces of shore were without snails. When we had localized a grazed patch we estimated the number of *L. saxatilis* in it. Surviving snails were, especially on rocky shores, found high up in the splash zone where the snails had probably

escaped the contaminated sea water. In addition, snails close to rill outflows survived in high numbers, probably as the flagellate did not tolerate brackish water.

### Saltö

Before 1988 the 2.3 km long west facing shore of this island had a continuous population of *L. saxatilis* (Janson & Sundberg, 1983) which was normally distributed over a 1 to 4 m wide zone of the shore. Estimates of population sizes (from August 1981) indicated that rocky cliffs had 200–2800 snails per meter of shore (mean 1100; juveniles <2 mm not counted), while boulder shores had 20–800 snails per meter of shore (mean 290). Snails per meter shore is used here rather than per m$^2$, because the densities vary with shore level.

We searched carefully for surviving snails along 10 transects on one boulder shore in September 1988 and found an average of 2.1 snails per meter of shore, which indicates a survival rate of around 1%. At this time juveniles born after May 1988 were also noticed (approximately 50 per meter of shore). We also searched for surviving snails along rocky pieces of shore (six sites, total length 200–300 m) but found only a handful of snails altogether. This indicates a much lower rate of survival in the exposed as compared to the boulder shores.

In September 1989 the boulder shores of Saltö (total length about 1.2 km) in most cases had tens of snails or more per meter of shore, while the rocky shores (total length about 1.1 km) were still to a large extent (7 of 12 sites visited) unpopulated, although occasionally we found tens of snails (or rarely hundreds) per meter of shore in local spots.

The situation was much the same in June 1990, although the local spots of rocky populations noted in 1989 had become more dense and more spread. In July 1992 all boulder shore populations had recovered to the normal (1981) densities of the snails ($>100$ m$^{-2}$). Several rocky shore populations of *L. saxatilis* had by then increased to normal densities too, but about half the stretches of exposed rocks on Saltö were still completely free, or nearly so, of snails of this species. Some of the emptied sites were instead inhabited by the littorinid snail, *Melarhaphe neritoides*. This species has rarely been found in Sweden before 1988 (Johannesson, 1992). It has planktonic larvae, and it seems possibly that larvae have dispersed from Britain or France. We found this species to be particularly common on vertical walls of exposed rocky shore, and in Saltö we estimated local densities of hundreds or even thousands per m$^2$ in many areas.

## Flata Kalveskär

Careful searching around 3/4 of the shoreline of this small ($80 \times 80$ m), flat and extremely exposed island in June 1988 revealed only 10 surviving snails of *L. saxatilis* altogether. All lived in the same site about 1 m above mean water and on a relatively wave protected vertical cliff face. In October of the same year we found hundreds of small snails close to the original site and in August 1989 there were thousands of individuals per meter of shore in this area. One year later (July 1990), snails, presumably originating from this spot, had spread up to 50 m along the shore, and occurred in densities of 10 to 100 per meter of shore. This suggests an along shore dispersal rate of approximately 5 m per month. We also observed that the most recently colonized areas of the island were completely dominated by subadult individuals. This suggests either that juveniles were more prone to disperse or, which seems more likely, that a few adults migrated and generated offspring at high rates in areas of low intraspecific competition.

*Melarhaphe neritoides* was only found in one spot on Flata Kalveskär, possibly due to a general lack of exposed vertical walls, and thus most parts of the shore of this island were devoid of snails. We have found no indications of any further recruitment of juveniles to the *M. neritoides* population, and this may be the explanation as to why this species did not fill up all the emptied spaces of shore over the years following the flagellate bloom.

## Bergstugan

Only one live individual of *L. saxatilis* was found in June 1988 over a 30 m stretch of shore in the north part of this small island ($180 \times 80$ m). In August 1988 likewise only one individual was located in the same area, while in August the next year, groups of a few tens of juvenile and adult snails were found at some sites while most of the area still lacked visible snails. However, only one year later (July 1990) we found normal densities of snails (hundreds to thousands per meter of shore) in the study area.

## Ursholmen

This island has a high, steep and very exposed rocky shore facing west. In 1987 the densities at all shore levels sampled (low - 0 to 1 m above mean water, mid - 2 to 3 m and high - 5 to 6 m) were hundreds to thousands per m$^2$. In October 1988, however, we found very few snails at low levels ($\leq 1$ m$^{-2}$) subnormal densities at mid levels (10–100 m$^{-2}$), but normal densities at high levels ($>100$ m$^{-2}$). In 1989 the low shore levels were also populated by fewer snails than normal, while in June 1990 about normal densities of snails ($>100$ m$^{-2}$) were recorded.

## Recolonization of intertidal skerries

### The situation before May 1988.

Sixty of the 105 intertidal skerries that we visited in 1984 and 1985 had population sizes of 100 snails of *L. saxatilis* or more, while 15 skerries lacked snails of this species altogether. Whether or not a skerry was densely populated appeared to be related to at least two external factors, namely the size of the skerry and exposure to wave action. The small & protected skerries never had more than 10 snails and on most of them *L. saxatilis* was completely absent (Fig. 2). Some of the large & protected and small & exposed skerries were also without snails, or had populations of less than 10. The large & exposed skerries, on the

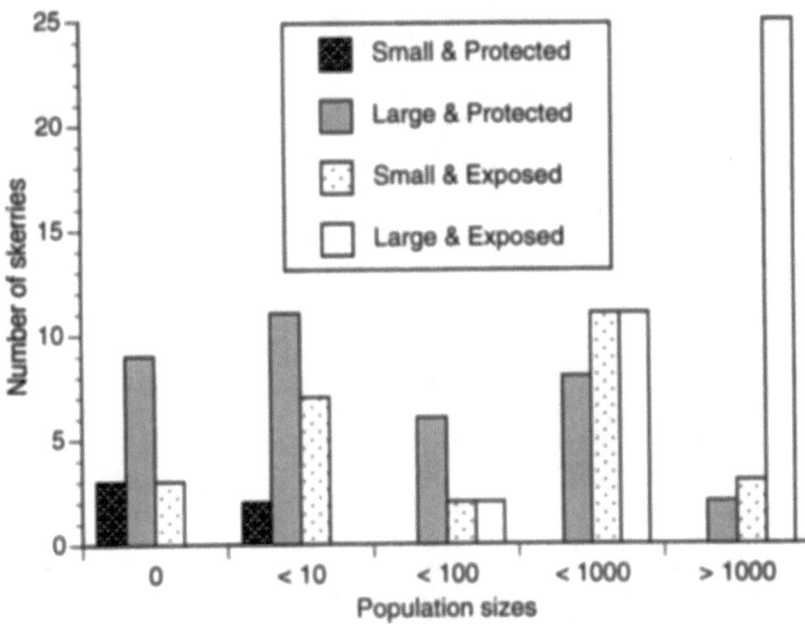

*Fig. 2.* Population sizes of *Littorina saxatilis* inhabiting 105 intertidal skerries in a Swedish archipelago in 1984-85, showing the marked differences in population sizes between skerries in the four defined categories (see Table 1 and text for definitions). All these populations were eliminated by the bloom of a microflagellate in May 1988.

other hand, were all populated, and all but two (2.5 and 15 m$^2$ in size) had population sizes of 100 or more. While only two of the 36 large & protected skerries had populations of over 1000 individuals, 25 of the 38 large & exposed populations were of this size (Fig. 2).

The large & protected skerries were often closer to an island than the large & exposed skerries (Table 1, $P = 0.0002$, Mann-Whitney U-test). However, close intertidal skerries were not more densely populated than remote ones. In fact, skerries with populations of more than 1000 were significantly more distant from islands than were skerries with less than 10 (Student-Newman-Keuls test; $P < 0.05$, following a one factor analysis of variance; $F_{4,40} = 2.74$, $P = 0.042$ in which square roots of distances were used to improve homogeneity of variances as indicated by Cochran's test, e.g. Underwood, 1981).

### The situation after May 1988

In June 1988, one month after the bloom of *Chrysochromulina*, we carried out the first estimate of population sizes at 17 skerries and found altogether only one individual of *Littorina saxatilis* (Table 2). We made additional counts in October 1988, September 1989, August 1990 and June 1992 of 21 skerries. Eleven of these were visited on all these occasions: the

remaining ten were visited two or three times (Table 2). The situation in October 1988, September 1989 and in August 1990 was much the same as in June 1988, i.e. most skerries were unpopulated and we found none with before-1988 densities of snails (Table 2). In June 1992 three of the 21 skerries were populated by thousands of snails while two had low densities, and 16 had no snails (all except one were populated before May 1988) (Table 2).

An additional 14 skerries that we knew were populated before May 1988 were visited in June 1992. Of these, 10 had no snails at all, three had ten snails or less, and only one had hundreds of snails. These skerries had not been visited shortly after the flagellate bloom in May-June 1988, so we could not be certain that the four populated ones had been completely snail free, although this seems most likely.

Of the 35 skerries (the 21 of Table 2 and the 14 mentioned above) the nine that were populated in June 1992 by one or more snails were, on average, closer to larger islands than the 26 that were completely unpopulated ($P = 0.008$, two-tailed Mann-Whitney U-skerry test). There was, however, no significant difference in skerry size between populated and unpopulated skerries ($P = 0.19$, two-tailed Mann-Whitney U-test).

178

Table 2. Population sizes of Littorina saxatilis on 21 intertidal rocky skerries before and after a bloom of a toxin-producing flagellate in May 1988 which killed practically all Littorina saxatilis at or slightly above mean water level. (*) denotes populations which were not sampled at that date.

| Category of skerry | Area of skerry (m²) | Height above mean water (m) | Distance to nearest island (m) | Population sizes | | | | | |
|---|---|---|---|---|---|---|---|---|---|
| | | | | Before bloom | After bloom | | | | |
| | | | | 1984 and 1985 | June 1988 | October 1988 | September 1989 | August 1990 | June 1992 |
| Large & protected | 2 | 0.3 | 15 | 2 | * | * | 0 | * | 0 |
| | 30 | 0.2 | 30 | 10≤N<100 | 1 | * | 0 | * | 0 |
| | 9 | 0.2 | 30 | 10≤N<100 | 0 | * | 0 | * | 0 |
| | 10 | 0.4 | 30 | 10≤N<100 | 0 | * | 0 | * | 0 |
| | 2 | 0.1 | 50 | 0 | 0 | * | 0 | * | 0 |
| | 5 | 0.1 | 40 | 100≤N<1000 | * | * | 0 | * | 1 |
| | 6 | 0.6 | 40 | ≥1000 | 0 | * | 0 | * | 10≤N<100 |
| | 60 | 0.6 | 100 | ≥1000 | 0 | * | 5 | * | ≥1000 |
| Small & exposed | 1 | 0 | 80 | 10≤N<100 | * | * | 0 | * | 0 |
| Large & exposed | 6 | 1 | 25 | ≥1000 | * | * | 1 | * | ≥1000 |
| | 2 | 0.1 | 100 | ≥1000 | 0 | 0 | 0 | 0 | 0 |
| | 8 | 0.5 | 180 | ≥1000 | 0 | 0 | 0 | 0 | 0 |
| | 13 | 0.3 | 170 | ≥1000 | 0 | 0 | 0 | 0 | 0 |
| | 40 | 0.4 | 160 | ≥1000 | 0 | 0 | 0 | 0 | 0 |
| | 2 | 0.4 | 220 | ≥1000 | 0 | 0 | 0 | 0 | 0 |
| | 2 | 0.4 | 290 | 100≤N<1000 | 0 | 0 | 0 | 0 | 0 |
| | 5 | 0.7 | 290 | 100≤N<1000 | 0 | 0 | 0 | 0 | 0 |
| | 2 | 0.5 | 5 | ≥1000 | 0 | 0 | 0 | 11 | ≥1000 |
| | 25 | 0.3 | 120 | ≥1000 | 0 | 0 | 0 | 0 | 0 |
| | 3 | 0.1 | 210 | 100≤N<1000 | 0 | 0 | 0 | 0 | 0 |
| | 4 | 0 | 100 | ≥1000 | 0 | 0 | 0 | 0 | 0 |

## Discussion

Despite its lack of an effective dispersal stage, such as a planktotrophic larva, Littorina saxatilis does indeed spread along shore lines and among island habitats, although, as expected, at slower rates than in species with a pelagic stage. The congeneric species L. littorea, for example, which has a pelagic larval stage of 5 to 8 weeks (Fretter & Graham, 1980), spread along the American east coast at an average rate of 34 km per year following the introduction of this species to Canada in the last century (Carlton, 1982).

The within shore dispersal rate of L. saxatilis found at Flata Kalveskär (5 m per month) was, however, higher than an earlier estimate from Saltö (Janson, 1983) which was <1 m per month within a normally dense population. The difference may suggest that disper-sal rates are higher in unpopulated than in populated areas.

In spite of the seeming handicap of lacking a pelagic dispersal stage it seems as if this does not prevent L. saxatilis from invading island habitats within an archipelago over a reasonably short time. Approximately 3% of the skerries in this study were recolonized per year, if we count those with more than a few founders present. This suggests perhaps that it will take 33 years to recolonize all of the skerries, assuming a constant rate of recolonization. However, as more remote skerries were recolonized at a lower rate, 3% per year may be an overestimate. On the other hand, it is possible that the rate of recolonization was quite low during 1988 and 1989 because, during this period, the mother populations of the islands were suffering from subnormal densities of snails.

Another way to estimate the magnitude of the dispersal rate is to use the information on the height of the skerries. The ground in this area rises at a constant rate due to a geological phenomenon (post-glacial elevation), and we may thus estimate the approximate time the intertidal skerries have been potential habitats for *L. saxatilis* from their height. The ground elevation rate is approximately 0.25 m per 100 y, (Loberg, 1980) and, because in Sweden *L. saxatilis* is not found below the barnacle zone, only intertidal skerries that extend into the barnacle zone (at approximately mean water) or higher are suitable habitats for *L. saxatilis*. The height over mean water of the islets of Table 2 ranged from 0.0 to 1.0 m, which means that they are all less than 400 years old, and all but one were colonized before the exceptional flagellate bloom of 1988. This suggests that the dispersal rate is of the magnitude indicated above, or greater. Certainly dispersal rates of *L. saxatilis* between breakwaters and other harbour constructions along the 60 km long sandy Belgian coast seem to be somewhat higher. Nearly all man-made substrates, except for the ones that were only one or two years old, were colonized by *L. saxatilis* (Johannesson & Warmoes, 1990).

Janson (1987) and Johannesson (1988) have argued that a direct development makes a colonization likely if a small founder group, possibly only one mated female, reaches a new habitat. It is evident from the observations after May 1988 of rapid population increases that even populations founded by very few individuals may expand at a high rate (for example, the island populations, and the three skerries that were recolonized over the period 1988 to 1992).

An intricate problem is how migrators disperse. Rafting has been shown as a probable way for several species of benthic marine invertebrates (Gerlach, 1977; Highsmith, 1985; O Foighil, 1989; Jokiel, 1990). However, the exposed shores of the skerries and islands in this study have very little rafting material available, because the strong wave action prevents growth of large seaweeds in the littoral zone. Floating with the aid of mucuos threads has been shown to be a possible dispersal mechanism for juveniles and small adults in a number of direct developing molluscs, including a *Littorina* sp. (Martel & Chia, 1991). It seems, however, less likely that a new colony of snails may be established through occasional dispersal of juveniles. Only mature females (carrying sperm or embryos) can, on their own, give rise to a new population. Johannesson & Warmoes (1990) suggested that in Belgium snails may be swept off one breakwater and then rolled to an adjacent one along the sandy bottom by waves and currents. The Belgian breakwaters may be very effective in catching dislodged snails as they run 400 m out from, and perpendicular to, the shore and are separated by only 200–500 m of sandy shore. This mechanism is obviously impossible in rocky coastal areas, and may explain the higher dispersal rate in Belgium compared to that in the Koster archipelago.

Floating in the surface film is a possible mechanism of dispersal for small individuals of intertidal organisms such as tanaids, amphipods and small bivalves (Highsmith, 1985), and in small snails (*Hydrobia*; e.g. Levinton, 1979). We have observed that small individuals of *Littorina* (1–2 mm) may remain floating for at least a couple of minutes if dislodged when they sit above the surface and have dry shells. While *Hydrobia* uses the foot attached to the surface film, the floating of *Littorina* is dependent on a dry shell, which seems less reliable.

Thus, although we acknowledge that we do not have any good suggestion of the actual mechanism of dispersal, we may conclude that *Littorina saxatilis* may spread and start new populations in adjacent (<1 km) island habitats over tens of years. Even if the habitats are small skerries, the populations may expand rapidly to a size of several thousand snails and, if so, the population of a single skerry may be large enough to prevent a random extinction. This seems to be particularly true for exposed cliff skerries. The populations of the protected ones were never so dense, and these may thus be more prone to stochastic forces of elimination. It seems probable that the turn-over rate of skerry populations in more sheltered areas is much higher than in exposed areas, which may explain why some small skerries lacked *L. saxatilis* completely.

## References

Carlton, J. T., 1982. The historical biogeography of *Littorina littorea* on the Atlantic coast of North America, and implications for the interpretation of the structure of New England intertidal communities. Malacol. Rev. 15: 146.

Fretter, V. & A. Graham, 1980. The prosobranch molluscs of Britain and Denmark. Part 5 - Marine Littorinacea. J. moll. Stud. Suppl. 7.

Gerlach, S. A., 1977. Mean of meiofauna dispersal. Mikrofauna Meeresboden 61: 89–103.

Highsmith, R. C., 1985. Floating and algal rafting as potential dispersal mechanisms in brooding invertebrates. Mar. Ecol. Prog. Ser. 25: 169–179.

Janson, K., 1982. Phenotypic differentiation in *Littorina saxatilis* Olivi (Mollusca, Prosobranchia) in a small area on the Swedish west coast. J. moll. Stud. 48: 167–173.

Janson, K., 1983. Selection and migration in two distinct phenotypes of *Littorina saxatilis* in Sweden. Oecologia (Berl.) 59: 58–61.

Janson, K., 1987. Genetic drift in small and recently founded populations of the marine snail *Littorina saxatilis*. Heredity 58: 31–37.

Janson, K. & P. Sundberg, 1983. Multivariate morphometric analysis of two varieties of *Littorina saxatilis* from the Swedish west coast. Mar. Biol. 74: 49–53.

Janson, K. & R. D. Ward, 1985. The taxonomic status of *Littorina tenebrosa* Montagu as assessed by morphological and genetic analyses. J. Conch. 32: 9–15.

Johannesson, K., 1988. The paradox of Rockall: why is a brooding gastropod (*Littorina saxatilis*) more widespread than one having a planktonic larval dispersal stage (*L. littorea*)? Mar. Biol. 99: 507–513.

Johannesson, K., 1989. The bare zone of Swedish rocky shores: why is it there? Oikos 54: 77–86.

Johannesson, K., 1992. Genetic variability and large scale differentiation in two species of littorinid gastropods with planktotrophic development, *Littorina littorea* (L.) and *Melarhaphe (Littorina) neritoides* (L.) (Prosobranchia: Littorinacea), with notes on a mass occurrence of *M. neritoides* in Sweden. Biol. J. linn. Soc. 47: 285–299.

Johannesson, K. & B. Johannesson, 1990. Genetic variation within *Littorina saxatilis* (Olivi) and *Littorina neglecta* Bean: Is *L. neglecta* a good species? Hydrobiologia 193 (Dev. Hydrobiol. 56): 89–97.

Johannesson, K. & T. Warmoes, 1990. Rapid colonization of Belgian breakwaters by the direct developer, *Littorina saxatilis* (Olivi) (Prosobranchia, Mollusca). Hydrobiologia 193 (Dev Hydrobiol. 56): 99–108.

Jokiel, P. L., 1990. Long-distance dispersal by rafting: reemergence of an old hypothesis. Endeavour, New Series 14: 66–73.

Levinton, J. S., 1979. The effect of density upon deposit-feeding populations: movement, feeding and floating of *Hydrobia ventrosa* Montagu (Gastropoda: Prosobranchia). Oecologia (Berl.) 43: 27–39.

Lewis, J. R., 1972. The ecology of rocky shores. The English Universities Press, London, 323 pp.

Loberg, B. 1980. Geologi. Material, processer och Sveriges berggrund. Norstedts, Stockholm, 302 pp.

Malone, C. R., 1965. Killdeer (*Charadrius vociferus* Linnaeus) as a means of dispersal of aquatic gastropods. Ecology 46: 551–552.

Martel, A., & F.-S. Chia, 1991. Drifting and dispersal of small bivalves and gastropods with direct development. J. exp. mar. Biol. Ecol. 150: 131–147.

O Foighil, D., 1989. Planktotrophic larval development is associated with a restricted range in *Lasaea*, a genus of brooding, hermaphroditic bivalves. Mar. Biol. 103: 349–358.

Rosenberg, R., O. Lindahl, & H. Blanck, 1988. Silent spring in the sea. Ambio 17: 289–290.

Sigurdsson, J. B., C. W. Titman & P. A. Davies, 1976. The dispersal of young post-larval bivalve molluscs by byssus threads. Nature 262: 386–387.

Underwood, A. J., 1981. Techniques of analysis of variance in experimental marine biology and ecology. Oceanogr. Mar. Biol. annu Rev. 19: 513–605.

*Hydrobiologia* **309**: 181–193, 1995.
*P. J. Mill & C. D. McQuaid (eds), Advances in Littorinid Biology.*
©1995 *Kluwer Academic Publishers.*

# A geographically-based study of shell shape in small rough periwinkles

K. J. Caley, J. Grahame & Peter J. Mill
*Department of Pure & Applied Biology, University of Leeds, Leeds LS2 9JT, UK*

*Key words:* shell shape, principal component analysis, discriminant analysis, North Atlantic, *Littorina*

## Abstract

A study using principal component analysis and discriminant analysis was carried out on shell shape variation in 3093 specimens of rough periwinkles, 2500 of which were below 5.5 mm in columella length, from around the North Atlantic. Using a combination of colour plus sculpture, and life history trait, the snails were classified by inspection and examination into *Littorina nigrolineata, L. arcana, L. saxatilis* and *L. neglecta*. Principal component analyses indicated that similar aspects of variation were important in the different taxa, but these were sometimes of differing levels of importance between *L. saxatilis* and *L. neglecta*. Crossvalidation in a discriminant analysis showed classification of shells larger than 5.5 mm to have at least an 88% accuracy. That of shells below 5.5 mm showed an accuracy of 49% in *L. arcana*, increasing to 54% in *L. saxatilis* and 63% in *L. neglecta*, with 76% accuracy for small *L. nigrolineata*. This last was a special case as only one site was sampled, therefore comparative data are not available. This geographically-based study reveals that *L. neglecta* is more homogeneous over its range than recently reported by other workers and shows greater differences from *L. saxatilis* than the latter does from either *L. nigrolineata* or *L. arcana*. Size effects do not account for these differences because *L. neglecta* is morphometrically distinct from both large and small *L. saxatilis*. Furthermore, small, mid-shore *L. saxatilis* classify with large high-shore *L. saxatilis* in discriminant analysis, not with *L. neglecta*. These results provide evidence that the taxon *L. neglecta* is more distinct than has sometimes been suggested.

## Introduction

One of the fundamental requirements in any biological programme is that species should be recognised. This involves definition, and a great deal of effort has been devoted to satisfactory biological definitions of 'species' (e.g. see Templeton (1989) for review). For present purposes the lack of wholehearted satisfaction with any of the definitions in use need not trouble us; it is sufficient to adopt a working idea related to the 'biological' concept as advocated by Mayr (1942, 1963) or the 'cohesion' concept of Templeton (1989). In either of these, the central fact is that there are groups of populations whose members actually or potentially interbreed, while not interbreeding with members of other groups. Therefore, there is gene flow between populations within a species, but (ideally) not between populations in different species. Nevertheless, both of these concepts must take into account the occurrence of

hybridisation between members of what are generally regarded as 'good' species.

Problems of definition aside, there is a requirement for recognition (by the biologist). Ultimately, decisions about species status must require investigation of the putative reproductive barrier, imperfect as it may sometimes be, but this is a later step, not a first one. The first step is to decide on what the likely groups are, and if this is not done adequately the result is likely to be confusion rather than enlightenment. The problem was succinctly put by Richards (1938): 'The division of living organisms into groups defined by correlated characters must come first, and the quantitative taxonomic analysis afterwards'. Although Richards was writing well before the genetic analysis of populations based on enzyme polymorphisms was possible, his remarks are still applicable.

The study of shape has long been regarded as an important technique in the identification of species

182

(e.g. Dimm, 1902; Campbell & Mahon, 1974; Tissot, 1988; Grahame & Mill, 1989, McNamee & Dytham, 1993) and in quantifying the nature of morphological variation within a species (e.g. Jolicoeur, 1959; Phillips *et al.*, 1973; Sacchi, 1980; Beaumont & Wei, 1991; McMahon, 1992). Shape studies in *Littorina* and other molluscs have shown that shell shape variation within a species reflects such environmental factors as wave exposure (Smith, 1979; Sacchi, 1980; Janson, 1982a, b; Grahame *et al.*, 1990; McMahon, 1992) and predation (Newkirk & Doyle, 1975; Heller, 1976; Johannesson, 1986; Palmer, 1990). Presumably some of this may be phenotypic, as demonstrated for thaiids by Palmer (1990) and inferred for littorinids by McMahon (1992). Among littorinids, it has been shown by several authors that there must be a genotypic basis as well (Newkirk & Doyle, 1975; Janson, 1982a; Etter, 1988; Boulding, 1990; Grahame & Mill, 1993).

Shell shape variation in the rough periwinkle group of species has been studied particularly extensively (Emson & Faller Fritsch, 1976; Daguzan, 1977; Raffaelli, 1979; Smith, 1981; Brandwood, 1982; Janson, 1982a, b; Janson & Sundberg, 1983; Atkinson & Newbury, 1984; Grahame & Mill, 1989, 1992; Dytham *et al.*, 1990; Mill & Grahame, 1995). Size variation has been said to be the cause of much of the variation exhibited in *Littorina saxatilis* (Olivi), especially along gradients of exposure (Sacchi & Torelli, 1973; Sundberg, 1988). Other authors have shown that factors such as foot area (Grahame & Mill, 1986) and changes in growth pattern with age (Van Marion, 1981) are also instrumental in defining shell shape.

It seems uncontroversial that there are at least three species in the 'rough periwinkle' complex: the ovoviviparous *Littorina saxatilis* and two oviparous species, *L. nigrolineata* Gray and *L. arcana* Hannaford Ellis. The ovoviviparous *L. neglecta*, originally described by Bean (1844), has recently been the subject of some debate (B. Johannesson & K. Johannesson, 1990; K. Johannesson & B. Johannesson, 1990; Reid, 1993) as to its status. B. Johannesson & K. Johannesson (1990) used principal components analysis to investigate shape of *L. saxatilis* and *L. neglecta* from several sites in Britain, Iceland, Sweden and Norway. They concluded that samples from a given shore were more alike (regardless of shore level of origin) than were samples from the same habitat but different shores, and considered that there was no consistent difference in shape between snails living in the barnacle zone and the littoral fringe. In their view, *L. neglecta* could not be distinguished from *L. saxatilis* on morphological

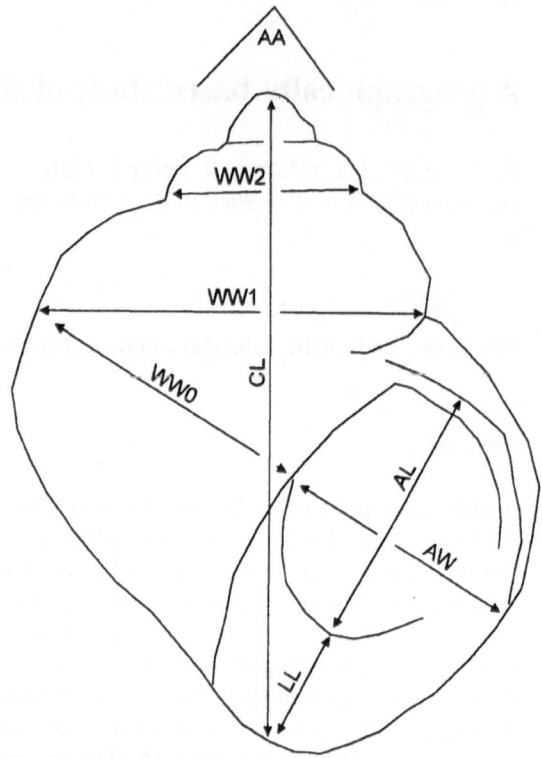

*Fig. 1.* Shell outline showing the variables measured. These were the apical angle (AA), columella length (CL), lip length (LL), aperture length (AL) and width (AW), width across the shell adjacent to aperture width (WW0), and width across the two whorls WW1 and WW2. These were chosen so as to provide a set of independently measured numerical variables summarising the shape of each shell measured. This shell is drawn from a specimen of *Littorina saxatilis*.

grounds, even in the same microhabitat. K. Johannesson & B. Johannesson (1990) then went on to consider enzyme polymorphisms, again with the conclusion that the taxa were virtually indistinguishable, and, moreover (and crucially), that gene flow was occurring between them. Therefore the suggestion has been made that '*L. neglecta*' is a barnacle-dwelling ecotype of *L. saxatilis*, in the same way as such ecotypes occur in the other taxa of the group (Reid, 1993). These conclusions conflict with other reports, which have indicated that the two taxa are distinct, both with respect to shell shape (Grahame *et al.* 1995) and enzyme polymorphisms (Heller, 1975; Wilkins & O'Regan, 1980).

In view of the uncertainty over the recognisability of *L. neglecta*, we set out here to investigate shape variation in small periwinkle shells from a variety of shores, principally in Britain. We seek to answer the question, is there a distinct shell form which both corresponds to the accepted definition of *L. neglecta* (*sensu*

*Table 1.* Samples used in the present analysis.

| Site and collector | Brit. Nat. Grid | Lat./Long. | Samples |
|---|---|---|---|
| St. Alban's Head, S. England (KJC) | SY/957758 | 50°N 2°W | *L. saxatilis* 25 |
| St. Govan's Head, S. Wales (KJC) | SM/976928 | 51°N 5°W | *L. saxatilis* 199; *L. neglecta* 87; *L. arcana* 58; *L. nigrolineata* 41 |
| Dale Peninsula + Grassholme, S. Wales (KJC) | SM/798058 + SM/599092 | 51°N 5°W | *L. saxatilis* 64; *L. neglecta* 137 |
| Scarborough, N.E. England (KJC) | TA/048895 | 54°N 0°W | *L. saxatilis* 74; *L. neglecta* 175 |
| Ravenscar, Robin Hood's Bay (N.E. England) (KJC) | NZ/979026 | 54°N 0°W | *L. saxatilis* 39; *L. neglecta* 151; *L. arcana* 57 |
| Ness Point + West Scar Robin Hood'sBay (N.E. England) (KJC) | NZ/960061 + NZ/955020 | 54°N 0°W | *L. saxatilis* 452; *L. neglecta* 422; *L. arcana* 109 |
| Wick of Skaw (Unst) (PJM) | HP/661163 | 60°N 0°W | *L. neglecta* 23 |
| Tunisia (site notdescribed) (NHM) | – | 35°N 10°W | *L. saxatilis* 12 |
| Spain (Baiona) (NHM) | – | 42°N 8°W | *L. saxatilis* 50; *L. neglecta* 49 |
| Belgium (Oostende) (NHM) | – | 51°N 2°E | *L. saxatilis* 53; *L. neglecta* 54 |
| Eire (Dooagh, Achill I.) (NHM) | – | 54°N 10°W | *L. saxatilis* 98; *L. neglecta* 100; *L. arcana* 74 |
| Iceland (Vik) (NHM) | – | 63°N 19°W | *L. saxatilis* 190 |
| U.S.A. (Appledore I.; Isles of Shoals, Maine) (NHM) | – | 44°N 69°W | *L. saxatilis* 99 |

*Table 2.* Coefficients forming the first two eigenvectors (PC1 and PC2) from principal component analyses on small shells (< 5.5 mm) from Robin Hood's Bay. The variables are according to rank, and relative importance of the vectors is indicated by percentages. Abbreviations: CL, columella length; LL, lip length, AL, aperture length; AW, aperture width; WWO, whorl width 0; WW1, whorl width 1; WW2, whorl width 2; AA, apical angle. The square brackets outside the Table indicate those vectors between which there are significant correlations (Table 3).

| PC1 | | | PC2 | | |
|---|---|---|---|---|---|
| *L. saxatilis* | *L. neglecta* | *L. arcana* | *L. saxatilis* | *L. neglecta* | *L. arcana* |
| 37% | 35% | 34% | 29% | 28% | 22% |
| AL 0.466 | WW2 0.445 | AL 0.534 | AA 0.599 | AL 0.605 | AA 0.681 |
| AW 0.398 | CL 0.437 | AW 0.49 | AW 0.305 | AW0.558 | WW0 0.373 |
| WW1 0.367 | WW1 0.408 | CL 0.282 | AL 0.232 | AA0.346 | AL 0.184 |
| WW2 0.298 | WW0 0.202 | WW1 0.259 | WW0 0.168 | CL 0.210 | AW 0.150 |
| WW0 0.229 | AL 0.000 | WW0 0.117 | LL 0.123 | WW1 −0.002 | LL 0.080 |
| CL 0.14 | AW −0.132 | WW2 0.015 | WW1 −0.153 | WW0 −0.071 | WW1 −0.193 |
| AA 0.096 | AA −0.389 | AA 0.014 | WW2 −0.442 | WW2 −0.113 | WW2 −0.316 |
| LL −0.566 | LL −0.485 | LL −0.560 | CL −0.482 | LL −0.376 | CL −0.444 |

*Table 3.* Spearman rank correlation coefficients for the first two vectors (PCs) from Table 2. Significant correlations (in bold) are followed by their probability level *P* (in italics).

| | | L. saxatilis | | L. neglecta | | L. arcana | |
|---|---|---|---|---|---|---|---|
| | | PC 1 | PC2 | PC 1 | PC2 | PC 1 | PC2 |
| *L. saxatilis* | PC1 | . | 0.143 | 0.310 | 0.595 | **0.833** | 0 |
| | | | | | | *0.01* | |
| | PC2 | | . | **−0.738** | 0.547 | 0.048 | **0.905** |
| | | | | *0.04* | | | *0.002* |
| *L. neglecta* | PC1 | | | . | −0.143 | 0.286 | −0.643 |
| | PC2 | | | | . | **0.762** | 0.357 |
| | | | | | | *0.03* | |
| *L. arcana* | PC1 | | | | | . | −0.095 |

Bean) and is of fairly widespread occurrence? Then, if there is such a form, how does it compare in distinctness with an accepted sibling species of *L. saxatilis*, namely *L. arcana*, and with *L. saxatilis* groups from different locations and habitats? Specimens of *L. nigrolineata* from St Govan's Head in south-west Wales have been included because barnacle-dwelling ecotypes of this species and of *L. arcana* have been reported from this location (Reid, 1993). However, while there are comparative data for other taxa, since they were collected from many shores, *L. nigrolineata* is represented by this collection only.

We return to the remark of Richards (1938) and stress that the initial sorting of material is crucial. There are obvious potential problems here – are the groups as defined merely artificial? Moreover, there is a danger of circularity – it is not surprising if objects sorted by shape turn out to be different in shape when this is explicitly analysed. In this study, we have tried to break this circularity by first sorting the shells on qualitative characters, and then using multivariate analyses on quantitative characters. Moreover, we are not simply interested in difference of shape as such (if it exists), but in quantifying the degree and nature of variation of shape in different groups.

## Material and methods

The object in this study was to explore variation in animals living on open rocky coasts of varying degrees of exposure. Therefore, while a range of shell forms is covered, the estuarine '*Littorina tenebrosa*' (considered to be a form of *L. saxatilis* – e.g. see Janson &

*Table 4.* Possible and observed numbers of correlated first and second vectors (PC1s and PC2s) in comparisons of the groups of small shells (< 5.5 mm columella length). Number of shores per group: 9 (*L. neglecta*), 13 (*L. saxatilis*), 4 (*L. arcana*)

| Correlation pair | Possible maximum (number of cells) | Observed results (number of cells) | Percentage obs./exp. |
|---|---|---|---|
| *L. neglecta/L. neglecta* | 76.5 | 66.0 | 86.3 |
| *L. saxatilis/L. saxatilis* | 162.5 | 97.0 | 59.7 |
| *L. arcana/L. arcana* | 14.0 | 8.0 | 57.1 |
| *L. neglecta/L. arcana* | 72.0 | 54.0 | 75.0 |
| *L. neglecta/L. saxatilis* | 234.0 | 129.0 | 55.1 |
| *L. saxatilis/L. arcana* | 104.0 | 47.0 | 45.0 |

Ward, 1985) has not been extensively sampled, being represented in shape by only one sample from Tunisia. Details of sample sites are given in Table 1. While they cover a wide latitudinal range, most sites are in the British Isles.

### Sorting and categorisation

There are two oviparous species, *Littorina nigrolineata* and *L. arcana*, the former was readily identified on the basis of its characteristic shell sculpture. For ovoviviparous winkles, work was commenced using material collected from various locations in the northern part of Robin Hood's Bay (see Table 1 for details). This location is particularly well known, with large sample sizes in the present study, and is convenient

*Table 5.* Coefficients forming the first three eigenvectors (PC1, PC2 and PC3) from principal component analyses on small shells (< 5.5 mm) using data from all shores. The variables are according to rank, and relative importance of the vectors is indicated by percentages. Abbreviations as in Table 2. The square brackets outside the Table indicate those vectors between which there are significant correlations (Table 6).

| PC 1 | | | | | | PC2 | | | | | | PC3 | | | | | |
|---|---|---|---|---|---|---|---|---|---|---|---|---|---|---|---|---|---|
| *L. saxatilis* | | *L. neglecta* | | *L. arcana* | | *L. saxatilis* | | *L. neglecta* | | *L. arcana* | | *L. saxatilis* | | *L. neglecta* | | *L. arcar* | |
| 41% | | 36% | | 36% | | 28% | | 28% | | 23% | | 15% | | 15% | | 20' | |
| WW1 | 0.461 | AA | 0.461 | AL | 0.518 | AA | 0.609 | AL | 0.495 | WW2 | 0.571 | WW0 | 0.787 | WW0 | 0.667 | AA | 0.64 |
| AL | 0.432 | AW | 0.398 | WW1 | 0.459 | AW | 0.367 | CL | 0.367 | WW0 | 0.543 | AA | 0.247 | AA | 0.357 | LL | 0.23 |
| AW | 0.346 | AL | 0.314 | AW | 0.391 | AL | 0.330 | WW1 | 0.352 | AA | 0.143 | WW1 | 0.191 | WW1 | 0.312 | WW0 | 0.11 |
| WW2 | 0.265 | LL | 0.293 | AA | 0.218 | LL | 0.023 | AW | 0.347 | WW1 | 0.114 | WW2 | 0.104 | LL | 0.066 | AL | 0.06 |
| CL | 0.258 | CL | −0.237 | WW0 | 0.187 | WW1 | 0.002 | WW0 | 0.230 | AL | −0.154 | LL | −0.027 | AL | −0.106 | AW | 0.04 |
| WW0 | 0.198 | WW0 | −0.242 | CL | 0.150 | WW0 | −0.036 | AA | 0.138 | CL | −0.199 | AL | −0.139 | WW2 | −0.216 | WW1 | −0.04 |
| AA | −0.036 | WW1 | −0.331 | WW2 | −0.054 | CL | −0.410 | WW2 | −0.126 | LL | −0.265 | CL | −0.297 | AW | −0.240 | WW2 | −0.32 |
| LL | −0.551 | WW2 | −0.470 | LL | −0.510 | WW2 | −0.465 | LL | −0.537 | AW | −0.461 | AW | −0.406 | CL | −0.459 | CL | −0.63 |

*Table 6.* Spearman rank correlation coefficients for the first three vectors (PCs) from Table 5. Significant correlations (in bold) are followed by their probability level P (in italics).

| | | | PC / species | | | | | | | | |
|---|---|---|---|---|---|---|---|---|---|---|---|
| | | | *saxatilis* | | | *neglecta* | | | *arcana* | | |
| | | | PC1 | PC2 | PC3 | PC1 | PC2 | PC3 | PC1 | PC2 | PC3 |
| | | PC1 | . | −0.07 | −0.29 | −0.26 | 0.69 | −0.29 | **0.74** *0.037* | −0.02 | −0.57 |
| | *saxatilis* | PC2 | | . | −0.10 | **0.90** *0.002* | 0.05 | 0.24 | 0.50 | −0.43 | 0.69 |
| | | PC3 | | | . | −0.29 | −0.38 | **0.90** *0.002* | −0.07 | **0.79** *0.021* | 0.45 |
| | | PC1 | | | | . | 0.10 | 0.00 | 0.31 | −0.52 | −0.60 |
| PC/ species | *neglecta* | PC2 | | | | | . | −0.31 | **0.74** *0.037* | −0.24 | −0.43 |
| | | PC3 | | | | | | . | 0.12 | 0.50 | **0.71** *0.047* |
| | | PC1 | | | | | | | . | −0.10 | 0.02 |
| | *arcana* | PC2 | | | | | | | | . | 0.00 |
| | | PC3 | | | | | | | | | . |

*Table 7.* Discriminant analysis crossvalidation table of small shells (< 5.5 mm in columella length). The numbers of shells classified to each category are followed by the percentage (in brackets) they form of the total for each sample (*n*).

| Group | Group | | | | |
| | *L. nigrolineata* | *L. arcana* | *L. saxatilis* | *L. neglecta* | Total shells for each sample (*n*) |
|---|---|---|---|---|---|
| *L. nigrolineata* | 32 (76.2%) | 0 (0.0%) | 4 (9.5%) | 6 (14.3%) | 42 |
| *L. arcana* | 13 (5.2%) | 123 (49.2%) | 40 (16.0%) | 74 (29.6%) | 250 |
| *L. saxatilis* | 136 (17.9%) | 120 (15.8%) | 407 (53.6%) | 96 (12.7%) | 759 |
| *L. neglecta* | 76 (5.4%) | 334 (23.9%) | 107 (7.7%) | 882 (63.1%) | 1399 |
| total classified | 257 | 577 | 558 | 1058 | 2450 |

because of the distribution of colour/sculpture morphs (Dytham *et al.*, 1990). The ovoviviparous animals fell into two quite clear and simply defined groups, namely with or without shell sculpturing. The sculpturing, when present, consisted of distinct sharp or rounded ridges and was indicative of *L. saxatilis*. Further discussion of the groups is deferred to the 'Results' section.

In the cases of animals from collections made for this study, identification included characterisation of female reproductive anatomy. Dissection was not possible with museum collections, and here identification relied on the labels of the original collectors, with subsequent evaluation of shell colour and sculpture by one of us (KJC). Shells used in all taxa were from as near a uniform size range as possible. Within the chosen range, most mature snails belonged to *L. neglecta*; snails of the other three taxa were generally immature. In all, 3093 shells from eleven different shores and covering most of the geographical range of *L. saxatilis* were analysed with respect to shell shape.

*Measurement and data handling*

The shells were positioned in the apertural view, and measured using a digitiser and microcomputer. The methods of measurement and nomenclature follow those used by Grahame & Mill (1992) (Fig. 1).

The variable apical angle was used untransformed, following Grahame & Mill (1989). Each linear measurement was divided by the geometric mean of all the linear variables for that shell, the resulting ratios were then transformed to base 10 logarithms. This procedure reduces all the linear data for each specimen to an apparent absolute size, so that individuals exhibit a variation in shape which would occur at that size (Reist, 1985). However, although this reduces the effect of size on the data, there are likely to be some residual effects of allometry. These were further countered by using similar sized specimens in the major analyses. Thus, shells were divided into those less than 5.5 mm in columella length ('small', *n* = 2450 specimens) and those greater than this height ('large', *n* = 643 specimens). The threshold measurement of 5.5 mm was equivalent to the largest *Littorina neglecta* specimen used in the present study. The data were separated by shore and taxon, but not by shore level; this provided 27 samples, i.e. one of *L. nigrolineata*, four of *L. arcana*, nine of *L. neglecta* and 13 of *L. saxatilis*.

*Data analyses*

Analyses were carried out using procedures in the Statistical Analytical System package (SAS Inst. Inc., 1990). The multivariate techniques used were principal component analysis (PCA) and discriminant analysis. The former uses no information on groups, though of course grouping is manipulated by the extent of the data set used. Discriminant analysis, in contrast, uses information on groups supplied by the investigator within the data set as a class variable (in our case, sample or taxon name).

*Table 8.* Discriminant analysis crossvalidation table of large shells (> 5.5 mm in columella length). The number of shells classified to each category is followed by the percentage (in brackets) they form of the total for each sample (*n*).

|  | Group | | |
|  | *L. arcana* | *L. saxatilis* | Total (*n*) |
| --- | --- | --- | --- |
| *L. arcana* | 40 | 5 | 45 |
|  | (88.9%) | (11.1%) |  |
| *L. saxatilis* | 56 | 542 | 598 |
|  | (9.4%) | (90.6%) |  |
| Total | 96 | 547 | 643 |

The first approach was to discover whether shape variation in the individual samples and groups showed any concordance. For this, an independent PCA was carried out on data sets consisting of measurements of shells less than 5.5 mm in columella length in each sample. PCA is a technique which estimates successive 'components' which maximise the variation in the original measurements (see Reyment *et al.*, 1984). The components are extracted in descending order of importance and each is uncorrelated with the others. For each component there is a vector of coefficients showing the extent of the relationship of each of the variables with that component – i.e., it is possible to see how much the variables have contributed to the component. If vectors with the same ordering of variables are calculated from analyses of different data sets, this indicates that the same sort of shape variation is occurring in the two sets of data.

In the current context, PCA is a way of answering the question, what shell variables most explain the variation observed in a given sample? Furthermore, if there is consistency in the shell variability between samples, the principal components should be similar among the PCAs for these separate samples. The likelihood of such consistency was examined by calculating correlation coefficients between all possible PC1 and PC2 combinations. We found that there was agreement between rank and parametric correlations, although the former are more appropriate to these data.

For comparisons within a taxon, take the case of 10 independent PCA calculations. Each will produce a series of vectors, PC1, PC2, etc. In any one analysis, vectors PC1 and PC2 will themselves be uncorrelated; either (but not both) may correlate with vector PC1 or with vector PC2 from any other analysis. The maxi-

mum number of positive correlations we might observe between all vectors PC1 and PC2 would then be $(20^2-20)/4 = 95$. The denominator here is 4 because of the assumption that a vector will correlate with only one of the pair available from each other analysis. For inter-taxon comparisons, the maximum possible number of positive correlations is more simply $(2N_1 \bullet 2N_2)/2$, where $N_1$ and $N_2$ are the number of analyses for the two taxa.

After exploring the variation within and between the samples independently, PCAs were carried out on the pooled data for each of the taxa *Littorina neglecta*, *L. saxatilis* and *L. arcana*, and correlations were sought between first, second and third principal components to estimate similarities within and between taxa.

We next ask the question, if groups can be defined, what is the reliability of classification using the measurement data? This was approached using discriminant analysis. In this analysis, a set of so called discriminant functions is estimated such that the variation between the declared groups is maximised. The set of functions is then used to classify one or a number of objects to see where the new example(s) best fit into the existing groups. In applying this to the 'small' shells (<5.5 mm), each individual shell was classified by using the discriminant functions estimated from all the remainder of the data set but not the shell in question (the 'cross validate' option in PROC DISCRIM). This was repeated for the 'large' shells, and finally these were classified with respect to the 'small' ones.

## Results

We found that in Robin Hood's Bay among the ovoviviparous animals the sculptured shells were from a population which attained a large size before brooding. The absolute minimum for a fertile female in this group was 3.6 mm, which was exceptional (only one such was observed). The more usual size for the onset of fertility was between 5 and 8 mm. These animals reached sexual maturity in the upper part of the shore, although small individuals were frequently found in the barnacle zone at Ness Point. Additionally at Ness Point this population showed the orange and dark brown banded morph of *Littorina saxatilis* described by Dytham *et al.* (1990). The smooth shells were from a population in which brooding first occurred at a much smaller size (1.8 mm was the smallest observed) and the great majority (about 75% of individuals) were found in the barnacle zone.

*Table 9.* Discriminant analysis of large shells: classification into small shells.

| Test data groups (large shells) | Calibration data groups (small shells) | | | | |
| --- | --- | --- | --- | --- | --- |
| | *L. arcana* (s) | *L. saxatilis* (s) | *L. nigrolineata* (s) | *L. neglecta* | Total (n) % |
| *L. arcana* | 27 | 16 | 2 | 0 | 45 |
| (1) | (60.0%) | (35.4%) | (4.4%) | (0.0%) | 100 |
| *L. saxatilis* | 45 | 278 | 272 | 3 | 598 |
| (1) | (7.5%) | (46.5%) | (45.5%) | (0.5%) | 100 |
| total (n) | 72 | 294 | 274 | 3 | 643 |

It is difficult to escape the conclusion that the brooding animals from Robin Hood's Bay do fall into two groups, that there is a set of characters which reliably distinguish them, and that these two groups are related to the ecological features of size at maturity and habitat. In fact, the groups correspond to *L. saxatilis sensu stricto* and *L. neglecta sensu* Bean. We found that of 452 shells which we attributed to *L. saxatilis*, four were ambiguous on the sculpture criterion – they could then only be placed in *L. saxatilis* on grounds of 'general appearance', i.e., an appreciation of shape. However, this number is only ≈1% of the sample. Likewise of 422 shells attributed to *L. neglecta* three (again ≈1%) were ambiguous. After initial experience at Robin Hood's Bay, it was found that at a variety of sites in Britain the same procedures could be carried out, with the recognition of a form corresponding to *L. neglecta*.

Table 2 shows the coefficients of the first two eigenvectors calculated from independent principal component analyses (PCAs) using data sets from Robin Hood's Bay where the shells are easily sorted into three categories on qualitative characters. For *L. saxatilis* and *L. arcana*, the two PC1 vectors are remarkably similar, as are the two PC2 vectors. The similarities are further indicated by Spearman rank correlations between the vectors (Table 3). For both these taxa on this shore, the most important aspects of shell shape variation are associated with the variables of aperture size contrasted with lip length and apical angle (PC1) and with the variables of columella length and whorl width 2 contrasted with apical angle and aperture width (*L. saxatilis*) or apical angle and whorl width 0 (*L. arcana*) (PC2). In contrast, the most important aspects of shell variation in *L. neglecta* (represented by PC1) is quite unlike PC1 for the other two taxa, but very like PC2 for *L. saxatilis* (correlation coefficient 0.738, see Table 3). There is also a high correlation between PC2 for *L. neglecta* and PC1 for *L. arcana*. PC3 in each species showed no correlation with any other vectors. It may also be noted that, within taxa, there are no correlations between vectors PC1 and PC2 – as indeed, by definition, there should not be (see Methods).

Next, we carried out independent PCAs on each of 26 data sets representing data for three taxa (i.e. excluding *L. nigrolineata*) on a range of shores. Correlations were sought between the resulting vectors for principal components 1 and 2. This analysis is summarised in Table 4 which shows the fraction of the possible maximum of significant correlations as a percentage. Because of the size of the data matrix in this instance, it was convenient to handle the calculation in a spreadsheet, limiting us to a Pearson product-moment correlation analysis. The lowest significant ($P = 0.05$) coefficients were individually checked using a Spearman rank correlation, with no loss of significance. No account is taken here of the possibility of type I errors – this is considered to be unimportant, since the object is not to decide on the reality of any given correlation, but to record the occurrence of high levels of correlation. The incidence of significant correlations is well in excess of that expected by chance at $P = 0.05$, namely about 5%. *L. neglecta* shows the greatest degree of consistency within taxa (86.3%), compared with the figures of 59.7% and 57.1% for *L. saxatilis* and *L. arcana* respectively. In the inter-taxon comparisons, the percentage correlation between *L. neglecta* and *L. arcana* is the greatest.

We next returned to the first approach, but using now the data pooled for the groups across all shores.

The coefficients for the first three principal components (vectors) for *L. saxatilis*, *L. neglecta* and *L. arcana* are given in Table 5. Here we refer to the first three vectors because, using the pooled data, we found similarities between the third vectors and others (*cf.* Tables 2 and 3 for Robin Hood's Bay). These vectors account for 84% of the total variation in *L. saxatilis* and for 79% in each of the other two taxa.

Relationships between the vectors in Table 5 were sought using Spearman rank correlations (Table 6). Again, within each taxon, the vectors (PCs) are uncorrelated. For each vector in each group, there is a very similar vector in one other group, and sometimes in both (e.g. for *L. neglecta*, PC3 is very similar to PC3 in both *L. saxatilis* and, to a lesser extent, *L. arcana*). The relationships between the first two vectors (PC1 and PC2) for the three taxa indicated in Tables 2 and 3 (Robin Hood's Bay) are evident in Tables 5 and 6 (all shores), as is the association of the first vector (PC1) for *L. saxatilis* with that for *L. arcana*. However, the strong association of the second vectors (PC2s) for *L. saxatilis* and *L. arcana* at Robin Hood's Bay is not evident when all the data are pooled.

The results of a discriminant analysis of the data are given in Tables 7 to 9. Table 7 shows the crossvalidation of small shells, i.e. those with a columella length below 5.5 mm. For each taxon the shells classify most reliably to their own taxon. The greatest reliability is shown by *L. nigrolineata* (76.2%) followed by that for *L. neglecta* (63.1%).

An analysis of the large specimens (necessarily involving only *L. saxatilis* and *L. arcana*) (Table 8) shows a very high proportion of self classification, as has been shown before for these two taxa (Grahame & Mill, 1989). As with all parts of the analysis, '*L. saxatilis*' here represents the combined populations of the *L. saxatilis* taxon which may be sympatric with or allopatric from those of *L. arcana*.

Table 9 shows the results of a discriminant analysis on the large specimens, using the small specimens as the calibration data set. Few *L. saxatilis* classify into *L. neglecta* (0.5%), but a large number (46.5%) do classify into small *L. saxatilis* and into small L. nigrolineata (45.5%). There is also some classification into small *L. arcana* (7.5%). In the case of *L. arcana*, large specimens show greater affiliation with small members of their own taxon (60.0%) than they do with any other taxon available. The greatest misclassification of *L. arcana* is with small *L. saxatilis* (35.4%); few large specimens of *L. arcana* misclassify into small

*L. nigrolineata* (4.4%), while none are misclassified into *L. neglecta*.

## Discussion

Hitherto, comparative analyses of shell characters in *Littorina neglecta* and *L. saxatilis* have depended upon comparatively small data sets (e.g. B. Johannesson & K. Johannesson, 1990), or have been predominantly qualitative in their approach (e.g. Heller, 1975). In this study, we have set out to explore shell shape in rough periwinkles using comparatively large data sets from a variety of shores and microhabitats on shores, and concentrating on small-size shells. Thus, we take into account the occurrence of juvenile *L. saxatilis* in the barnacle zone, and the occurrence of the 'neglecta' form in rock crevices above the barnacle zone (*cf.* B. Johannesson & K. Johannesson, 1990). As littorinids may migrate up and down the shore (Newell, 1958; Raffaelli, 1978; Hannaford Ellis, 1985; Takada, 1992), great care must be taken when analysing shell variation in these snails. We explicitly address the question of whether there might be consistent differences such as to separate *L. neglecta* from *L. saxatilis*.

A problem in multivariate morphometrics is the extent to which apparent differences shown by the analyses are differences of size rather than of taxonomically useful or ecologically meaningful shape characters. The first principal component (PC1) of a PCA is usually interpreted as a 'general size factor' if the coefficients within it are all positive and similar in magnitude, approximating the expression $(N)^{-1/2}$, where N is the number of shell characters being analysed (Jolicoeur & Mosimann, 1960; Sundberg, 1988). PC2 is then interpreted as describing 'general shape' (Jolicoeur, 1963), with subsequent principal components describing decreasing amounts of variation in the data set. In our 'small shell' analyses, carried out after using a size transformation and relying on specimens below 5.5 mm in columella length, the coefficients in the first principal component vectors are not all positive, of similar value and approximately $(N)^{-1/2}$ (Table 5). Therefore, we consider that we have at least substantially reduced the size problem and that the variation revealed is principally to do with shape.

In Robin Hood's Bay, the ovoviviparous winkles fall into two groups when characterised by the qualitative shell characters of shell colour and sculpture. This makes this shore a useful starting point for the work described here. Having sorted the small shells into three groups, partly on the characters of the reproductive tract (differentiating *L. arcana* from *L. saxatilis* and *L. neglecta*) and on qualitative shell characters, it then can be shown that between two of these groups the pattern of variation in shell metric characters is very similar. Between these first two and the third group, this is less so. This conclusion rests on the observation that the most important principal component ($\geq$ 34% of variation explained) is apparently the same in the groups corresponding to *L. saxatilis* and *L. arcana*, but is a different component in the group *L. neglecta* (Tables 2 and 3). The first *L. neglecta* component (35% of variation explained) is not unique, but corresponds to the second component identified for *L. saxatilis* (29% of variation explained). We have tried to avoid the potential circularity of sorting on shape and analysing shape by using qualitative characters. The latter provided excellent discrimination at Robin Hood's Bay, with only about 1% of shells of ovoviviparous snails unclassifiable as *L. saxatilis* or *L. neglecta* using shell colour and sculpture. This small fraction cannot have significantly affected the analyses. Moreover, since the first principal component is different in *L. neglecta*, the groups (however distinguished) show different shape properties between *L. neglecta* and the other two taxa.

Using a similar qualitative approach, shells from all the sites sampled for this survey (a total of 2113, i.e. 68% of those used) could be sorted into groups corresponding to the taxa named herein, including *L. neglecta*. For the remaining shells, we used the identifications ascribed by the collector or authority involved, but only if the principal investigator (KJC) was content with the identification. In common with other workers, we have found that the sets of qualitative characters separating *L. saxatilis* from *L. neglecta* are not necessarily consistent between shores – as is found for the qualitative characters used for any of the rough periwinkles. This is even true for the supposedly characteristic sculpturing of *L. nigrolineata*. Thus, populations on The Smalls rock in the southern Irish Sea are very thin-shelled and have sharp ridges (J. Grahame & P. J. Mill, unpublished observations). We are not surprised, therefore, to find that sets of characters useful in distinguishing *L. saxatilis* from *L. neglecta* on one shore may not be useful on another – the point is that distinction can be made for any given shore among those which have been studied.

PCA calculations carried out on these groups using pooled data within each group yielded essentially the same vectors of coefficients for the first and second principal components (Tables 5 and 6, *cf*. Tables 2 and 3). Of course the large Robin Hood's Bay samples are making a contribution here, but the relative sample sizes involved are 33% (*L. saxatilis*), 35% (*L. neglecta*) and 37% (*L. arcana*), and it may be supposed that the other samples must be in good agreement or the close similarity between the two analyses would not exist. This is confirmed by the finding that there is good agreement between the vectors (PCs) when all the samples are independently compared (Table 4). The first principal component (and therefore the most important aspects of shell variation) appear to be substantially the same in *L. saxatilis* and *L. arcana*, with a correlation coefficient of 0.74 (Table 6). The most important variables here are whorl width 1 and aperture length, which vary inversely with lip length and apical angle (*L. saxatilis*) or lip length and whorl width= 2 (*L. arcana*) (Table 5). Therefore, in these two taxa the most important aspects of shell variation in small shells seem to be much the same. However, in *L. neglecta*, the most important aspects of shell variation are apical angle and aperture width varying inversely with whorl width 2 and whorl width 1. PC1 in *L. neglecta* is evidently the same as PC2 in *L. saxatilis*, and this is a particularly interesting finding because it indicates that, whereas the same sorts of shell variation are identifiable in these two taxa (which is unsurprising), the relative importance of the different combinations of variables differs between them. Thus, the most important aspects of shell variation in small *L. saxatilis* are not the same as those in comparably sized *L. neglecta*. It is also noticeable that this particular ordering of variables is not represented in *L. arcana*, i.e. there is no significant correlation between PC1 (*L. neglecta*) or PC2 (*L. saxatilis*) with any of the *L. arcana* vectors (Table 6).

If the apparent shape differences between *L. neglecta* and *L. saxatilis* were mostly attributable to variation in size, as was suggested by Sundberg (1988) for variation in *L. saxatilis sensu stricto*, this might be taken as evidence that *L. neglecta* is a form of *L. saxatilis* (B. Johannesson & K. Johannesson, 1990; K. Johannesson & B. Johannesson, 1990; Reid, 1993). We might then expect similar vectors (PCs), in a similar order, to be calculated when data from shells of comparable sizes are analysed. This does not happen, and consideration of the nature and ordering of PCs in our analyses indicates distinctness of the three groups *L. saxatilis*, *L. arcana* and *L. neglecta* on grounds of shell shape differences.

In terms of the incidence of correlation between principal components calculated from independent data sets, *L. neglecta* is more like *L. arcana* than like *L. saxatilis* (Table 4). This table also indicates that *L. neglecta* is the most homogeneous of the groups for which comparative data are available. The apparent homogeneity within *L. nigrolineata* (indicated by the level of crossvalidation; Table 7) is presumably due to the fact that only one sample was included in the analysis. It is also clear that large *L. saxatilis* classify well into small shells of their own taxon, but almost as many misclassify into *L. nigrolineata* (Table 9). Only 7.5% misclassify into *L. arcana*, while misclassification into *L. neglecta* at 0.5% is trivial.

## Conclusion

A geographically-based analysis shows that *Littorina neglecta* apparently represents a homogeneous and discrete group over its range in Europe (see also Fish & Sharp, 1985). Shell shape intermediates between this taxon and *L. saxatilis* (defined as those animals misclassifying in a discriminant analysis) are no more common than those between *L. arcana* and *L. saxatilis*, indeed, they are rarer (Table 7). Furthermore, shell shape intermediates are more common between *L. neglecta* and *L. arcana* than between the two ovoviviparous taxa (Table 7). It has also been shown here that differences between *L. saxatilis* and *L. neglecta* are not merely size related; the most important shape eigenvector in *L. neglecta* is evidently the same as the second most important one in *L. saxatilis* (Tables 2 & 5). Thus the data presented here show that, on a wide range of shores, a group of periwinkles shells can be identified on qualitative grounds which are then found after analysis of shape to be quite as distinct as any other such group, and which correspond to *L. neglecta* *sensu* Bean.

Reid (1993) suggested that the observation of differing zonation patterns of *L. saxatilis* and *L. neglecta* and the shell shape variation between them might be explained if *L. saxatilis*, being relatively intolerant of exposure, had its potential vertical range bisected under conditions of exposure. Then the species might be surviving as different ecotypes in two refugia: a population of large shelled forms high on the shore, and a population of small shelled forms ('*L. neglecta*') low on the shore among barnacles. B. Johannesson & K. Johannesson (1990), in contrast considered that shells of *L. neglecta* could not be distinguished from

the variable shells of *L. saxatilis*, and that there were no grounds for specific status for *L. neglecta* on morphological grounds. In a subsequent paper (K. Johannesson & B. Johannesson, 1990) they used data on polymorphic enzymes to infer gene flow between populations of the two 'species', finding it to be apparently so extensive as to make the 'neglecta' form an ecotype of *L. saxatilis*. We cannot here address the question of gene flow, which remains for future consideration. However, we do report that our data show the widespread existence of a discrete group corresponding to *L. neglecta* on a variety of shores – notably at Robin Hood's Bay, where *L. saxatilis* is abundant, and the two taxa in fact overlap in distribution (K. J. Caley, unpublished observations). *L. neglecta* does not merge into *L. saxatilis* in terms of shell shape; there is real variation between the populations. *L. neglecta* is as discrete a group as is any of the groups of shells we have considered. This situation might be maintained by disruptive selection in some way generating shell polymorphism (the 'ecotype' solution), or by a reproductive barrier (the 'species' solution). We consider that this may only be resolved by genetic analyses of material which take into account the reality of shape variation between the populations, and therefore the question is still open.

## Acknowledgments

KJC wishes to thank the SERC in conjunction with the Natural History Museum (London) for the CASE award studentship which made this research possible. The authors are grateful to D. G. Reid and K. & B. Johannesson for much discussion.

## References

Atkinson, W. D. & S. F. Newbury, 1984. The adaptations of the rough winkle *Littorina rudis* to desiccation and to dislodgement by wind and waves. J. anim. Ecol. 53: 93–105.

Bean, W., 1884. A supplement of new species. In C. Thorpe (ed.), British Marine Conchology, being a descriptive catalogue, arranged according to the Lamarckian System, of the saltwater shells of Great Britain. Edward Lumley, London.

Beaumont, A. & J. H. C. Wei, 1991. Morphological and genetic variation in the Antarctic limpet *Nacella concinna* (Strebel, 1908). J. moll. Stud. 57: 443–450.

Boulding, E. G., 1990. Are the opposing selection pressures on exposed and protected shores sufficient to maintain genetic differentiation between gastropod populations with high intermigration rates? Hydrobiologia 193 (Dev. Hydrobiol. 56): 41–52.

192

Brandwood, A., 1982. Intraspecific variation and distribution of *Littorina rudis* (Maton) in the Fleet – a coastal lagoon in Dorset. Proc. Dors. Nat. Hist. and Arch. Soc. 104: 165–167.

Campbell, N. A. & R. J. Mahon, 1974. A multivariate study of variation in two species of rock crab of the genus *Leptograspus*. Aust. J. Zool. 22: 417–425.

Daguzan, J., 1977. Analyse biometrique du dimorphisme sexual chez quelques Littorinidae (Mollusques, Gasteropodes, Proso-branches). Haliotis 6: 17–40.

Dimm, A. C., 1902. Quantitative study of the effect of environment upon the forms of *Nassa obsoleta* and *Nassa trivittata* from Cold Spring Harbour, Long Island. Biometrika 2: 24–43.

Dytham, C., J. Grahame & P. J. Mill, 1990. Distribution, abundance and shell morphology of *Littorina saxatilis* and *L. arcana* at Robin Hood's Bay, North Yorkshire. Hydrobiologia 193 (Dev. Hydrobiol. 56): 233–240.

Emson, R. H. & R. J. Faller Fritsch, 1976. An experimental investigation into the effect of crevice availability on abundance and size-structure in a population of *Littorina rudis* (Maton): Gastropoda: Prosobranchia. J. exp. mar. Biol. Ecol 23: 285–297.

Etter, R. J., 1988. Asymmetrical developmental plasticity in an intertidal snail. Evolution 42: 322–334

Fish, J. D. & L. Sharp, 1985. The ecology of the periwinkle *Littorina neglecta* Bean. In P. J. Moore & R. A. Seed (eds), The Ecology of Rocky Coasts. Hodder & Stoughton, Lond.: 143–156.

Grahame, J. & P. J. Mill, 1986. Relative size of foot of two species of *Littorina* on a rocky shore in Wales. J. Zool., Lond. 208: 229–236.

Grahame, J. & P. J. Mill, 1989. Shell shape variation in *Littorina saxatilis* and *L. arcana* – a case of character displacement? J. mar. biol. Ass. U.K. 69: 837–855.

Grahame, J. & P. J. Mill, 1992. Local and regional variation in shell shape of rough periwinkles in southern Britain. In J. Grahame, P. J. Mill & D. G. Reid (eds), Proceedings of the 3rd International Symposium on Littorinid Biology. The Malacological Society of London, London: 99–106.

Grahame, J., P. J. Mill & A. C. Brown, 1990. Adaptive and non-adaptive variation in two species of rough periwinkle (*Littorina*) on British shores. Hydrobiologia 193 (Dev. Hydrobiol. 56): 223–231.

Grahame, J. & P. J. Mill, 1993. Shell shape variation in rough periwinkles: genotypic and phenotypic effects. In J. C. Aldrich (ed.), Quantified phenotypic responses in morphology and physiology (Proceedings of the Twenty seventh European Marine Biology Symposium). JAPAGA, Ashford, Ireland: 25–30.

Grahame, J., P. J. Mill, S. Hull & K. J. Caley, 1995. *Littorina neglecta* Bean: ecotype or species? J. nat. Hist.

Hannaford Ellis, C. J., 1985. The breeding migration of *Littorina arcana* Hannaford Ellis, 1978 (Prosobranchia: Littorinidae). Zool. J. linn. Soc. 84: 91–96.

Heller, J., 1975. The taxonomy of some British *Littorina* species, with notes on their reproduction (Mollusca: Prosobranchia). Zool. J. linn. Soc. 56: 131–151.

Heller, J., 1976. The effects of exposure and predation on the shell of two British winkles. J. Zool., Lond. 179: 201–213.

Janson, K., 1982a. Phenotypic differentiation in *Littorina saxatilis* Olivi (Mollusca, Prosobranchia) in a small area on the Swedish west coast. J. moll. Stud. 46: 167–173.

Janson, K., 1982b. Genetic and environmental effects on the growth rate of *Littorina saxatilis*. Mar. Biol. 69: 73–78.

Janson, K. & P. Sundberg, 1983. Multivariate morphometric analysis of two varieties of *Littorina saxatilis* on the Swedish west coast. Mar. Biol. 74: 49–53.

Janson, K. & R. D. Ward, 1985. The taxonomic status of *Littorina tenebrosa* as assessed by morphological and genetic analysis. J. Conch. 32: 9–15.

Johannesson, B., 1986. Shell morphology of *Littorina saxatilis* Olivi: the relative importance of physical factors and predation. J. exp. mar. Biol. Ecol. 102: 183–195.

Johannesson, B. & K. Johannesson, 1990. *Littorina neglecta* Bean: a morphological form within the variable species *L. saxatilis* (Olivi)? Hydrobiologia 193 (Dev. Hydrobiol 56): 71–87.

Johannesson, K. & B. Johannesson, 1990. Genetic variation within *Littorina saxatilis* (Olivi) and *Littorina neglecta* Bean: is *L. neglecta* a good species? Hydrobiologia 193 (Dev. Hydrobiol. 56): 89–97.

Jolicoeur, P., 1959. Multivariate geographical variation in the wolf *Canis lupus* L.. Evolution 13: 283–299.

Jolicoeur, P., 1963. The multivariate generalisation of the allometry equation. Biometrics 19: 497–499.

Jolicoeur, P. & J. Mosimann, 1960. Size and shape variation in the painted turtle; a principal components analysis. Growth 24: 339–354.

McMahon, R. F., 1992. Microgeographic variation in the shell morphometrics of *Nodilittorina unifasciata* from south-western Australia in relation to wave exposure of shore. In J. Grahame, P. J. Mill & D. G. Reid (eds), Proceedings of the 3rd International Symposium on Littorinid Biology. The Malacological Society of London, London: 107–118.

Mayr, E., 1942. Systematics and the origin of species from the viewpoint of a zoologist. Columbia University Press, New York, 334 pp.

Mayr, E., 1963. Animal species and evolution. Oxford University Press, London, 797 pp.

Mill, P. J. & J. Grahame, 1995. Shape variation in the rough periwinkle *Littorina saxatilis* on the west and south coasts of Britain.

McNamee, S. & C. Dytham, 1993. Morphometric discrimination of the sibling species *Drosophila melanogaster* (Meigen) and *D. simulans* (Sturtevant) (Diptera: Drosophilidae). Syst. Entomol. 18: 231–236.

Newell, G. E., 1958. The behaviour of *Littorina littorea* (L.) under natural conditions and its relation to position on the shore. J. mar. biol. Ass. U.K.. 37: 229–239.

Newkirk, G. F. & R. W. Doyle, 1975. Genetic analysis of shell-shape variation in *Littorina saxatilis* on an environmental cline. Mar. Biol. 30: 227–237.

Palmer, A. R., 1990. Effect of crab effluent and scent of damaged conspecifics on feeding, growth and shell morphology of the Atlantic Dogwhelk *Nucella lapillus*. Hydrobiologia 193 (Dev. Hydrobiol. 56): 155–182.

Phillips, B. F., N. A. Campbell & B. R. Wilson, 1973. A multivariate study of geographical variation in the whelk *Dicathais*. J. exp. mar. Biol. Ecol. 11: 27–69.

Raffaelli, D., 1978. Factors affecting the population structure of *Littorina neglecta*. J. moll. Stud. 44: 223–230.

Raffaelli, D., 1979. The taxonomy of the *Littorina saxatilis* species-complex, with particular reference to the systematic status of *Littorina patula* Jeffreys. Zool. J. linn. Soc. 65: 219–232.

Reid, D. G., 1993. Barnacle-dwelling ecotypes of three British *Littorina* species and the status of *Littorina neglecta* Bean. J. moll. Stud. 59: 51–62.

Reist, J. D., 1985. An empirical evaluation of several univariate methods that adjust for size variation in morphometric data. Can. J. Zool. 63: 1429–1439.

Reyment, R. A., R. E. Blackith & N. A. Campbell, 1984. Multivariate morphometrics, 2nd Edition. Academic Press, London, 233 pp.

Richards, O. W., 1938. The formation of species. Methods of studying the early stages of evolutionary divergence in animals. In G. R. de Beer (ed.), Evolution. Essays on aspects of evolutionary biology presented to Professor E. S. Goodrich on his seventieth birthday. Clarendon Press, Oxford, 350 pp.

Sacchi, C. F., 1980. Ricerche sulle variazioni di mole in *Littorina saxatilis* (Olivi) e sul loro significato ecologico. Boll. Mus. Civ. Venezia 31: 51–67.

Sacchi, C. F. & A. M. Torelli, 1973. Présence, variabilité et cycle biotique de *Littorina saxatilis* (Olivi) (Gastropoda, Prosobranchia) dans la lagune de Venise. Atti. Soc. Peloritana di scienze fisiche matem. e naturali 19: 181–188.

SAS Institute Inc., 1990 SAS/STAT® Users Guide, Version 6, fourth edition, volumes 1 and 2. SAS Institute, Cary N.C., 1739 pp.

Smith, J. E., 1981. The natural history and taxonomy of shell variation in *Littorina saxatilis* and *L. rudis*. J. mar. biol. Ass. U.K. 61: 215–241.

Smith, S. M., 1979. *Littorina rudis* var. *scotia* and its adaptation to the extreme environment of Rockall (Mollusca: Gastropoda). Porcupine Newsletter 1: 138–139.

Sundberg, P., 1988. Microgeographic variation in shell characters of *Littorina saxatilis* (Olivi) a question mainly of size? Biol. J. linn. Soc. 35: 169–184.

Takada, Y., 1992. The migration and growth of *Littorina brevicula* on a boulder shore in Amakusa, Japan. In J. Grahame, P. J. Mill & D. G. Reid (eds), Proceedings of the 3rd International Symposium on Littorinid Biology. The Malacological Society of London, London: 277–279.

Templeton, A. R., 1989. The meaning of species and speciation: a genetic perspective. In D. Otte & J.A. Endler (eds), Speciation and its Consequences. Sinnauer Associates Inc., Mass.: 3–27.

Tissot, B. N., 1988. Geographic variation and heterochrony in two species of cowries (genus *Cypraea*). Evolution 42: 103–117.

Van Marion, P., 1981. Intra-population variation of the shell of *Littorina rudis* (Maton) (Mollusca: Prosobranchia). J. moll. Stud. 47: 99–107.

Wilkins, N. P. & D. O'Regan, 1980. Generic variation in sympatric sibling species of *Littorina*. Veliger 22: 355–359.